I0055883

New Theories and Predictions on the
Ozone Hole and Climate Change

New Theories and Predictions on the
Ozone Hole and Climate Change

Qing-Bin Lu

University of Waterloo, Canada

World Scientific

NEW JERSEY · LONDON · SINGAPORE · BEIJING · SHANGHAI · HONG KONG · TAIPEI · CHENNAI

Published by

World Scientific Publishing Co. Pte. Ltd.

5 Toh Tuck Link, Singapore 596224

USA office: 27 Warren Street, Suite 401-402, Hackensack, NJ 07601

UK office: 57 Shelton Street, Covent Garden, London WC2H 9HE

Library of Congress Cataloging-in-Publication Data
Lu, Qing-Bin.
  New theories and predictions on the ozone hole and climate change / Qing-Bin Lu, University of Waterloo, Canada.
    pages cm
  Includes bibliographical references and index.
  ISBN 978-9814619448 (hardbound : alk. paper) -- ISBN 9814619442 (hardbound : alk. paper)
  1. Chlorofluorocarbons--Environmental aspects. 2. Global warming. 3. Ozone layer depletion.
4. Climatic changes. I. Title.
  QC879.7.L8 2015
  551.51'42--dc23
                                    2015007719

British Library Cataloguing-in-Publication Data
A catalogue record for this book is available from the British Library.

Copyright © 2015 by World Scientific Publishing Co. Pte. Ltd.

All rights reserved. This book, or parts thereof, may not be reproduced in any form or by any means, electronic or mechanical, including photocopying, recording or any information storage and retrieval system now known or to be invented, without written permission from the publisher.

For photocopying of material in this volume, please pay a copying fee through the Copyright Clearance Center, Inc., 222 Rosewood Drive, Danvers, MA 01923, USA. In this case permission to photocopy is not required from the publisher.

Dedication

This book is dedicated to my father (Zhuo Lu), who had been encouraging me to make contribution to science and the world before he passed away on August 2nd, 2001.

Preface

The idea to write this book originated from an invitation by a scientific editor of World Scientific Publishing Co., who had read the preprint of my paper that was first published in the open-accessible physics electronic preprint archive (*arXiv*) [Lu, 2012b] and then published in *International Journal of Modern Physics B (IJMPB)* [Lu, 2013]. The preprint drew some immediate attention from the public, including the above-mentioned editor. This encouraged me to accept the editor's invitation and submit a book proposal.

The editors of both the book and *IJMPB* took serious and independent peer-review processes of my work. In total, I received the reports of five reviewers, who gave very positive comments using words "interesting", "convincing", "solid" and "valuable" to describe my work. The reviewers also raised the suggestion that I should give more thorough comparisons between conventional theories and the new theories proposed in my papers. I was encouraged by all the comments! These were indeed helpful to me in deciding the content of this book.

In regard to big environmental and climate problems, I have detected three main views. The first view is that the Earth is quite a complicated system and therefore it is almost impossible to write any scientific theory that can predict the Earth's change with some accuracy. That is, the environmental and climate changes are essentially unpredictable.

The second view is the assumed consensus that the sciences behind environmental and climate changes, such as the ozone hole and global surface temperature change, are well settled. They are referred to as the photochemical theory of O_3 depletion and the IPCC-supported CO_2 theory of global warming (each receiving a Nobel Prize recognition), which are the so-called mainstream theories. This opinion is held by numerous scientists who have been involved either in developing the theories or in making the WMO/IPCC Reports, and by some politicians.

Another extreme in the second view is the argument that the observed warming in the late half of the 20th century is primarily caused by natural rather than anthropogenic factors.

The third view is that some scientists believe that the cause of global warming is unknown, i.e., no principal cause whether man-made or natural can be ascribed to the observed temperature changes.

Although the mainstream theories appear dominant in the opinion of the public media and the policy makers, there is a fairly long list of prominent scientists (especially physicists) who have openly questioned the validity of such theories, especially the conclusions in IPCC Reports. To name some of them, the list includes Drs. Ivar Giaever (the 1973 Nobel laureate in Physics), Freeman Dyson, Steven Koonin, William Happer, Robert Austin, Nir Shaviv, Richard Lindzen, Hans von Storch, Judith Curry, Claude Allègre, Peter Stilbs, Wibjörn Karlén, Fred Singer, Roy Spencer, Henk Tennekes, and Antonio Zichichi.

Outsiders of atmospheric and climate research are aware of the problems with the IPCC-used climate models more than with the photochemical models of the ozone hole. This is probably due to the alarming consequence of IPCC and some climate researchers as well as politicians, which has drawn tremendous attention of society and the general public. As a matter of fact, there are no less discrepancies between observations and state-of-the-art photochemical models of ozone depletion. There is still a lack of solid evidence to place not only the Kyoto Protocol but the Montreal Protocol on firm scientific foundations within the current context of atmospheric chemistry and climate models. This is evident by the fact that the predictions rarely agreed with the observations.

Atmospheric chemistry has mainly addressed the photochemical processes in the atmosphere associated with the sunlight, which can be seen by our eyes and easily conceivable. However, many physical and chemical processes initiated by invisible particles such as electrons and charged particles are not unfamiliar to physicists. In addition to solar radiation, there are also well-known cosmic rays in the atmosphere. The cosmic-ray-driven electron-induced reaction can play an important or dominant role in some important processes in the atmosphere.

There are also assumptions implicitly included in climate models. In many cases, they have actually failed to be validated with observations. The failure of models to explain real world observations often indicates that the assumptions in the models are most likely incorrect. Among the problems, the assumption of an incomplete saturation in greenhouse effect of non-halogen gases (CO_2, CH_4 and N_2O) contradicts the observations over the periods of not only the past 16 years but prior to that, as revealed by the well-observed data of global surface temperatures and polar stratospheric ozone and temperatures.

I have learned from the history of science that successful predictions are usually an indicator of the validity of a model, and a model of greater simplicity and stronger predictive capability is likely the one closer to the truth. And it is also my belief that it is possible to solve a seemingly complex problem with a succinct theory if the latter captures the essentials of the problem.

This book therefore aims to show how basic physical principles can be applied to solve vital and challenging problems concerning the ozone hole and global climate change. The emphasis in the text is on the *conceptual physical models* underlying the important issues, rather than on complicated atmospheric and climate models that rely heavily on the use of supercomputers. Also for reasons of focus I have omitted the short-term atmospheric and climate effects, such as ENSO and volcanic eruptions, and the pollutants (*e.g.*, black carbon) that have no absorption in the atmospheric 'window' and their effect is subject to a threefold or larger uncertainty. Thus, by no means do the new models reviewed in this book aim to give a perfect match with observed data in every detail.

To help the readers better understand the sciences underlying the ozone hole and climate changes, this book first gives a balanced introduction to basic physical and chemical processes induced by both electromagnetic (photon/light) radiation and charged particle (cosmic ray) radiation in the Earth's atmosphere in Chapter 2. A review of the interactions of electrons with molecules is given in Chapter 3. Brief descriptions of the conventional models of ozone hole and climate change are given in Chapters 3 and 6, respectively, before the new theories and observations are presented in Chapters 4, 5, 7, and 8.

Different from many atmospheric chemistry and physics texts, this book treats electron-molecule interactions as an important topic (Chapters 2 and 4). Substantial observations from both laboratory and atmospheric (balloon- and space-based) measurements are discussed in Chapters 4 and 5, respectively. Particularly, some topics that seem unfamiliar to atmospheric scientists but are known to molecular/surface physicists, such as questions as to why electrons having zero energy can make dissociation of molecules and how molecules can stick to ice surfaces at polar stratospheric temperatures of 180-200 K, are fairly robustly addressed by experimental observations in Chapter 4. This was partially motivated by comments/criticisms I received from some researchers (including the most prominent atmospheric chemists). Moreover, in-depth analyses of observed datasets in terms of the new theory of the ozone hole are given in Chapter 5. The new theory has predicted and well reproduced observed 11-year cyclic variations of polar O_3 loss and associated stratospheric cooling, and also clearly placed the Montreal Protocol on a firmer scientific ground. These observations have shown the superior simplicity (*with one parameter only*), explaining and predictive capabilities of the new theory for the O_3 hole.

Equally intriguingly, questions about the natural contribution to climate change and the evidence of the complete saturation in warming effect of non-halogen gases (CO_2, CH_4 and N_2O) are addressed by fairly solid observations in Chapters 7 and 8, respectively. In particular, the present physical model with *no parameters* well reproduces the observed data of global surface temperature since 1950, including the stopping in global warming over the past 16 years. A continued global cooling trend is therefore predicted in the coming decades under the international regulations to phase out CFCs, HCFCs, and HFCs.

Notably, the present theory of ozone hole includes *only one parameter*, while the new theory of climate change *does not include any parameter*. These are in marked contrast to state-of-the-art complex atmospheric and climate computer models with multiple parameters used in WMO and IPCC Reports. Most strikingly, excellent agreements of the new theories with observations *within 10% uncertainties* are shown (in

Chapters 5 and 8). The book ends in Chapter 9 with a summary of main results and conclusions, together with a brief discussion of the implications of the new theories and observations for the society and policy makers.

This book is self-contained and unified in presentation. It should be useful as an advanced book for graduate students and ambitious third- or fourth-year undergraduates in physics, chemistry, environmental and climate sciences. It also includes useful information for non-expert readers and policy makers who wish to have an overview of the sciences behind atmospheric ozone depletion and global climate change (who might skip to read Chapter 9 directly).

Qing-Bin Lu

Acknowledgments

I am extremely grateful to the following scientific teams for making the data used in my previous studies and in this book available: NASA TOMS and OMI Teams (especially Dr. P. K. Bhartia and co-workers), NASA UARS's CLAES Team (Dr. A. E. Roche and co-workers), the British Antarctic Survey's Ozone Team (Dr. J. D. Shanklin), the University of Oxford's MIPAS team (Dr. A. Dudhia), the Argentine GAW Station Belgrano Team (Dr. M. Gil and co-workers), the Sun-Earth's Climate Team (Dr. N. Krivova and co-workers), the Bartol Research Institute's Neutron Monitor Team (Dr. J. W. Bieber and co-workers), the UK Met Office Hadley Centre (Dr. P. D. Jones and co-workers), the US NOAA National Climatic Data Center (NCDC) and Global Monitoring Division, the Royal Observatory of Belgium's Solar Influences Data Analysis Center (SIDC), the Swiss PMOD / WRC team (Dr. C. Fröhlich), and the US NASA's ACRIM team (Dr. R. C. Willson).

I thank my colleagues, especially the Vice President in Research (Professor George Dixon), the Dean of Science (Professor Terry McMahon), and the former Provost (Professor Geoff McBoyle) at the University of Waterloo for their encouragement and support over the years. I also thank my research group members for their understanding and support during the writing of this book.

I particularly thank Mrs. Anna Perry and my daughter Linda Lu (an undergraduate in science) for their careful proofreading, constructive and helpful criticisms of chapters of this book. I also thank Linda's help in preparing the illustrations. The help from the Publisher, Dr. Chandra Nugraha, is also appreciated.

I am also grateful to my family's long-term enthusiastic support, to Ning, Linda, Joseph and Timothy.

List of Abbreviations

ACRIM	Active Cavity Radiometer Irradiance Monitor
AOGCM	coupled atmosphere-ocean general circulation model
BAS	British Antarctic Survey
CFC	chlorofluorocarbon
CLAES	Cryogenic Limb Array Etalon Spectrometer
CMIP5	Coupled Model Intercomparison Project Phase 5
CR	cosmic ray
CRE	cosmic-ray-driven electron-induced reaction
DA	dissociative attachment
DEA	dissociative electron attachment
DET	dissociative electron transfer
ECS	equilibrium climate sensitivity
EECl	equivalent effective chlorine
EESC	equivalent effective stratospheric chlorine
ENSO	El Niño Southern Oscillation
ERF	effective radiative forcing
ESD	electron stimulated desorption
ESDIAD	electron stimulated desorption ion angular distribution
GCM	general circulation model
GCR	galactic cosmic ray
GH	greenhouse
GHG	greenhouse gas
GMST	global mean surface temperature
HALOE	Halogen Occultation Experiment
HCFC	hydrochlorofluorocarbon
HFC	hydrofluorocarbon

IMG	Interferometric Monitor for Greenhouse gases
IPCC	Intergovernmental Panel on Climate Change
IR	infrared
IRIS	Infrared Interferometric Spectrometer
MLS	Microwave Limb Sounder
NIR	negative ion resonance
OLR	outgoing longwave radiation
OMI	Ozone Monitoring Instrument
PFC	perfluorocarbon
PMOD	Physikalisch Meteorologisches Observatorium Davos
PSC	polar stratospheric cloud
RCP	Representative Concentration Pathway
RMIB	Royal Meteorological Institute of Belgium
SATIRE	Spectral And Total Irradiance Reconstructions
SPE	solar proton event
SSI	Solar Spectral irradiance
SSN	sunspot number
TCR	transient climate response
TDS	thermal desorption spectroscopy
TNI	transient negative ion
TOA	top of the atmosphere
TOMS	Total Ozone Mapping Spectrometer
TSI	total solar irradiance
UARS	Upper Atmosphere Research Satellite
UHV	ultrahigh vacuum
WMGHG	well-mixed greenhouse gas
WMO	World Meteorological Organization
ΔF	radiative forcing
α_c	climate sensitivity factor
$\alpha_c^{\ s}$	solar climate sensitivity factor
β	total climate amplification factor
λ_c	equilibrium climate sensitivity factor
$\lambda_c^{\ s}$	solar equilibrium climate sensitivity factor
ΔT_s	global mean surface temperature change
$\Delta T_s^{\ t}$	global transient temperature change
κ	ocean heat uptake efficiency

Contents

Chapter 1

Basic Physics and Chemistry of the Earth's Atmosphere

1.1 Introduction

This chapter gives an introduction to the basic physical and chemical processes in the Earth's atmosphere. Although there are a number of excellent atmospheric chemistry and physics textbooks [*e.g.*, Brasseur *et al.*, 1999; Jacob, 1999; Andrews, 2000; Salby, 2012], most of them focus only on photochemical processes induced by electromagnetic waves (photons) from sunlight, which can be seen by our eyes. However, there are also many invisible particles such as electrons and other charged particles either originating from outer space (*e.g.*, cosmic rays) or produced from the ionization of atoms and molecules in the atmosphere [Johnson, 1990; Jackman, 1991]. In this Chapter, an introduction to the Earth's atmosphere is given in Sec. 1.2, followed by a description of various radiation sources (mainly solar radiation and cosmic-ray radiation) in the atmosphere in Sec. 1.3. Subsequently, photon interactions with atmospheric molecules are described briefly in Sec. 1.4, while Sec. 1.5 is devoted to a description of atmospheric ionization. Then, a brief review of ion chemistry in the atmosphere is given in Sec. 1.6. Finally, a summary is given in Sec. 1.7.

1.2 The Earth's atmosphere

The Earth's atmosphere is vital to sustaining life on its surface. This atmosphere consists of a mixture of ideal gases; the major constituents are molecular nitrogen (N_2) and oxygen (O_2) by volume. But the minor gases, in particular, water vapor (H_2O), carbon dioxide (CO_2) and ozone (O_3), are also crucial to the ecological system of the Earth.

The Sun is the primary energy source for the Earth. The Sun is a radiating star. Different planets in the Solar system receive various solar radiation intensities, inversely proportional to the squares of their distances from the Sun. The actual intensity of sunlight that reaches the surface depends also on the thickness and composition of the atmosphere of the Earth. While solar radiation makes life on Earth possible, the incoming higher-energy ultraviolet (UV) radiation must be absorbed by the Earth's atmosphere (especially by the ozone molecules), protecting us from the harmful aspects of solar radiation. The Earth's nearly ideal positioning in the Solar system and its unique atmospheric constituents render us the benefits of proximity to the Sun without being baked or dried like Venus or Mars.

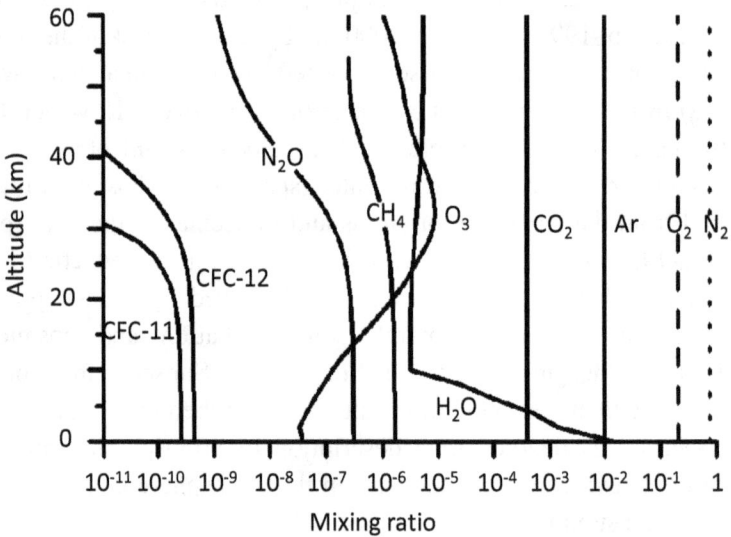

Fig. 1.1. Typical vertical concentration distributions of chemical constituents in the atmosphere. Based on data from Smith and Adams [1980] and Brasseur *et al.* [1999].

1.2.1 *Atmospheric compositions*

The Earth's atmosphere is retained by gravity. As shown in Fig. 1.1, the atmosphere (dry air) is composed of about 78.1% N_2, 20.9% O_2, 0.93% argon (Ar) gas, 0.04% CO_2 (percent by volume), and trace amounts of other gases. A variable percent, 0.001-7%, of water vapor is also present in the atmosphere.

1.2.2 *Atmospheric layers*

The Earth's atmosphere at altitudes below 100 km can be divided into several layers from lower to higher altitudes: *troposphere, stratosphere, mesosphere* and *thermosphere*. With increasing altitudes, the atmospheric temperature decreases in the troposphere, increases in the stratosphere, decreases again in the mesosphere, and increases again in the thermosphere. A typical structure of atmospheric temperature versus altitude below 100 km is shown in Fig. 1.2.

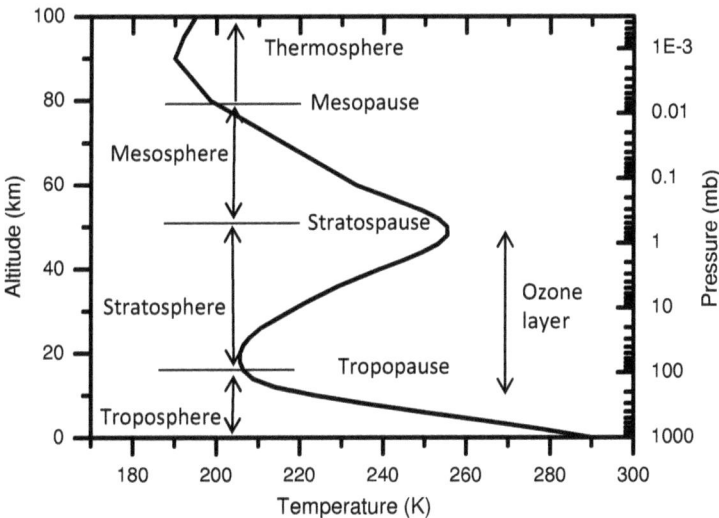

Fig. 1.2. Typical vertical structure of atmospheric temperature in the atmosphere. Based on data from Brasseur *et al.* [1999].

The *troposphere* is the region extending from Earth's surface to between 8-10 km near the polar regions and 16-18 km in tropical regions with some seasonal variations. In the troposphere, the atmospheric temperature decreases with rising altitude. This layer contains about 80% of the entire atmosphere's mass, and it is the place most daily weather occurs that is observed from the ground. Commercial airlines typically fly at altitudes of about 10 km (30 kft). The *tropopause* is the highest troposphere where the temperature reaches a minimum and the air is almost completely dry.

Extending from the tropopause to about 50 km is the *stratosphere* in which the temperature increases with altitude. Ozone and oxygen in the stratosphere absorb much of the UV radiation from the Sun. This is an important protecting layer to living creatures on Earth. The *strotopause* is the top of the stratosphere where maximum temperature is reached.

Next, the *mesosphere* is the region extending from 50 km to 80-85 km, in which the temperature falls again with altitude. There are visible meteors between 65 and 120 km above the Earth, disintegrating at altitudes of 50-95 km; millions of meteors enter Earth's atmosphere daily. The top of the mesosphere is referred to as the *mesopause* with an altitude of about 80 km where the temperature reaches a second minimum,

Above the mesopause, the *thermosphere* is the region extending from 80-85 km to more than 500 km, where the temperature increases rapidly with altitude. The *International Space Station* orbiting the Earth is located at the altitude of 330-400 km in the thermosphere.

Lastly, the *ionosphere* overlaps the thermosphere as it extends from about 80 km to 480 km. This layer contains electrons, ions and neutral molecules and is the inner edge of the magnetosphere. The magnetosphere is the region around Earth influenced by the geomagnetic field. This region is used to reflect radio signals over long distances and it is where auroral displays occur.

The troposphere is generally called the *lower atmosphere*; the stratosphere and mesosphere are called the *middle atmosphere*; and the thermosphere is called the *upper atmosphere*.

1.2.3 *Atmospheric temperature profile*

The variation of atmospheric temperature with altitude, shown in Fig. 1.2, can be roughly explained by some physical mechanisms. In the upper atmosphere, absorption of solar radiation at short wavelengths (0-100 nm) is an efficient process, resulting in the effective photoionization and photodissociation of O_2 and N_2 molecules (Fig. 1.3). Thus, the atmospheric temperature increases drastically. Below 100 km, all the neutral gas particles, the electrons and the ions in the plasma are in equilibrium with an identical temperature, while above 100 km the electron temperature becomes significantly higher than the ion temperature (and also the atmospheric temperature) due to the lower pressure (the less frequent electron-ion collisions) and the large amount of energy carried by the photoelectrons.

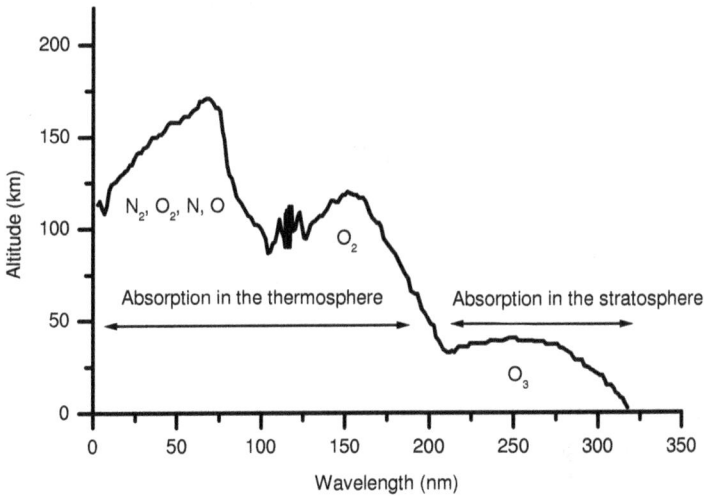

Fig. 1.3. The absorption spectrum for solar radiation in the Earth's atmosphere. Based on data from Smith and Adams [1980].

In the stratosphere, the increase of atmospheric temperature with altitude is largely due to the existence of the ozone layer, which peaks in the lower stratosphere (see Fig. 1.1). Ozone molecules (O_3) are produced by the well-known photochemical reactions involving the absorption of

UV solar radiation by molecular oxygen (O_2) in the stratosphere. However, the absorption of UV solar radiation by ozone molecules also leads to the destruction of the ozone and causes atmospheric heating. The equilibrium ozone profile depends on the ozone production and destruction chemical processes, which are determined by UV photon / cosmic ray flux, atomic and molecular oxygen number density, and the number densities of other atmospheric gases including pollutants. As a result, the temperature rises throughout the stratosphere and reaches a maximum at the stratopause. The decreasing temperature in the mesosphere is a result of the relatively low ozone concentration. The observed cluster ions formed due to the relatively high gas pressure and low temperature may also contribute to the temperature structure in the mesosphere.

The mechanisms accounting for the temperature profile in the lower atmosphere are more complicated. At first, the temperature in the troposphere is expected to decrease with rising altitude as the total gas density decreases and the distance from the longwave thermal radiation source (the Earth's surface) increases. But the resultant convection also in turn modifies the temperature profile. Thus, the observed temperature shows a much less rapid decrease in the troposphere, which is determined by the heating of longwave radiation from the Earth's surface and the conduction and convection. The altitude of tropopause at which the temperature reaches a minimum also depends on seasonal and latitude variations. At high latitudes, the effect of the strong ionization caused by cosmic rays may affect the temperature structure in the upper troposphere and the lower stratosphere; the detailed information is still a subject of investigation.

Below we will focus more on the physics and chemistry of the lower and middle atmosphere, having relevance to stratospheric ozone depletion and global climate change.

1.3 Radiation in the atmosphere

1.3.1 *Solar radiation*

The Sun emits radiation across most of the electromagnetic spectrum, including X-rays, extreme ultraviolet (EUV), ultraviolet (UV), visible light, infrared (IR), and even radio waves. However, X-rays and extreme ultraviolet (EUV) make up only a very small amount of the total solar power output. The spectrum of the solar electromagnetic radiation impacting the Earth's atmosphere covers a wavelength range of 100 nm to about 1 mm. This spectrum can roughly be divided into three regions in increasing order of wavelengths: UV light ranging from 100 to 400 nm, largely absorbed by molecular oxygen (O_2) and ozone (O_3) in the Earth's atmosphere (Fig. 1.3); visible light from 380 to 780 nm, visible to the naked eye; and IR light from 700 nm to about 1 mm.

1.3.1.1 *Solar activities*

Sunspots are observed as relatively dark areas on the radiating 'surface' (photosphere) of the Sun. This is because sunspot areas are cooler than the surrounding photosphere. Sunspots occur where the strong Sun's magnetic field inhibits the convection of hot gases from the surface of the Sun and cools the photosphere. Solar activity is based on the number of sunspots and groups of sunspots visible on the Sun. Thus, the sunspot number is an index of solar activity. The sunspot cycle has a periodicity of about 11 years on the average.

Solar wind boils continuously off the Sun and produces an interplanetary magnetic field. The solar wind is composed by mass of about 80% protons, 18% α particles, and traces of heavier charged particles [Burch, 2001]. The discontinuous magnetic fields carried by the solar wind, the so-called scattering centers, deflect some less energetic galactic cosmic rays away from the Earth, which would otherwise enter our atmosphere. During the active phase of the solar cycles, the solar wind is at its highest intensity.

Solar proton event (SPE) is a surge of subatomic particles from the Sun. It is defined as an average solar proton flux ≥10 particles/

($cm^2 \cdot$steradian$\cdot$s) with all proton energies >10 MeV in three consecutive 5-minute periods. It is now generally believed that SPE events arise most likely from a coronal mass ejection (CME), which is an explosive ejection of huge amounts of matter and embedded magnetic fields from the solar corona as a result of an occasional intense magnetic disturbance in the Sun. Usually a CME event originates in a magnetically active region around visible sunspots. A large CME event can blast a huge number of charged particles into space. While a fast CME penetrates a solar wind that moves slower, it produces interplanetary shock waves. The latter can then produce showers of high energy particles in the Earth's atmosphere [Smart and Shea, 1997]. These particles strike the atmosphere from all directions and have the same interactions with atoms and molecules in the air as galactic cosmic ray particles.

Solar flare is a sudden release of electromagnetic energy and particles from a relatively small volume of the Sun [Smart and Shea, 1997]. The amount of energy and matter released during a solar flare is much smaller than that released during a CME. Few solar flares are directed at the earth, and particles associated with a solar flare may have energy not sufficient to increase radiation levels in the Earth's atmosphere. Thus, their effect on the atmosphere is limited.

1.3.1.2 *Solar Cycle*

The *solar cycle* is the periodic variations in the sun's activity, including changes in the levels of solar radiative output, changes in the number of sunspots and changes in other solar indexes (solar flares, ejection of solar material, *etc.*). The main periodic variation that is referred to as the solar cycle has a mean duration of about 11 years, which may span 9-12 years.

Only the 11-year and closely related 22-year cycles are clearly observed. The 11 year cycle is the so-called the *Schwabe cycle*. Most significantly, the number of sunspots has a gradual increase and a more rapid decrease over an average period of 11 years, ranging from 9 to 12 years. Differential rotation of the Sun's convection zone versus latitude consolidates magnetic flux tubes, increasing their magnetic field intensity and causing them buoyant. As they rise through the solar atmosphere they partially inhibit the convective flow of energy, cooling their region of the photosphere and thus producing sunspots. The sun's photosphere radiates more actively when there are more sunspots. Satellite measurements have confirmed the direct relationship between the solar activity (sunspot) cycle and solar luminosity.

The 22 year cycle is the so-called *Hale cycle*. During each 11-year cycle, the magnetic field of the Sun reverses. Hence the magnetic poles return to the same state after two 11-year cycles.

The *total solar irradiance* (TSI), averaged at 1361–1362 W/m^2 received at the outer surface of Earth's atmosphere, has been measured to vary by approximately 0.1% (that is about 1.36 W/m^2) from solar maximum to solar minimum during the 11-year solar cycles [IPCC, 2013]. In addition to this variation during solar cycles, there are also possible long-term (secular) changes in solar irradiance over decades or centuries. However, satellite-based total solar irradiance measurements which started in 1978 have shown that the average solar constant has had only small variations since then.

1.3.2 *Cosmic ray radiation*

Cosmic rays (CRs) are high-energy charged particles. They originate from outer space (outside the Solar system) and strike the Earth from all directions. Upon penetrating the Earth's atmosphere, galactic cosmic rays produce showers of secondary particles that can impact the atmospheric processes significantly. In order to evaluate the impact of cosmic rays, we need to know the basic physical properties of cosmic ray radiation, including its elemental composition, energy spectrum, and temporal variations.

1.3.2.1 *Discovery and early research*

In 1909, Theodor Wulf developed a device called electrometer to measure the production rate of ions inside a closed chamber, and used it to show an elevated level of radiation at the top of the Eiffel Tower than at the ground level. Then, Victor Hess used Wulf electrometers with a higher accuracy and balloons to measure the ionization rate with various altitudes up to 5.3 km in 1912. He found that the ionization rate increased with height, and also ruled out the Sun as the radiation source by making measurements in a balloon rising during a nearly total eclipse. Hess explained his observation as due to a kind of radiation entering the Earth's atmosphere from above. For his discovery of cosmic rays, Hess was awarded the Nobel Prize in Physics in 1936.

However, the term "cosmic rays" was coined by Robert Millikan in the 1920s. He proposed that the primary cosmic rays were energetic photons (γ rays), which were produced in interstellar space as by-products of the fusion of hydrogen atoms into the heavier elements, and that secondary electrons were produced in the atmosphere by Compton scattering of γ rays. But soon after that, J. Clay found that the cosmic ray intensity varied with latitude, indicating that the primary cosmic rays are deflected by the Earth's magnetic field and must therefore be charged particles not photons. Clay's observation was later confirmed by many other researchers. During the period from 1930 to 1945, a variety of studies confirmed that the primary cosmic rays are mostly protons, and the secondary radiation produced in the atmosphere is primarily electrons, photons and muons. In 1948, observations with nuclear emulsions carried by balloons to near the top of the atmosphere showed that approximately 10% of the primaries are helium nuclei (α particles) and 1% are heavier nuclei of the elements such as carbon, iron, and lead.

In 1937, Pierre Auger studied the phenomenon of atmospheric ionization by cosmic rays and concluded that high-energy primary cosmic-ray particles interact with air nuclei high in the atmosphere, initiating a cascade of secondary interactions that ultimately yield a shower of electrons and photons that reach the ground level.

The term "cosmic rays" usually refers to galactic cosmic rays (GCRs), which originate from outside the solar system. However, this term also includes other classes of energetic particles in space, including solar energetic particles produced by energetic events on the Sun (called solar cosmic rays), and particles accelerated in interplanetary space.

1.3.2.2 *Primary cosmic ray composition and energy distribution*

Primary cosmic rays, originating outside the Solar system, consist of roughly 90% protons and 9% α particles (He^+, helium nuclei) with the remainder (1%) being high-energy electrons and the nuclei of heavier elements.

Primary cosmic rays have energies ranging from 1 MeV (1 MeV= 1.6×10^{-13} J) to 3×10^{20} eV (48 J) with the peak of the energy distribution at about 0.3 GeV (4.8×10^{-11} J). At 48 J, the ultrahigh energy cosmic rays have energies about 40 million times the highest particle energy that can be achieved by human-made accelerators (the Large Hadron Collider).

It is generally believed that primary cosmic rays mainly obtain their energy from supernova explosions. There is significant evidence that cosmic rays are accelerated as the shock waves from these explosions travel through the surrounding interstellar gas.

1.3.2.3 *Secondary cosmic rays and solar cosmic rays*

When cosmic rays strike the Earth's atmosphere, they collide with molecules (mainly oxygen and nitrogen) of the upper atmosphere, producing a cascade of secondary particles, known as an air shower. These secondary particles include pions, which quickly decay to produce muons, neutrinos, γ rays, and electrons and positrons as well. The latter are produced by muon decay and γ ray interactions with atmospheric molecules, as shown in Fig. 1.4. Muons are able to penetrate the atmosphere and reach the surface of the Earth, with an average intensity of ~100 /m²·s.

Solar cosmic rays (also called solar energetic particles, SEP) are energetic particles produced during the explosive energy release at the Sun and accelerated in the interplanetary space. They impact the atmosphere sporadically, with a positive correlation to solar activity. But only a small fraction of SEPs with energy around several GeV results in cascades in the atmosphere, because of their steep energy spectrum. Another kind of energetic particles is that of magnetospheric electrons which can precipitate into the atmosphere. Although they are absorbed in the upper atmosphere, the X-rays produced by these energetic electrons can penetrate down to the altitude of about 20 km. Electron precipitation is often in anti-phase with solar activity in the 11-year solar cycle.

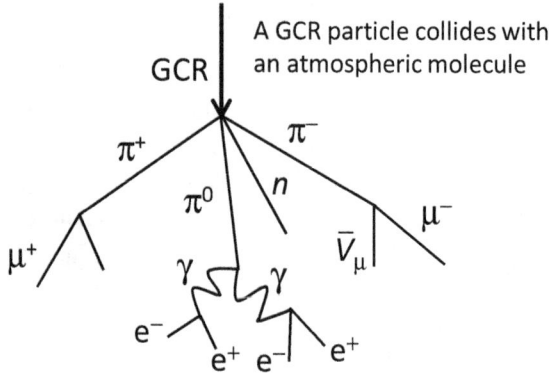

Fig. 1.4. Schematic illustration of a GCR collision with an atmospheric molecule.

1.3.2.4 *Cosmic ray flux in Earth's atmosphere*

The flux of incoming CRs at the top of the atmosphere is dependent on the solar wind, the geomagnetic field, and the energy of the cosmic rays. There are several different features of the cosmic ray flux in the Earth's atmosphere: Cosmic ray intensities vary with latitude (because of geomagnetic properties), with altitude (because of Earth's atmosphere), and with the time in the sun's magnetic activity (solar cycle).

First, the geomagnetic field deflects GCRs from their original propagation directions, giving rise to the observation that the GCR flux depends on geomagnetic location (latitude, longitude, and azimuth angle). Moving a charged particle in the geomagnetic field depends on the particle rigidity $R=cp/ze$, where c is the speed of light, e is the charge of an electron, p and z are the momentum and ionic charge number of the particle respectively. For particles with equal rigidities, they move in a similar way in a given magnetic field. Each geomagnetic latitude may approximately be characterised by a cutoff rigidity, R_C, while particles with a rigidity smaller than R_C cannot reach this latitude. Therefore only particles with a very high energy (>14 GeV) can reach equatorial regions, while low-energy particles can only arrive at high latitudes. As a consequence, the polar regions and high latitudes receive more penetrating GCR (and SEP) particles, as shown in Fig. 1.5.

Second, low-energy incoming GCR particles are absorbed in the atmosphere, while those with high energies above 1.0 GeV generate secondary particles through collisions with atomic nuclei in air (Fig. 1.4). Fig. 1.6 shows that energetic GCRs initiate nuclear-electromagnetic cascades in the atmosphere, causing a maximum in secondary particle intensity at the altitude of 15-18 km, the so-called Pfotzer maximum. Below this altitude, the cosmic rays have been significantly absorbed by the atmosphere and the ionization rate decreases. The typical positive or negative ion concentration observed is 5×10^3 cm^{-3} in the lower stratosphere, and drops to about 10^3 cm^{-3} in the troposphere [Smith and Adams, 1980].

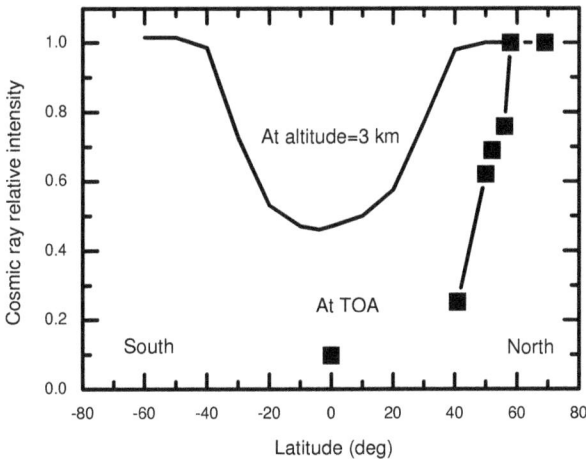

Fig. 1.5. Cosmic-ray intensity as a function of latitude: the solid line reproduces data obtained at ~3 km from the sea level (after Pomerantz [1971]); the solid squares are for the primary cosmic ray flux at the top of the atmosphere (TOA) for northern latitudes (after Van Allen and Singer [1950]). Adapted from Lu and Sanche [2001a].

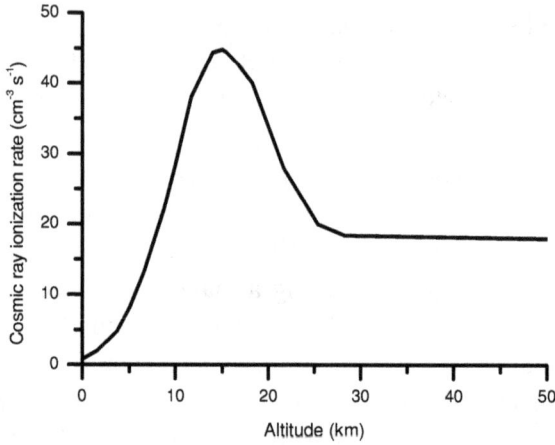

Fig. 1.6. Cosmic-ray ionization-rate variation as a function of altitude (after Hayakawa [1969], and Cole and Pierce [1965]).

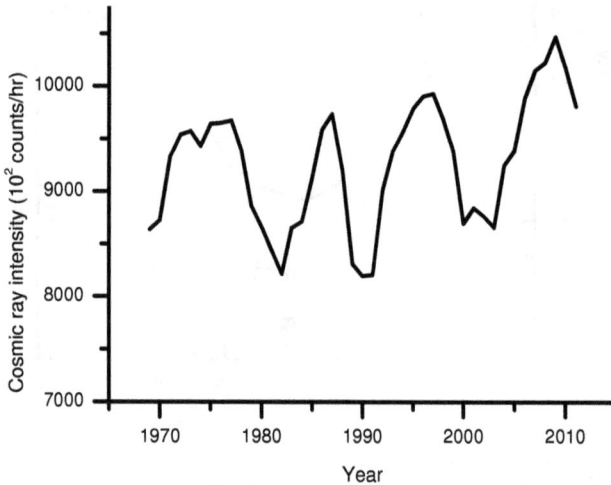

Fig. 1.7. Observed time-series cosmic ray intensities measured at McMurdo (77.9° S, 166.6° E), Antarctica during 1979-2012. Adapted from Lu [2010a, 2013].

Third, the solar cycle modulation of cosmic rays is well known and is shown in Fig. 1.7. The original theory was put forward by Parker [1965], whose basic concept is described as follows. The solar wind (the plasma of ions and electrons blowing out from the solar corona) flows nearly radially outward from the Sun and carries the heliospheric magnetic field. The cosmic rays being charged particles are affected by the interplanetary magnetic field embedded in the solar wind and therefore have difficulty reaching the inner solar system. This effect decreases the flux of cosmic rays at lower energies (≤ 1 GeV) by about 90%. Since the strength of the solar wind is not constant over the 11 year solar cycle, the intensity of cosmic rays at Earth's atmosphere is anti-correlated with solar activity. When solar activity is high, GCR flux is low, and vice versa. This effect is most significant at high latitudes; at equatorial altitudes, there is little variation in GCR rate as the geomagnetic field repels most low-energy GCR particles that might otherwise enter the atmosphere during low solar activity. In addition, the cosmic ray flux averaged over the solar cycle may have small variations with time in the decadal or centurial timescales, corresponding to the variation in total solar irradiance.

The combined effects of all of the factors mentioned contribute to the flux of cosmic rays at Earth's atmosphere. For 1 GeV particles, the rate of arrival on the Earth surface is about 10,000 /m^2·s.

1.3.3 *Terrestrial radiation*

In addition to solar radiation and cosmic rays, the heat (longwave thermal) radiation from the Earth surface also contributes to atmospheric radiation. Knowledge of this terrestrial radiation is crucial to the understanding of the *greenhouse effect* of the atmosphere, which will be discussed in Chapters 6 and 8.

1.4 Photon interactions with atmospheric molecules

The different parts of the electromagnetic radiation spectrum have very different types of interactions with atoms and molecules. The basic physical processes include *scattering, photoabsorption, photoexcitation* and *photoionization*. The probability of cross-section for each interaction depends on the energy of the photon and on the chemical composition of the target, as illustrated in Fig. 1.8.

Photons can be *absorbed* by atoms or molecules if the energy gap ΔE between two discrete quantum states of the atoms or molecules matches the photon energy $h\nu$, i.e.

$$\Delta E = h\nu = hc/\lambda, \qquad (1.1)$$

where h is the Planck constant, and ν (λ) the frequency (wavelength) of the electromagnetic radiation.

1.4.1 *Photoexcitation*

Photoexcitation is the first step in the photophysical and photochemical processes where the molecule is elevated to a state of higher energy, an excited state

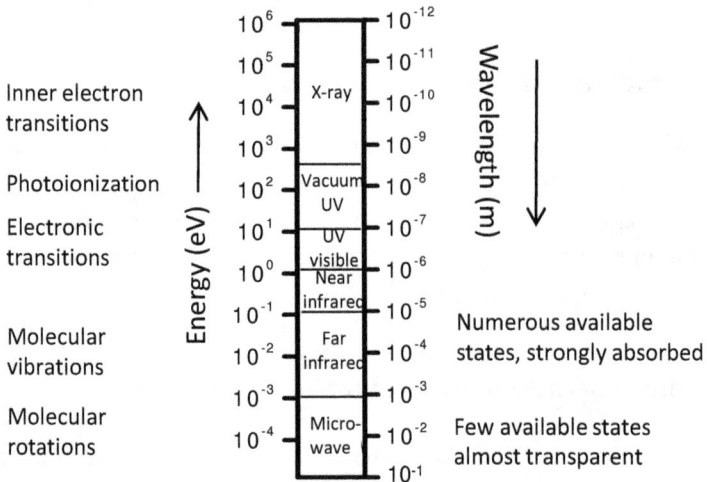

$$h\nu + AB \rightarrow AB^*. \qquad (1.2)$$

Fig. 1.8. Schematic illustration of interactions of radiation with atoms and molecules.

In molecules, there are three types of quantum energy states (levels): electronic, vibrational and rotational. Correspondingly, the absorption of a photon by a molecule may result in three types of transition (Fig. 1.8)

- electronic transitions $\Delta E=1.7-10.8$ eV (for photons in near infrared, visible, and ultraviolet wavelengths);
- vibrational transitions $\Delta E=0.04-0.4$ eV (for photons in infrared wavelengths);
- rotational transitions $\Delta E=0-0.04$ eV (for photons in microwave or millimeter wavelengths).

The energy differences between two vibrational or rotational states (levels) in molecules are much smaller than those of two electronic states (levels). Thus, the energy required for an electronic transition (excitation) is much higher than those for a vibrational or rotational transition. Besides, electronic excitations may involve changes in the vibrational and rotational energy, while vibrational excitations are usually accompanied by a change in the rotational energy.

The energy levels of a molecule are quantized, and transitions can occur only when the molecule possesses a transient dipole oscillating at the frequency of the electromagnetic radiation (the photon). Theoretically the transition probability is proportional to the square of the transient dipole moment, which is dependent on the wavefunctions of the two energy levels and on the dipole moment operator.

For rotational transitions to occur, a molecule must have a permanent electric dipole. Therefore molecules such as CO and H_2O possess rotational spectra, whereas molecules likes N_2, O_2, and CO_2 do not.

Vibrational transitions can occur only when there are changes in the dipole moment of the molecule during the molecular vibrations. Thus, N_2 and O_2 do not possess vibrational transitions. CO_2 does not have a permanent dipole, but it still possess allowed vibrational transitions as a transient dipole is generated by its asymmetric stretch and bending motions.

The absorption and emission of infrared radiation by atmospheric molecules are of special significance. The heat trap of the terrestrial radiation in the IR region by some atmospheric gases (the so-called *greenhouse gases*) leads to warming of the Earth's lower atmosphere and surface. The main greenhouse gases include H_2O, CO_2, CH_4, N_2O, O_3, and halogenated molecules such as chlorofluorocarbons (CFCs). An important example is the infrared absorption of CFCs, which have strong absorption in the so-called "atmospheric window" between 8 and 13 μm through which the majority of the Earth's blackbody radiation is emitted into outer space. In contrast, far more abundant IR-absorbing atmospheric gases such as CO_2 and H_2O do not absorb strongly in the atmospheric window. The greenhouse effect of these gases will be explored in Chapters 6 and 8.

In the case of electronic transitions, the required non-zero transient dipole moment is a result of charge redistribution generated during the electronic transitions. The electronic transition energy is typically comparable to or larger than that required to break a chemical bond. Thus, electronic transitions often lead to photochemical reactions, *e.g.*, photodissociation, which will be discussed in next section.

Absorption normally takes place in the electronic ground state, though vibrational and/or rotational states may already be excited due to thermal interactions. For studies of photochemical processes, the most important is the first excited electronic state because it has a sufficient long lifetime for further reactions. Excitations into higher states are possible, but they quickly decay into the first excited state by intramolecular or intermolecular interactions.

There are two different kinds of excited electronic states. If the excited electron and the ground state electron are still paired with different spin directions, that is, the sum S of spins of electrons in the molecule/atom (electrons are paired) is zero (S=0), the excited state is called "singlet excited state". In contrast, if the excited electron is unpaired with the ground-state electron (the spin direction of the excited electron is reversed to become identical to that of the ground state electron), the excited state is called "triplet excited state". This can occur via intersystem crossing. Except for rare molecules (*e.g.*, O_2), the ground state of molecules is usually a singlet state (S_0). The radiative decay from the lowest singlet excited state S_1 to S_0, which is called fluorescence emission, is spin-allowed and fast. The lifetimes of singlet excited states are typically short in nanoseconds. Differently, the radiative decay from the lowest triplet excited state T_1 to S_0, called phosphorescence emission, is spin-forbidden and much slower. The triplet states are therefore much long-lived and the associated photochemical reactions are readily observed.

Photochemistry is the study of chemical reactions that proceed with the absorption of light by atoms or molecules. A famous example is the photosynthesis of vitamin D with sunlight. A chemical reaction takes place only when a molecule is provided with the required activation energy, which can be provided in the form of electromagnetic energy. In a photochemical reaction, light provides the activation energy; it involves electronic reorganization initiated by electromagnetic radiation. Photochemical reactions are typical several orders of magnitude faster than thermal reactions; reactions as fast as 10^{-15} to 10^{-9} seconds are often observed.

Once a molecule (AB) is promoted into an excited electronic state (AB*) by absorption of a visible or UV photon, the excited state AB* can proceed with a number of processes:

1. AB* in a singlet or triplet excited state can decay into a lower electronic state, accompanying the emission of fluorescence or phosphorescence (AB* $\rightarrow$ AB + $h\nu'$).

2. AB* can also be deactivated (called quenched) by another species via energy transfer (AB* + CD $\rightarrow$ AB + CD*).

3. AB* can also be deactivated (quenched) by another species via electron transfer (AB* + CD $\rightarrow$ AB$^+$ + CD$^-$ or AB* + CD $\rightarrow$ AB$^-$ + CD$^+$).

4. AB* can undergo photoionization (AB* $\rightarrow$ AB$^+$ + e$^-$) or photodissociation (AB* $\rightarrow$ A + B*).

5. AB* can also proceed with chemical reaction with a reactant (AB* + CD $\rightarrow$ EF).

1.4.2 *Photodissociation*

The most common photochemical event of relevance to atmospheric chemistry is photodissociation (also called *photolysis*), which is a chemical reaction in which a molecule is broken down by photons. It is defined as the interaction of one or more photons with a target molecule. Any photon with sufficient energy can affect the chemical bonds of a molecule. Visible light or electromagnetic radiations at shorter wavelengths, such as UV light, x-rays and γ-rays, are usually involved in such reactions.

The two most important photodissociation reactions in the troposphere are the formation of the hydroxyl radical (OH$^\bullet$) and ozone. The first example of reaction is

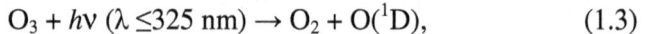

$$O_3 + h\nu \ (\lambda \leq 325 \text{ nm}) \rightarrow O_2 + O(^1D), \tag{1.3}$$

which generates an excited oxygen atom to react with a water molecule, resulting in the formation of hydroxyl radicals

$$O(^1D) + H_2O \rightarrow 2 \ OH^\bullet. \tag{1.4}$$

The OH$^\bullet$ radical plays a central role in atmospheric chemistry as it acts as a detergent via initiating the oxidation of hydrocarbons in the atmosphere.

The second example of reaction is

$$NO_2 + h\nu \ (\lambda \leq 400 \text{ nm}) \rightarrow NO + O(^3P), \tag{1.5}$$

which produces a ground-state oxygen atom that can react with an oxygen molecule to form an ozone molecule

$$O(^3P) + O_2 + M \rightarrow O_3 + M, \tag{1.6}$$

where M is a third body. This is the dominant source for formation of ozone in the troposphere.

Photodissociation of NO_2 can occur in the near UV region with a near unity of quantum yield, which is defined as the probability that this process will occur upon adsorption of a photon. But the quantum yield for the NO_2 photodissociation drops rapidly to zero as the wavelength increases to above 400 nm (the photon energy becomes too low to break the NO-O chemical bond).

The formation of stratospheric O_3 is also caused by photodissociation via the well-known Chapman reactions [Chapman, 1930]

$$O_2 + hv \ (\lambda \leq 242 \text{ nm}) \rightarrow O + O, \tag{1.7}$$

$$O + O_2 + M \rightarrow O_3 + M. \tag{1.8}$$

Since the O-O bond energy in O_2 is large (500 kJ mol^{-1}), O_2 can only be broken by more energetic UV photons with wavelengths shorter than 242 nm. This process leads to the formation of ozone in the stratosphere.

Also, photodissociation of O_3 can occur and result in the destruction of tropospheric and stratospheric ozone

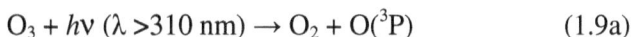

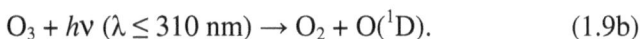

$$O_3 + hv \ (\lambda > 310 \text{ nm}) \rightarrow O_2 + O(^3P) \tag{1.9a}$$

$$O_3 + hv \ (\lambda \leq 310 \text{ nm}) \rightarrow O_2 + O(^1D). \tag{1.9b}$$

The O_3 molecule possesses a relatively small chemical bond energy (~1.1 eV), and therefore the photolysis of O_3 can even occur readily upon the absorption of visible light. The resultant oxygen atom can be in the ground state $O(^3P)$, mainly for absorbed wavelengths $\lambda > 310$ nm, or in the first excited state $O(^1D)$, for $\lambda \leq 310$ nm. In addition, there are other processes that can lead to the destruction of the stratospheric ozone layer, which will be discussed in Chapter 3.

1.4.3 *Photoionization*

Photoionization is the physical process in which an incident photon ejects one or more electrons from an atom, ion or molecule. Not every photon which encounters an atom or ion will photoionize it. The probability of photoionization is related to the photoionization cross-section, which depends on the energy of the photon and the target. For photon energies below the ionization threshold, the photoionization cross-section is near zero. With intense coherent light (e.g., pulsed lasers), multi-photon ionization may occur, but this is extremely rare in the natural atmosphere.

There are three kinds of photoionization processes: photoelectric effect, Compton scattering and pair/triplet production. The most famous one is the photoelectric effect, whose explanation for photoelectron emission from a metal surface led to the awarding of the Nobel Prize in Physics to Albert Einstein in 1921. This interaction involves the photon and a tightly bound orbital electron of an atom of the absorber. The photon gives up its energy entirely; the electron is ejected with a kinetic energy equal to the energy of the incident photon less the binding energy E_B of the electron in the atomic orbital prior to emission

$$E^e(KE) = h\nu - E_B. \tag{1.10}$$

This formula defines the photoelectric effect. Thus, the photo-ejected electron, known as photoelectron, carries information about its initial state. Photons with energies less than the electron binding energy may be absorbed or scattered but will not photoionize the atom or ion at whatever intensity below the threshold for multiple-photon ionization.

In photoelectric effect, the hole in the original electron shell (orbital) may be filled by an electron from an outer orbital. The difference in energy is emitted as either a photon of characteristic x-rays or an Auger electron.

Compton scattering is a collision between a photon of zero rest mass and an outer orbital electron of an atom. The atomic electron can be considered as a free electron, since the photon energy is much larger than the electron binding energy. After the collision, there is a fast electron and a photon of reduced energy. The scattered photon and the electron can undergo further interactions, resulting in more excitations and ionizations, until their energies are completely lost.

When the photon energy is very high (≥ 1 MeV) (*e.g.*, from cosmic ray radiation), both the photoelectric and Compton effect have lower interaction probabilities with increasing photon energy, and pair formation becomes dominant. The latter leads to the formation of an electron-position pair, in which the photon energy is distributed by the kinetic energies of the two particles as well as the atomic nucleus that receives part of the photon momentum. The resultant particles have typically high energies and can cause further excitations and ionizations along their paths.

All the above three processes result in the generation of energetic electrons that then lose energy by further excitations and ionizations of atoms and molecules, producing a large number of electrons and ions in the air (medium). Finally, the kinetic energy of electrons is completely consumed and they are captured by atoms or molecules in the air (medium).

1.5 Atmospheric ionization

In the upper atmosphere, solar radiation is the main source of ionization, while galactic cosmic rays are the main source of ionization in the stratosphere and troposphere. In addition, solar cosmic rays (also called solar energetic particles, SEP) and magnetospheric energetic electrons can also cause ionization in the upper and middle stratosphere.

1.5.1 *Ionization in the upper atmosphere*

Atmospheric ionization was first demonstrated in the upper atmosphere. It was first discovered by radio wave reflection from the ambient plasma. In 1901, Guglielmo Marconi performed an impressive experiment. He received a radio signal in St. John's, Newfoundland, that had been transmitted from Land's End, England, about 2900 km away. Because radio waves were thought to travel in straight lines, radio communication over large distances on Earth had been assumed to be impossible. Marconi's successful experiment suggested that Earth's atmosphere in some way substantially affects radiowave propagation. Marconi's discovery stimulated intensive studies of the upper atmosphere. Then, the existence of electrons in the upper atmosphere was established by experimental studies in the 1920s.

For each electron present in the upper atmosphere, there must be a corresponding positively charged ion. The electrons in the upper atmosphere result mainly from the photoionization of molecules, caused by solar radiation. As discussed above, photoionization occurs when a molecule absorbs a photon, causing the ejection of an electron from the molecule. The requirement is that the absorbed photon must have an energy sufficient to remove the electron at a bound orbital of the molecule. Thus, photons from sunlight with energies sufficient to cause ionization have usually wavelengths in the high-energy region of the ultraviolet. Most photons with these wavelengths are absorbed and filtered out by the upper atmosphere.

The photoionization process and its effects on the Earth's atmosphere are complicated and elusive. The impacts of photoionization are rather dependent on the regions in the atmosphere. Indeed, the Earth's upper atmosphere is conventionally divided into several distinct 'regions' in altitudes, according to the temperature variation and the ionization density.

The ionosphere has several characteristic regions of ionization: D region (60-90 km), E region (90-150 km) and F region (above 150 km), with free electron densities typically 10^3, 10^5 and 10^6 cm^{-3}, respectively. The primary atmospheric constitutes for ionization are N_2, O_2 and oxygen atom.

In F region, ionization is predominantly induced by EUV and X-rays to produce electrons and ions, and the main ions are O^+ and NO^+, with some N_2^+ ions. In E region, ionization is mainly induced by the 80 to 102.7 nm part of the EUV spectrum. The major observed ions in E region are O_2^+ and NO^+, with some initial N_2^+ and O^+ ions.

Ionization in D region is the most complex and least understood in the ionosphere. However, understanding of the ionization in D region will be important to explore the even more complicated and less understood physics and chemistry in the stratosphere and upper troposphere.

It was once concluded that no significant ionization existed below 80 km as the free electron density decreased sharply with decreasing altitudes and no radio reflection could be detected. This conclusion was then known to be incorrect by the 1970s. It is now known that the sharp decrease in free electron density below this altitude is due to the rapid recombination of electrons with the water clusters which are efficiently formed in this region. Some of the free electrons can also rapidly be attached to form ions that do not efficiently reflect radio waves and therefore were not detected. This was well confirmed up to the 1970s [Smith and Adams, 1980].

In D region, the primary source of ionization is the photoionization of O_2 and N_2 by soft X-rays (0.2-0.8 nm) and of NO by Lyman-α (121.6 nm)

$$O_2 + h\nu \ (0.2\text{-}0.8 \ nm) \rightarrow O_2^+ + e^-, \tag{1.11}$$

$$N_2 + h\nu \ (0.2\text{-}0.8 \ nm) \rightarrow N_2^+ + e^-, \tag{1.12}$$

$$NO + h\nu \ (121.6 \ nm) \rightarrow NO^+ + e^-. \tag{1.13}$$

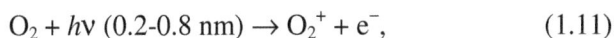

In addition, ionization by precipitating magnetospheric electrons and by galactic cosmic rays also plays some roles

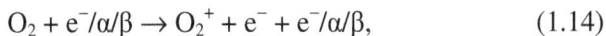

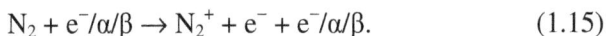

$$O_2 + e^-/\alpha/\beta \rightarrow O_2^+ + e^- + e^-/\alpha/\beta, \tag{1.14}$$

$$N_2 + e^-/\alpha/\beta \rightarrow N_2^+ + e^- + e^-/\alpha/\beta. \tag{1.15}$$

1.5.2 *Ionization in the stratosphere and troposphere*

In the stratosphere and upper troposphere (at altitudes below 50 km), solar radiation sufficiently energetic to photoionize the ambient gas has been absorbed in the atmosphere above it and cosmic rays are the only source of ionization. Near the Earth's surface, energetic particles emitted from radioactive decay (mainly from radon) are the primary source of ionization. Additionally, local ionization may be created by lightning.

The Earth's atmosphere is continuously bombarded from all directions by high-energy charged particles of GCRs. Striking the atmosphere, cosmic ray particles collide with air molecules and create cascades of secondary particles, as shown in Fig. 1.4. The GCR-air collisions are primarily due to Coulomb interactions of GCR particles with orbital electrons of air molecules, ejecting energetic electrons and producing positive ions (leaving behind electron-ion pairs). The ejected electrons usually are energetic sufficiently to cause similar multiple ionizing events. The cosmic ray particles lose a small fraction of their energies in each atomic collision and must experience a large number of collisions before their energies are completely consumed. Rarely extremely high-energy cosmic ray particles (mainly ions) can collide with the nucleus of air molecules to generate nuclear reactions. The resultant nucleus is highly unstable, undergoing radioactive decays by various channels and emitting further secondary (air nuclear) particles.

Among the secondary particles produced in GCR-air interactions, the most important one is the neutron, which penetrates deep into the Earth's atmosphere because of its charge neutrality. The neutron creates further ionizing events along its path and contributes to a major part of the atmospheric ionization in the stratosphere and upper troposphere.

In addition to cosmic rays as the main source of ionization in the lower atmosphere, the transient solar energetic particle (SEP) events (or solar cosmic rays), which are associated with eruptions on the Sun's surface, can occasionally occur and last for several hours to days with widely varying intensity (during high geomagnetic activities). The interaction mechanisms of SEP and air are the same as those of GCRs and air described above. The atmospheric ionization caused by a SEP also varies with altitude and geomagnetic field. On rare occasions, energetic electrons can also reach the upper stratosphere, and solar proton events or polar cap absorption events can also produce very intense ionization events into the middle stratosphere.

Similar to the photoionization in the upper atmosphere described in the preceding Section, primary and secondary cosmic rays ionize the nitrogen and oxygen molecules in the atmosphere, which leads to a number of chemical reactions, as further discussed later. Also, CRs are responsible for the continuous production of a number of unstable isotopes such as beryllium-10 and carbon-14 in the Earth's atmosphere.

1.6 Ion chemistry in the atmosphere below 100 km

Ion chemistry in the atmosphere below 100 km has been reviewed extensively in the literature [Ferguson *et al.*, 1979; Smith and Adams, 1980; Torr, 1985; Viggiano and Arnold, 1995]. Only a brief description will be given here.

1.6.1 *D-region (50-80 km)*

In D-region, dissociation of molecules can occur in atmospheric ionization events. Thus, the primary initial ions are N_2^+, O_2^+, and NO^+, with less O^+ and N^+. The N_2^+ ion is unstable and rapidly converted to O_2^+ via the following charge exchange reaction

$$N_2^+ + O_2 \rightarrow O_2^+ + N_2. \tag{1.16}$$

This process results in O_2^+ and NO^+ as the major positive ions in the D region. In the lower D region (near the stratopause), however, the ion composition changes from the simple atomic and molecular ions to cluster ions by the following reactions

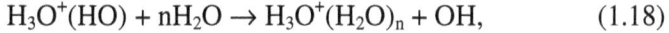

$$O_2^+(H_2O) + H_2O \rightarrow H_3O^+(HO) + O_2, \qquad (1.17)$$

$$H_3O^+(HO) + nH_2O \rightarrow H_3O^+(H_2O)_n + OH, \qquad (1.18)$$

and

$$NO^+(H_2O)_{n+1} + H_2O \rightarrow H_3O^+(H_2O)_n + HNO_2. \qquad (1.19)$$

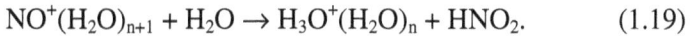

The resultant hydrated hydronium ions $H_3O^+(H_2O)_n$ (also called water cluster ions) are quite stable and the dominant ions. This change is due to the increased ambient pressure.

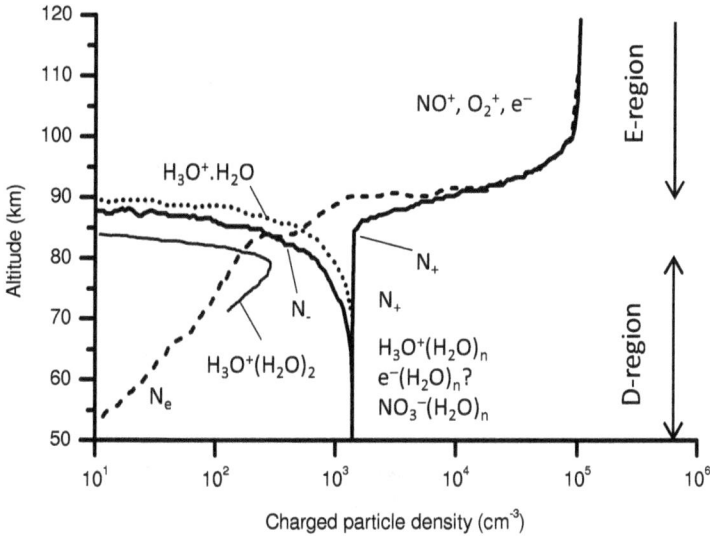

Fig. 1.9. The profiles of the free electron density N_e, the negative ion density N_- and positive ion density N_+ in D-region of the atmosphere. Adapted from Smith and Adams [1980]. Note that transient electrons trapped in water clusters, $e^-(H_2O)_n$, were not observable in conventional experimental methods prior to 1987, and were not discussed in the literature.

An interesting feature of the D region is also the occasional appearance of the so-called 'noctilucent clouds' in the upper D region (at or near the mesopause), despite the very minor atmospheric constituent of water vapor in these altitudes [Smith and Adams, 1980]. It has been shown that the ions in these clouds are heavily clustered with water molecules (with up to 20 H_2O molecules) and that the $H_3O^+(H_2O)_n$ clusters are prominent. The cluster ions play an important role as nucleation sites for droplet (aerosol) formation.

The negative ion composition in the D region is rather poorly understood and highly speculative. In the upper D region, free electrons are the dominant negatively-charged species, and the loss of electrons is believed to be controlled by dissociative recombination processes

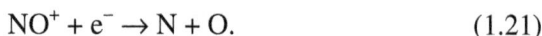

$$O_2^+ + e^- \rightarrow O + O, \tag{1.20}$$

$$NO^+ + e^- \rightarrow N + O. \tag{1.21}$$

With decreasing altitudes, however, free electrons are gradually replaced by negative ions until below 60 km (Fig. 1.9). The primary negative ion in the D region is mainly O_2^- which results from a three-body electron attachment process to O_2, as will be discussed in Sec. 2.4 in Chapter 2. Atomic O^- is the much minor negative ion of the D region, formed by *dissociative electron attachment* (DEA) to O_3

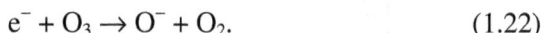

$$e^- + O_3 \rightarrow O^- + O_2. \tag{1.22}$$

This reaction has a small rate constant of 9×10^{-12} cm^3 s^{-1} at 300 K [Van Doren *et al.*, 2003].

The O_2^- ion is unstable and readily converted into the more stable ions through a series of ion-molecule reactions with the D-region minor molecular neutral constituents. The reactions eventually lead to the formation of more complex anions such as CO_3^-, NO_2^- and NO_3^-. Since NO_3 has a very large electron affinity of 3.937 eV [Miller, 2003], NO_3^- is very stable and is the most abundant negative ion in the D region. These transformation reactions are, to a large extent, controlled by the minor neutral constituents such as O, O_3, NO and H, which are the limiting factors. The ions are further converted into hydrated negative ions. Thus, the processes can be expressed as

$$O_2^-, O^- \rightarrow NO_3^-(H_2O)_m, CO_3^-(H_2O)_m. \qquad (1.23)$$

And then the loss of ionization in the ion-ion plasma results from the ion-ion neutralization process:

$$H_3O^+(H_2O)_n + NO_3^-(H_2O)_m / CO_3^-(H_2O)_m \rightarrow \text{products.} \qquad (1.24)$$

1.6.2 *Stratosphere and troposphere (below 50 km)*

Since cosmic rays are highly energetic particles, they ionize all gaseous molecules with approximately the same efficiency. Thus, the primary ions are N_2^+, O_2^+, N^+, and O^+, resulting from the ionization of the most abundant atmospheric gases N_2 and O_2.

As mentioned above, the water cluster ions $H_3O^+(H_2O)_n$ are the dominant positive ions in the lower D-region (near the stratopause). In the stratosphere and troposphere, these cluster ions are formed rapidly, on the order of 10^{-3} s, and are quite stable [Viggiano and Arnold, 1995]. In the stratosphere, the number n peaks around 3 to 5, while n increases in the troposphere where the H_2O concentration is large. The conversion time to form the cluster ions $H_3O^+(H_2O)_n$ is much shorter than the ion residence time in 100 to 10000 s. Thus, the process leading to the $H_3O^+(H_2O)_n$ ions is so rapid that they can be considered the starting ions for subsequent positive ion chemistry.

In the middle and lower stratosphere (near the tropopause), acetonitrile-containing ions $H^+(CH_3CN)_m(H_2O)_n$ and acetone-containing ions, $H^+((CH_3)_2CO)_m(H_2O)_n$ are also believed to exist. However, it is worthwhile to note that the concentration of H_2O is approximately 1000 times the concentrations of the other trace gases. Thus, the water cluster ions $H_3O^+(H_2O)_n$ are probably the most important positive ions in most part of the stratosphere and troposphere.

In the lower troposphere near ground level, the amine-containing ions are the dominant ions. In particular, protonated ammonia ions, $H^+(NH_3)_m(H_2O)_n$, become very abundant. In addition, other important species include the water cluster ions containing pyridine and ammonia derivatives.

In the stratosphere, the free electron density is very low and most of the negative charge is in the form of negative ions, as shown in Fig. 1.9. There is a sharp reduction in electron density in the decreasing altitude range of 80 to 50 km, whereas the density of negative ions rises to be approximately equal to that of positive ions (at about 2×10^3 cm^{-3}).

For the general Earth's atmosphere, the primary negative ion species is O_2^- resulting from the three-body electron attachment to O_2. The reaction rate coefficient for this electron attachment is very small. However, O_2 is the much more abundant atmospheric gas than any other gas that has a significant electron affinity to attach an electron. Thus, O_2^- is the primary negative ion species in the general atmosphere.

With decreasing altitudes, the primary ions are quickly converted into more complex anions CO_3^- and NO_3^- and their hydrates $CO_3^-(H_2O)_m$ and $NO_3^-(H_2O)_m$ on the order of 10^{-3} s [Viggiano and Arnold, 1995]. The $CO_3^-(H_2O)_m$ ions are further converted into $NO_3^-(H_2O)_m$ by reaction with nitrogen oxides. This conversion occurs on the order of 1 s in the stratosphere and considerably faster in the troposphere. Below 50 km, there are only two families of negative ions, NO_3^-- and HSO_4^--containing ions. The NO_3^--containing cluster ions are the dominant species for most of this altitude region, except at altitudes between 30 and 40 km where the HSO_4^--containing ions are dominant. In the upper stratosphere, the NO_3^--containing ions observed are mainly in the form of $NO_3^-(H_2O)_n$, while in the mid-stratosphere and below they can be in the form of $NO_3^-(HNO_3)_m(H_2O)_n$ with m=1~3 formed by ligand switching reactions with HNO_3. But due to the presence of large concentrations of water vapor, the equilibrium will tend to the right side of the following reaction

$$NO_3^-(HNO_3)_m \Leftrightarrow NO_3^-(H_2O)_n. \qquad (1.25)$$

Thus, the hydrated NO_3^- ions are believed to be the dominant negative ions in most part of the stratosphere and troposphere [Smith and Adams, 1980].

The HSO_4^--containing ions are mainly in the form of $HSO_4^-(H_2SO_4)_m(HNO_3)_n$, where the number m is relatively large with a peak at about 3 when HSO_4^- core ions are dominant at 30-40 km, and m=0 and n=1 or 2 mainly when NO_3^- core ions are dominant. It is formed by the following reactions

$$NO_3^-(HNO_3)_m + H_2SO_4 \rightarrow HSO_4^-(HNO_3)_n + HNO_3, \quad (1.26a)$$

$$HSO_4^-(HNO_3)_n + mH_2SO_4 \rightarrow HSO_4^-(H_2SO_4)_m(HNO_3)_n. \quad (1.26b)$$

However, again due to the relative high concentrations of water vapor in the lower atmosphere, the equilibrium of the following reaction would tend to shift to the right side

$$HSO_4^-(acid)_n + nH_2O \Leftrightarrow HSO_4^-(H_2O)_n + acid. \quad (1.27)$$

This returns the HSO_4^- hydrates to be the dominant HSO_4^--containing species.

Although the NO_3^-- and HSO_4^--containing water cluster ions are the most abundant species in the stratosphere and troposphere, other less abundant negative ions such as Cl^-, HCO_3^- and NO_2^- as well as their hydrates are also observed. In addition, the core ions of $CH_3SO_3^-$ (methyl sulfonate) and $C_3H_3O_4^-$ (malonate), resulting from the deprotonation of methanesulfonic and malonic acids also exist near the ground. Additional ligands such as HCl, H_2O, and HOCl are also observed to be clustered with both NO_3^- and HSO_4^- core ions.

Although it appears that NO_3^- hydrates are the terminal negative ions of the lower atmosphere below 50 km, the reaction paths leading to NO_3^- are somewhat uncertain. The proposed effective NO_3^- formation reactions include the following [Ferguson *et al.*, 1979]

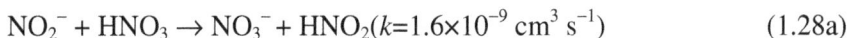

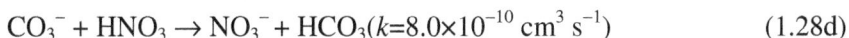

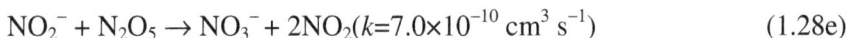

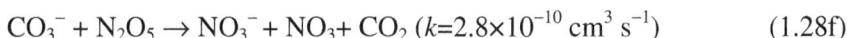

$$NO_2^- + HNO_3 \rightarrow NO_3^- + HNO_2 (k=1.6\times10^{-9}\ cm^3\ s^{-1}) \quad (1.28a)$$

$$Cl^- + HNO_3 \rightarrow NO_3^- + HCl\ (k=1.6\times10^{-9}\ cm^3\ s^{-1}) \quad (1.28b)$$

$$Cl^- + N_2O_5 \rightarrow NO_3^- + ClNO_2\ (k=9.4\times10^{-10}\ cm^3\ s^{-1}) \quad (1.28c)$$

$$CO_3^- + HNO_3 \rightarrow NO_3^- + HCO_3 (k=8.0\times10^{-10}\ cm^3\ s^{-1}) \quad (1.28d)$$

$$NO_2^- + N_2O_5 \rightarrow NO_3^- + 2NO_2 (k=7.0\times10^{-10}\ cm^3\ s^{-1}) \quad (1.28e)$$

$$CO_3^- + N_2O_5 \rightarrow NO_3^- + NO_3 + CO_2\ (k=2.8\times10^{-10}\ cm^3\ s^{-1}) \quad (1.28f)$$

The NO_3^- ion, once formed, is very stable, and its reactions with any neutral atmospheric trace gases are not expected to occur. In the 1970s, the reaction NO_3^- with ozone was proposed as a possible sink for stratospheric ozone by Ruderman *et al* [1976]

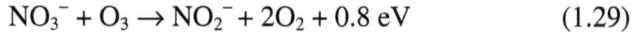

$$NO_3^- + O_3 \rightarrow NO_2^- + 2O_2 + 0.8 \text{ eV} \tag{1.29}$$

The attempt was to explain the observed small solar cycle variation of the ozone concentration. However, the rate constant of the reaction was measured to be less than 10^{-13} cm^3 s^{-1}, too small for the reaction to be significant in the stratosphere [Fehsenfeld *et al.*, 1976].

Table 1.1. Reaction rate constants for reactions of anions with O_3 at 300 K. Based on William *et al.* [2002].

Reaction	Products	Rate constant k (cm^3 s^{-1})
$SF_6^- + O_3 \rightarrow$	$O_3^- + SF_6$	2.2×10^{-10}
$O^- + O_3 \rightarrow$	$O_3^- + O$	1.4×10^{-9}
	$O_2^- + O_2$	3.0×10^{-10}
	$e^- + 2O_2$	$<5.0 \times 10^{-12}$
$O_2^- + O_3 \rightarrow$	$O_3^- + O_2$	1.3×10^{-9}
$OH^- + O_3 \rightarrow$	$O_3^- + OH$	1.3×10^{-9}
	$HO_2^- + O_2$	1.0×10^{-10}
	$O_2^- + HO_2$	3.0×10^{-11}
$NO_2^- + O_3 \rightarrow$	$NO_3^- + O_2$	1.8×10^{-10}
	$O_3^- + NO_2$	2×10^{-12}
$CO_4^- + O_3 \rightarrow$	$O_3^- + CO_2 + O_2$	4.3×10^{-10}
	$CO_3^- + 2O_2$	3.0×10^{-11}

William *et al.* [2002] studied the reactions of negative ions with O_3 at 300 K in a selected-ion flow tube coupled to a novel ozone source. It was shown that the reactions proceed primarily through charge transfer and oxygen atom transfer. They found that O^-, O_2^-, and OH^- are highly reactive with O_3, with large reaction rate constants that are approximately equal to the thermal energy capture rate constant. The negative ions NO_2^-, CO_4^-, SF_6^- and PO_2^- are less reactive, reacting at approximately 20-50% of the thermal capture rate. The NO_3^-, CO_3^-, PO_3^-, CF_3O^-, F^-, Cl^-, and Br^- anions are found to be unreactive with rate constants

$<5\times10^{-12}$ cm^3 s^{-1}, except the Γ ion, which was observed to cluster with O$_3$ with a rate constant of approximately 1×10^{-11} cm^3 s^{-1}. All of these anions are found to be unreactive with O$_2$ with rate constants $< 5\times10^{-13}$ cm^3 s^{-1} within the detection limits of their experiments. The listed reaction rate constants are given in Table 1.1 [William *et al.*, 2002].

1.6.3 *Ultrafast reactions with prehydrated electrons*

It must be noted that the existence of ultrashort-lived electrons trapped in water clusters could not be observed until the advent of femtosecond (fs, 1fs=10^{-15} s) time-resolved laser spectroscopy in 1987. Thus, it is not surprizing that the presence and reactivity of prehydrated and hydrated electrons, e$^-$(H$_2$O)$_n$, as major radicals of radiolysis of water, are completely missing in the conventional understanding of the negative ion chemistry of D-region and below (reviewed in the preceding Section), though it has generally been agreed that the stable hydrated hydronium ions H$_3$O$^+$(H$_2$O)$_n$ (another major radiolysis product of water) are the dominant positive ions. Note that ultrafast electron transfer reactions of molecules involving the trapped electron can effectively occur within timescales of sub-picosecond to a few picoseconds (10^{-13} to 10^{-12} s) [Lu, 2010a, 2010c], orders of magnitudes faster than the reactions leading to the formation of metastable or stable positive and negative ion species (clusters). This point is critical to understanding of atmospheric reactions of molecules, especially halogen-containing molecules, with the presence of water/ice under ionizing radiation, which will be discussed in Chapters 4 and 5.

1.7 Concluding remarks

It is well known that there are photons from the eye-seeing sunlight, which initiate photochemical processes in the atmosphere. On the other hand, there also exist many invisible particles such as electrons and other charged particles arising from cosmic rays in the atmosphere. The presence of the latter can also lead to many atmospheric processes (reactions), which are relatively not well known.

In particular, it is generally agreed that the charged particle chemistry of the stratosphere and troposphere is not yet well understood, and the negatively charged particle chemistry is even more speculative than the positively charged particle chemistry of these regions [Smith and Adams, 1980; Torr, 1985]. It is interesting to remark that the positive ion chemistry of the stratosphere and troposphere is quite similar to the well-established radiation physics and chemistry of liquid water under ionizing radiation. In both cases, the hydrated hydronium ions $H_3O^+(H_2O)_n$ are the dominant positive ion species [Hart and Anbar, 1970; Buxton *et al.*, 1988; Lehnert, 2008]. However, the current knowledge about the negatively charged particle chemistry of the atmosphere is very different, where the reactivity of prehydrated and hydrated electrons, $e^-(H_2O)_n$, other major radicals of radiolysis of water [Hart and Anbar, 1970; Lu, 2010a, 2010c], is completely missing in the current chemistry of the stratosphere and troposphere. The ultrafast reactions of *short-lived or ultrashort-lived radicals* generated from radiolysis of water with much less abundant trace molecules may play an important or key role in some atmospheric processes. This will be explored in Chapters 4 and 5.

Chapter 2

Interactions of Electrons with Atmospheric Molecules

2.1 Introduction

As discussed in Chapter 1, there are numerous electrons produced from photoionization in the upper atmosphere and cosmic-ray ionization in the stratosphere. Therefore, it is important to understand not only photochemical processes but also electron-induced processes of molecules, which have relevance to the atmosphere. In this Chapter, a description of the basic processes of electron interactions with molecules is given in Sec. 2.2. It is followed by a description of *negative ion resonances* of molecules in Sec. 2.3, in which an important electron-molecule reaction, the so-called *dissociative (electron) attachment* (DA/DEA), is emphasized. Subsequently, examples of anion resonances in interactions of important atmospheric molecules with electrons are given in Sec. 2.4. This chapter ends with a brief summary in Sec. 2.5.

2.2 Electron interactions with molecules

Similar to photon interactions with molecules, electrons resulting from atmospheric ionization can have strong interactions with molecules, including elastic and inelastic electron scattering, electron capture, rotational, vibrational and electronic excitations, ionization and dissociation of molecules. As a free electron collides with a molecule, a number of processes may occur. If the electron loses a part of its kinetic energy to the excitation of internal degrees of freedom of the molecule, the process is called the inelastic collision. If there is no energy transfer to the internal motion of the molecule, the process is called the elastic collision. In an elastic collision, the electron may lose some energy due to momentum transfer. But since this energy loss is proportional to the mass ratio of the electron to the molecule, it is far smaller than the energy loss due to an excitation of internal molecular degrees of freedoms and is therefore negligible.

The scattering processes can be expressed as

$$e^- + AB \rightarrow AB + e^- \text{ (elastic scattering)} \tag{2.1}$$

$$e^- + AB \rightarrow AB^* + e^- \text{ (inelastic scattering)} \tag{2.2}$$

In inelastic scattering, AB may become a vibrationally or electronically excited state AB*.

The electron-molecule interaction may lead to the ionization of the molecule in several different channels, producing ions and electrons

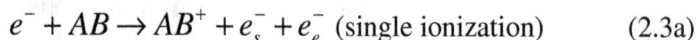

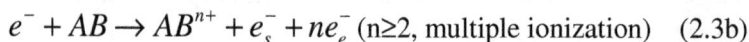

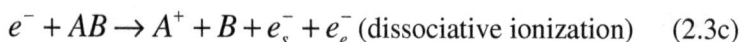

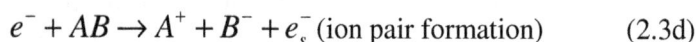

$$e^- + AB \rightarrow AB^+ + e_s^- + e_e^- \text{ (single ionization)} \tag{2.3a}$$

$$e^- + AB \rightarrow AB^{n+} + e_s^- + ne_e^- \text{ (n≥2, multiple ionization)} \tag{2.3b}$$

$$e^- + AB \rightarrow A^+ + B + e_s^- + e_e^- \text{ (dissociative ionization)} \tag{2.3c}$$

$$e^- + AB \rightarrow A^+ + B^- + e_s^- \text{ (ion pair formation)} \tag{2.3d}$$

where e_s^- stands for the scattered electron and e_e^- the ejected electron. Moreover, the ions or neutral fragments may be in excited states.

Electron impact may also lead to dissociative excitation of molecules

$$e^- + AB \rightarrow A + B^* + e_s^- \text{ (dissociative excitation)} \tag{2.4}$$

where the fragment atoms or molecules may be electronically, vibrationally, and/or rotationally excited.

Taking N_2 as an example, we can see that electron impact can lead to interesting results. Chemists have traditionally considered N_2 as a gas particularly difficult to dissociate, partially due to the large N-N bond energy (9.8 eV) and the absence of repulsive potential surfaces that can be reached in the thermochemical experiments involving only the ground state of N_2. However, a rather different picture of N_2 dissociation was obtained from electron impact experiments. N_2 can in fact be dissociated easily into atomic states and can be ionized by inelastic collision with electrons having energies less than the ionization potential (15.6 eV) of N_2. The electron impact can lead to simple dissociation

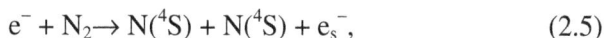

$$e^- + N_2 \rightarrow N(^4S) + N(^4S) + e_s^- , \qquad (2.5)$$

or dissociative excitation

$$e^- + N_2 \rightarrow N^* + N + e_s^- , \qquad (2.6)$$

or dissociative ionization

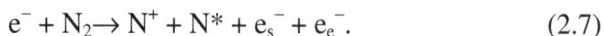

$$e^- + N_2 \rightarrow N^+ + N^* + e_s^- + e_e^- . \qquad (2.7)$$

All the processes (2.5)-(2.7) can take place effectively in electron impact with N_2 [Christophorou *et al.*, 1984].

   In addition to the above processes, electron-molecule interactions also include *dissociative electron attachment* (DEA) of molecules to low-energy electrons, which will addressed in next Section.

2.3 Negative ion resonances in electron-molecule interactions

In the context of electron collisions with atoms and molecules, resonance scattering is an interesting and important phenomenon, especially at low electron energies (0-20 eV) [Schultz, 1973a, 1973b]. Resonances in electron-molecule scattering have been well observed since the 1960s and extensively reviewed [Schultz, 1973b]. Their existence was first inferred from characteristic peaks in electron scattering cross sections of gas-phase atoms and molecules. Resonance scattering involves the formation of a *transient negative ion* (TNI) state, called a negative ion resonance (NIR), of the atom or molecule, which is a discrete state of the target plus electron. TNI is formed only at specific electron energy (the resonance energy), at which the incident electron is temporarily trapped in the vicinity of the target. The electron is trapped at a quasi-bound or virtual atomic or molecular orbital with a typical lifetime of 10^{-15} s to within 10^{-10} s

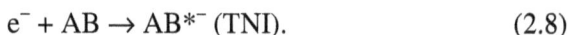

$$e^- + AB \rightarrow AB^{*-} \text{ (TNI)}. \qquad (2.8)$$

Negative ion resonances of molecules are traditionally classed into two categories, *shape resonances* and *Feshback resonances*. For a shape resonance, the energy of the TNI is above that of the parent state (either electronic ground or excited) from which it derives. In contrast, the energy of the TNI is below that of the parent state for a Feshback resonance.

If the incident electron is temporarily trapped at a previously unoccupied orbital of the ground-state neutral atom or molecule, the TNI is called the *single-particle shape resonance*. In this case, the incident electron is trapped in a potential well that results from the interaction between the electron and the neutral molecule in its electronic ground state (the 'parent' state). When the electron approaches the atom or molecule with a large collisional angular momentum, the electron feels a repulsive centrifugal potential. As this repulsive potential is combined with the static attractive interaction between the electron and the atom or molecule, a potential barrier is created. It is the shape of this potential barrier that results in the transient trapping of the electron. In this type of resonances, the potential-energy surface / curve of the TNI is above that of the neutral molecule. Thus, the TNI can decay via autodetachment into its parent state with emission of a free electron. The decay may also lead to the formation of vibrationally excited states of the molecule.

Shape, single-particle TNI resonances usually lie at low energise (0-4 eV) [Schultz, 1973b], and they decay by autodetachment preferentially to their parent state with lifetimes ranging from 10^{-15} s (fs) to 10^{-10} s, or by dissociative attachment if energetically allowed. Shape resonances often occur for diatomic molecules such as N_2 and O_2 as well as polyatomic molecules.

Another type of shape resonances is the *core-excited shape resonances*. They are similar to the single-particle shape resonances, except that they involve an electronically excited state rather than the electronic ground state of the neutral molecule. The TNI has an energy above the energy of the parent state, the electronically excited neutral molecule, and it decays by autodetachment or if energetically allowed by dissociative attachment.

Similar to the case of shape resonances, there are two types of Feshback resonances: *vibrational Feshback resonances* (also called nuclear-excited Feshback resonances) and *core-excited Feshback resonances*. The former involve the initial coupling of the incident electron's kinetic energy to the vibrational excitation of the neutral molecule. The energy of the TNI is below that of the parent state that is a vibrationally excited but electronic ground state of the target molecule. The latter has a positive electron affinity, and hence the attached electron can be trapped in a bound state of the parent molecule. Note that a vibrational Feshback resonance cannot decay into its parent state, though the decay into the vibrational ground state of the neutral molecule (the nonparent state) is possible. However, the latter involves a change in configuration, and therefore the TNIs in Feshback resonances often have long lifetimes and their cross sections are large. For many polyatomic molecules, the cross sections of vibrational Feshback resonances are even larger than 10^{-14} cm^2 with their maximum values at thermal electron energies.

Core-excited Feshbach resonances are very similar to vibrational Feshbach resonances, except that the former occur when the incident electron is transiently trapped in a bound state of the electronically excited state of the molecule. The energy of the TNI is below that of the parent state, that is, the electronically excited neutral molecule has a positive electron affinity. The lifetimes of core-excited Feshbach resonances are also typically long.

Once a TNI of a molecule is formed, there are a variety of decay channels. The electron in a TNI can autodetach with a finite lifetime τ_a (related to the width Γ of the negative ion resonance), resulting in the neutral molecule in a vibrationally excited state (resonance inelastic scattering) or the ground state (resonance elastic scattering). Alternatively, the TNI can dissociate into a stable negative ion and a neutral species. This is called the *dissociative (electron) attachment* (DA or DEA). The main decay processes of TNI include

$$AB^{*-} \rightarrow AB^* \, (AB) + e_s^- \text{ (autodetachment)} \tag{2.9}$$

$$AB^{*-} \rightarrow A + B^- \text{ (dissociative electron attachment)} \tag{2.10}$$

$$AB^{*-} \rightarrow AB^- + \Delta E \text{ (stabilization)} \tag{2.11}$$

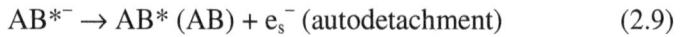

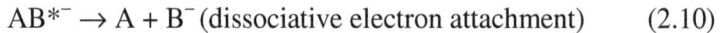

The last channel (stabilization) can occur only for the neutral molecule (the parent state) with a positive electron affinity. During its lifetime, the TNI may transfer its energy to another molecule (e.g., by collisions) or to the surrounding environment, and therefore becomes stabilized.

The autodetachment lifetime τ_a varies over a wide range from less than one vibration period (10 fs) to the μs-ms scale for larger molecules such as SF_6 and many perfluorinated compounds, depending on the resonance energy and the molecular size. These metastable negative ions are typically formed at narrow resonances near zero eV where the dissociation channel cannot operate.

The DEA channel typically occurs on a time scale of 10 fs to a few ps. This channel can occur only if (1) the lifetime of the TNI is comparable to the vibrational period of the molecule, (2) the TNI is dissociative in the Franck-Condon region, and (3) at least one of thermodynamically stable negatively-charged fragments of the molecule exists (the fragment has a positive electron affinity). Satisfying these conditions, the TNI may dissociate into a stable negative ion and a neutral fragment in either the ground state or the excited state. DEA, generally competitive with autodetachment, is very effective for many molecules, and this process will be addressed in more detail in next Section.

2.3.1 *Dissociative electron attachment (DEA) in the gas phase*

DEA occurs when a low-energy (0-20 eV) free, unbound electron resonantly attaches to a molecule to form a transient anion state, which then dissociates into a neutral and an anionic fragment. The physical process of DEA has been reviewed comprehensively by Schultz [1973b], Christophorou *et al.* [1984], Illenberger [1992], Sanche [1995] and Chutjian *et al.* [1996]. The DEA process can be expressed as

$$e^- + AB \rightarrow AB^{*-} \rightarrow A + B^-. \tag{2.12}$$

This reaction can be illustrated by the Born-Oppenheimer potential energy diagram as schematically shown in Fig. 2.1, which displays the potential energy curves (PECs) of the neutral molecule (AB) and of the TNI resonance (AB^{*-}). Since many features of TNI resonances of a molecule are associated with the nuclear motions of the atoms in the molecule itself, we will introduce the case of diatomic molecules for simplicity. For a diatomic molecule, only the nuclear separation R varies as the molecule vibrates, and the width Γ of the resonance state is itself a function of the variable R.

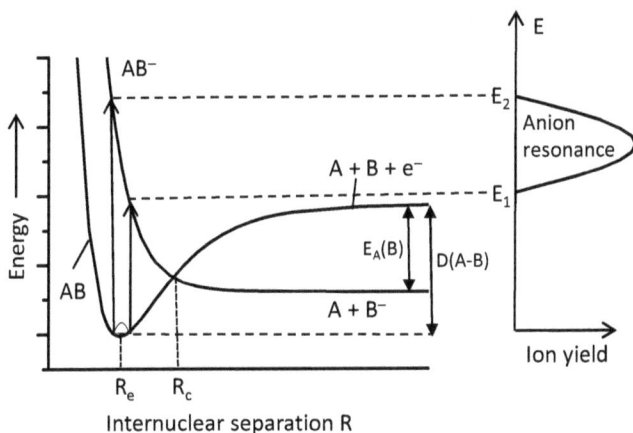

Fig. 2.1. Schematic diagram illustrating dissociative electron attachment (DEA) of a low-energy free electron to a molecule AB.

Since the nuclear motion of a molecule is much slower than the electronic transition, the molecule wave function can be expressed as a product of electron coordinate r and nuclear coordinate R, $\psi(r, R)\Theta(R)$. As shown in Fig. 2.1, electronic transitions between two electronic levels, which take place much faster than the nuclear motion in a molecular vibration, can be represented on a potential energy surface diagram by a vertical coupling. The Franck-Condon overlap factors can approximately be calculated for the vibrational-state distribution as overlap integrals in a single variable R.

According to the Frank-Condon principle, transition from the continuum state (AB + e⁻) to AB*⁻ are only possible within the energy range of E_1 to E_2. Once the TNI is formed, it may either dissociate into A and B⁻ or decay into the neutral molecule with re-emission of the attached electron. The two potential energy curves cross at an internuclear separation R=R_c. At R>R_c, the autodetachment of the electron is energetically not allowed.

Based on the expression derived by O'Malley [1966, 1967], the DEA cross section σ(E) in its simplest form can be expressed as

$$\sigma(E) = \sigma_0(E)P(E), \tag{2.13}$$

and the electron attachment probability $\sigma_0(E)$

$$\sigma_0(E) = \frac{h^2 g}{2m_e} \left[\frac{\Gamma_a}{E\Gamma_d} \right] |\chi|^2. \tag{2.14}$$

Here, $|\chi|^2$ is the Franck-Condon (FC) factor, Γ_a the autodetachment width, Γ_d the dissociation width, and g a statistical factor. P(E) is the survival probability P of the TNI (AB*⁻) state against autodetachment

$$P(E) = \exp\left(-\int_{R(E)}^{R_c} \frac{\Gamma_a(R)dR}{\hbar v(R)} \right), \tag{2.15}$$

where for a TNI AB*⁻, R(E) is the internuclear distance at energy E at which the electronic transition occurs, R_c the A-B distance at which the AB*⁻ and AB potential energy curves cross, and v(R) is the classical velocity of the dissociating anion.

Eqs (2.13)-(2.15) are approximate solutions to O'Malley's original formulation, and are instrumental to understanding the DEA process. However, more accurate solutions can be obtained from other theoretical methods, which include all parameters that must be incorporated into Eq. (2.15). For example, the R-matrix formalism has been developed to calculate the DEA cross sections by LeDourneuf *et al.* [1979] and Fabricant [1990, 2007]. In the R-matrix approach, one does not need to assume analytical expressions of the energy level shift and width functions; instead, the complete energy dependence is contained in the expression for the continuity of the logarithmic derivative of the electronic wave function at the R-matrix radius. Excellent results of numerical calculations of DEA for molecules with resonances near zero eV have been obtained by Fabrikant and co-workers [Fabrikant, 1994, 2007; Wilde *et al.*, 1999].

The DEA may not always take place directly via the purely repulsive potential energy surface as shown in Fig. 2.1. In fact, vibrational or electronic rearrangements or predissociation in the TNI may occur prior to its dissociation [Oster *et al.*, 1989]. These could result in different energy partition among the dissociation fragments. However, the energy position and width of the dissociated ion field essentially reflects the attachment energy in the Frank-Condon region.

The energy relationship for dissociation electron attachment is given from Fig. 2.1 by

$$E= D(A\text{-}B) - E_A(B) + E^* = \Delta H_0 + E^*, \qquad (2.16)$$

where $D(A\text{-}B)$ is the bond dissociation energy in the neutral molecule AB, $E_A(B)$ the electron affinity of B, E^* the excess energy of the process, and E in the electron energy range of E_1 to E_2. Here, $\Delta H_0 = D(A\text{-}B) - E_A(B)$ is the minimum heat of reaction for the process. Thus, the total excess energy for a given incident electron energy is given by

$$E^*=E-\Delta H_0. \qquad (2.17)$$

In polyatomic molecules, E^* is usually shared among the different degrees of freedom (translational and internal motion energy of the fragments).

2.3.2 *DEAs in liquid and solid phases*

It has been well established that DEAs occur in all states of matter, including gas phase, liquid phase and solid phase, and the medium introduces significant changes in the energetics of their formation, as well as in their other physical properties such as their lifetimes and decay channels [Christophorou *et al.*, 1984; Illenberger, 1992; Sanche, 1995].

With regard to the energetics of electron interactions in various media, the polarization energy plays a critical role. Negative ion resonances are generally shifted to lower electron energies due to the polarization energy of the medium, and DEA cross sections can be modified significantly, dependent on the DEA resonance energies in the gas phase [Sanche *et al.*, 1995; Fabrikant *et al.*, 1997; Lu and Sanche, 2003; Wang and Lu, 2007; Wang *et al.*, 2006, 2008a, 2008b].

2.4 Examples of negative ion resonances of atmospheric molecules

2.4.1 *Nitrogen (N₂)*

In gaseous N_2, the electron scattering cross sections exhibit a *shape* resonance of symmetry $^2\Pi_g$ around 2.3 eV, which decays by autodetachment into vibrational states of N_2 (Fig. 2.2)

$$e^-(\sim 2.3\ eV) + N_2(^1\Sigma_g) \rightarrow N_2^-(^2\Pi_g) \rightarrow N_2(v=1,2,\ldots) + e^- \qquad (2.18)$$

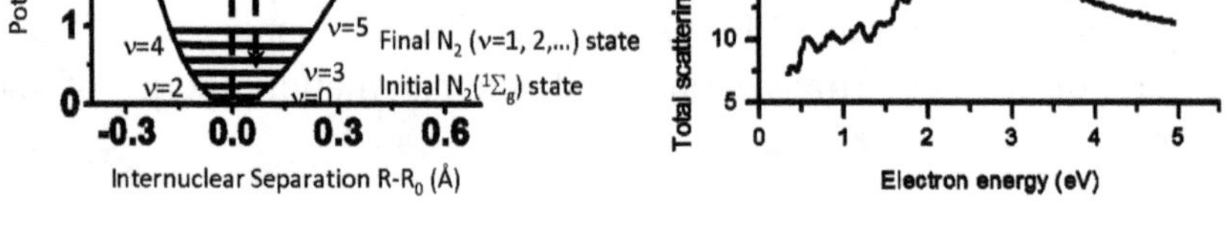

Fig. 2.2. (a) Potential energy curves for $N_2(^1\Sigma_g)$ and $N_2^-(^2\Pi_g)$ states involving in electron attachment to N_2. (b) The absolute total cross section for scattering of electrons on N_2 (Shultz, 1973b).

DEA at this electron energy does not occur. The total vibrational cross section of this resonant scattering, leading to excitations of the v=1 to 5 levels of N_2, is very large, which is about 5×10^{-16} cm^2 in the gas phase [Shultz, 1973b].

An interesting observation which may have implications for atmospheric processes is that electron trapping to N_2-adsorbed H_2O ice occurs effectively via electron transfer from the $N_2^- (^2\Pi_g)$ shape resonance at ~ 1.0 eV to a preexisting trap in ice

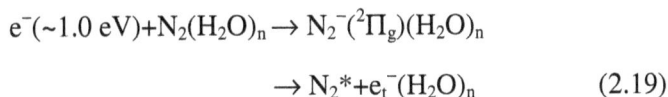

$$e^-(\sim1.0 \text{ eV})+N_2(H_2O)_n \rightarrow N_2^- (^2\Pi_g)(H_2O)_n$$

$$\rightarrow N_2^*+e_t^-(H_2O)_n \qquad (2.19)$$

The electron is then solvated in a deep trap by relaxation of the medium. The absolute cross section for this process is measured to be 5.5×10^{-16} cm^2 [Lu *et al.*, 2002]. This value is similar to the sum of the vibrational excitation cross sections due to decay of the $N_2^- (^2\Pi_g)$ resonance at ~2.3 eV in the gas phase, indicating a highly efficient electron transfer. The results will be presented in detail in Chapter 4. This phenomenon is likely to be significant for charging (electron trapping) of ice particles or clouds in the stratosphere due to the abundances of N_2 and H_2O clusters as well as electrons produced by cosmic rays.

In addition to the $N_2^- (^2\Pi_g)$ resonance at ~ 2.3 eV, the unstable N^{-*} (^3P) ion was observed, though the stable N^- ion does not exist. This is attributed to DEA of N_2 at 9.8-10.1 eV, leading to the formation of shorted-lived N^{-*} (^3P) ion which then decays to the ground-state N (^4S) state:

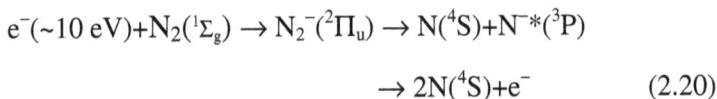

$$e^-(\sim10 \text{ eV})+N_2(^1\Sigma_g) \rightarrow N_2^- (^2\Pi_u) \rightarrow N(^4S)+N^{-*}(^3P)$$

$$\rightarrow 2N(^4S)+e^- \qquad (2.20)$$

The total cross section for this DEA process is $(0.8-2.5)\times10^{-18}$ cm^2 [Spence and Burrow, 1979; Huetz *et al.*, 1980].

2.4.2 *Oxygen (O₂)*

Similar to a N_2 molecule, molecular oxygen O_2 has two well-known negative ion resonances. The first one is the nondissociative attachment of electrons at energies near 0 eV to O_2. In contrast to N_2, O_2 has a small positive electron affinity of 0.451 eV [Miller, 2003]. The electronic ground state of O_2 is the triplet $^3\Sigma$ state, while the lowest TNI O_2*^- state has the configuration of $^2\Pi_g$. Fig. 2.3 shows the potential energy curves of the electronic ground-state O_2 ($^3\Sigma$) molecule and the $O_2^-(^2\Pi_g)$ ion. The resonant electron attachment produces a vibrationally excited TNI state $[O_2*^-(^2\Pi_g)]$ at vibrational levels (v≥4), whose lifetime is about ~10^{-10} s at low pressures (≤100 Torr). The $O_2*^-(^2\Pi_g)$ may decay into a neutral O_2 molecule via autodetachment of the electron or into a long-lived $O_2^-(^2\Pi_g)$ state via a three-body collision. These processes are expressed as

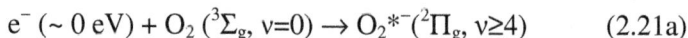

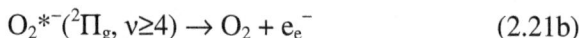

$$e^- (\sim 0 \text{ eV}) + O_2\ (^3\Sigma_g,\ v{=}0) \rightarrow O_2*^-(^2\Pi_g,\ v{\geq}4) \qquad (2.21a)$$

$$O_2*^-(^2\Pi_g,\ v{\geq}4) \rightarrow O_2 + e_e^- \qquad (2.21b)$$

$$O_2*^-(^2\Pi_g,\ v{\geq}4) + M \rightarrow O_2^-(^2\Pi_g,\ v{<}4) + M + \Delta E. \qquad (2.21c)$$

The formation of O_2^- is often described by an one-step process:

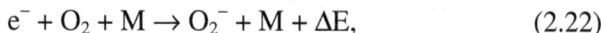

$$e^- + O_2 + M \rightarrow O_2^- + M + \Delta E, \qquad (2.22)$$

with a three-body attachment constant k=1.9×10^{-30} cm^6 s^{-1} for M=O_2 and k=1.0×10^{-31} cm^6 s^{-1} for M=N_2 at temperature T=300 K [Phelps, 1969].

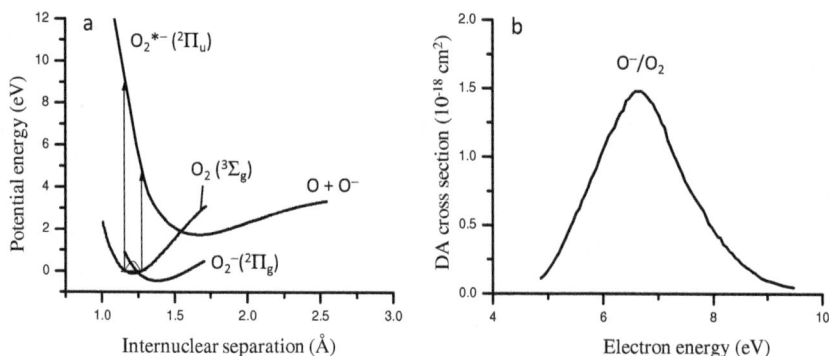

Fig. 2.3. (a) Approximate potential energy curves for O_2 ($^3\Sigma_g$), $O_2^-(^2\Pi_g)$ and $O_2*^-(^2\Pi_u)$ [Schulz, 1973b]; (b) The cross section for O^- formation by dissociative attachment in O_2, based on data from Christophorou *et al.* [1965].

Electron attachment in the energy range of 4.4-10 eV is associated with an electronically excited $O_2*^-(^2\Pi_u)$ state that then proceeds to autodetach or dissociate:

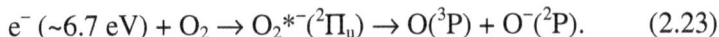

$$e^- (\sim 6.7 \text{ eV}) + O_2 \rightarrow O_2*^-(^2\Pi_u) \rightarrow O(^3P) + O^-(^2P). \qquad (2.23)$$

The DEA cross section peaking at 6.7 eV is about 1.4×10^{-18} cm^2. The potential energy curve of the $O_2*^-(^2\Pi_g)$ state involved in the DEA is also shown in Fig. 2.3. Since D(O-O)=5.16 eV, and $E_A(O)$=1.46 eV [Miller, 2003], the energy threshold to form the O$^-$ ion is 3.6 eV.

2.4.3 Nitrogen monoxide (N$_2$O)

Below 4 eV, there are two DEA resonances of N$_2$O, leading to the formation of O$^-$ and N$_2$* (Fig. 2.4)

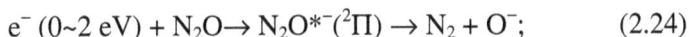

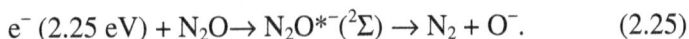

$$e^- (0 \sim 2 \text{ eV}) + N_2O \rightarrow N_2O*^-(^2\Pi) \rightarrow N_2 + O^-; \qquad (2.24)$$

$$e^- (2.25 \text{ eV}) + N_2O \rightarrow N_2O*^-(^2\Sigma) \rightarrow N_2 + O^-. \qquad (2.25)$$

The DEA cross section of the first reaction is strongly temperature dependent. It is due to the attachment of a nearly 0 eV electron to a vibrationally excited N$_2$O* molecule, forming a *vibrational Feshback resonance* N$_2$O*$^-(^2\Pi)$. The neutral N$_2$O is linear, while the anion N$_2$O$^-$ is bent and its potential energy depends significantly on the bond angle. Thus, the temperature dependence of the DEA cross section is mainly due to the excitation of the bending mode of vibration, and arises from the bond-angle dependence of the energy separation between the electronic ground states of N$_2$O and N$_2$O*$^-$. The threshold energy for dissociative attachment to an electron, derived from the NN–O dissociation energy of 1.67 eV and the electron affinity of the O atom of 1.46 eV, is 0.21 eV only.

The cross section of the second DEA resonance peaks at 2.25 eV and is temperature independent. It has a cross section of $\sim 9 \times 10^{-18}$ cm^2, and is ascribed to the second N$_2$O$^-$ state [Tronc *et al.*, 1977].

It is interesting to note that N$_2$O is known not only as a greenhouse gas but as a scavenger of the hydrated electron (e_{aq}^-) and a source of OH radical in radiation chemistry and biology [Hart and Anbar, 1970; Lehnert, 2008]. The reaction of N_2O with e_{aq}^- has been widely used to convert e_{aq}^- to OH radicals ($OH^\bullet$) in radiolysis of water. The reduction of N_2O by e_{aq}^- produces the N_2O^- TNI, which rapidly dissociates to give N_2 and O^- (Reaction 2.24 or 2.25). The O^- immediately reacts with a H_2O molecule to give an $OH^\bullet$ radical and OH^-

$$O^- + H_2O \rightarrow OH^\bullet + OH^- \tag{2.26}$$

Reactions 2.24-2.26 are very well known to radiation chemists and biologists [Hart and Anbar, 1970; Lehnert, 2008].

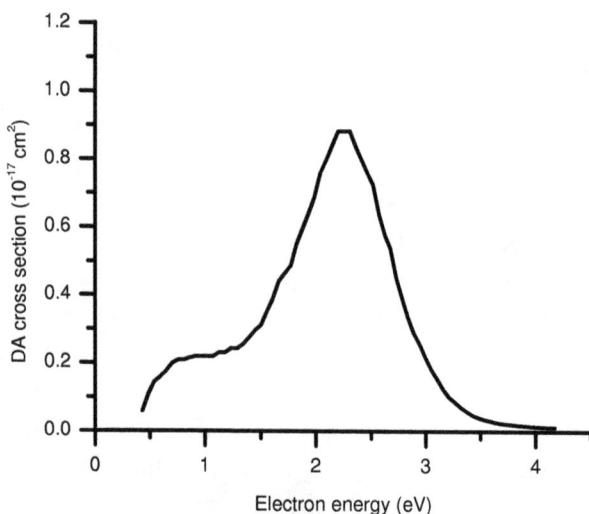

Fig. 2.4. The cross section for O^- formation by dissociative attachment in N_2O. Based on data from Christophorou *et al.* [1984].

2.4.4 *Nitrogen dioxide (NO₂)*

Three fragment negative ions were observed in DEA of NO_2: O^-, O_2^-, and NO^-, in decreasing order of abundance. The threshold energies for the formation of these ions are 1.61, 4.03 and 3.11 eV, respectively, consistent with the thermochemical onsets for the processes [Abouaf *et al.*, 1976]

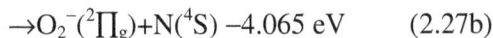

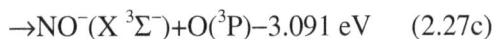

$$e^-+NO_2 \rightarrow NO_2{}^{*-} \rightarrow O^-(^2P)+NO(^2\Pi)-1.650 \text{ eV} \qquad (2.27a)$$

$$\rightarrow O_2{}^-(^2\Pi_g)+N(^4S)-4.065 \text{ eV} \qquad (2.27b)$$

$$\rightarrow NO^-(X\ ^3\Sigma^-)+O(^3P)-3.091 \text{ eV} \qquad (2.27c)$$

2.4.5 *Halogen-containing molecules*

There is a particularly long history of studies of DEAs of halogenated molecules such as chlorofluorocarbons (CFCs) and HCl/HBr since the 1950s [*e.g.*, Hickam and Berg, 1958; Curran, 1961; Bansal and Fessenden, 1972; Christophorou, 1976; Illenberger *et al.*, 1979; Peyerimhoff and Buenker, 1979; Christophorou *et al.*, 1984; Oster *et al.*, 1989]. Till the late 1970s, it had been well observed that DEA of gaseous halogen-containing molecules to low-energy free electrons near zero eV is an extremely efficient process. The measured DEA cross sections at near zero eV are up to the order of 10^{-14} cm^2, four orders of magnitude higher than their cross sections of photodissociation peaking at near 7.6 eV for CFCs [Bansal and Fessenden, 1972; Christophorou, 1976; Oster *et al.*, 1989].

For example, DEA of CCl$_4$ to low-energy free electrons near zero eV is an extremely efficient process (Fig. 2.5)

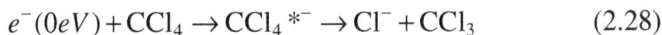

$$e^-(0eV)+CCl_4 \rightarrow CCl_4{}^{*-} \rightarrow Cl^- +CCl_3 \qquad (2.28)$$

The observed rate constant for this reaction in the gas phase ranges from 2.5×10^{-7} cm^3 s^{-1} to 4.1×10^{-7} cm^3 s^{-1} [Mothes *et al.*, 1972], corresponding to an attachment cross section of $\sim 2 \times 10^{-14}$ cm^2 [Christophorou *et al.*, 1984].

DEA resonances are indeed the dominant process observed upon impact of low-energy electrons with halogen-containing molecules, including organic CFCs, and inorganic HCl/HBr and ClONO$_2$. The DEAs of these important molecules in gas phase, condensed phase and adsorbed on ice surfaces, as well as their implications for atmospheric ozone depletion, will particularly be discussed in Chapters 4 and 5.

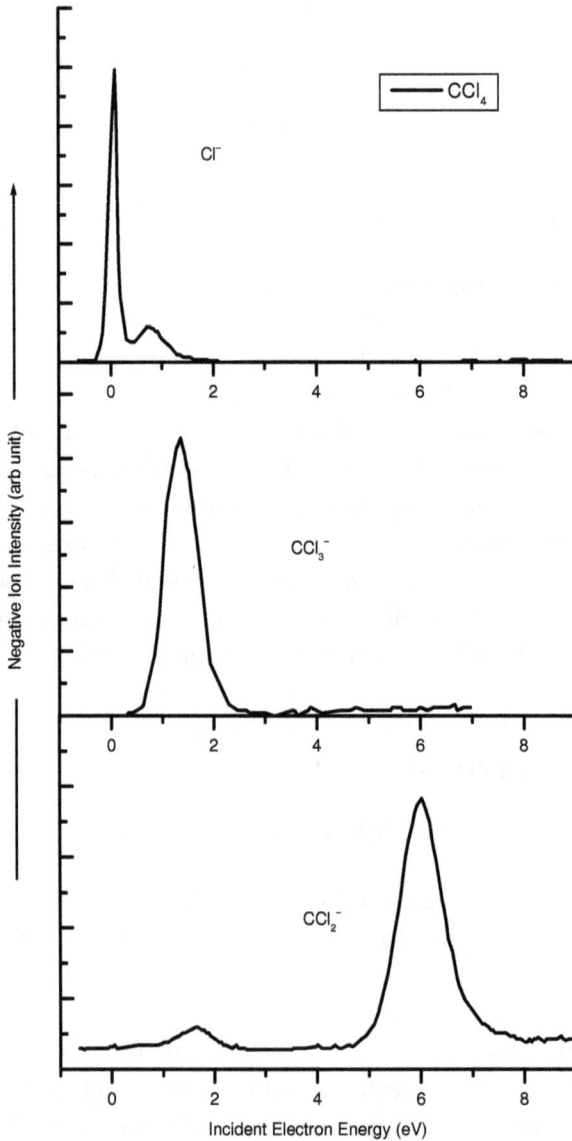

Fig. 2.5. Negative ion yield from dissociative attachment of electrons to CCl_4. Based on data from Oster *et al.* [1989].

2.5 Concluding remarks

Electron interactions with molecules must be considered as part of the physics and chemistry in the Earth's atmosphere. In particular, electron-induced charging and adsorption on solid (ice) surfaces and electron-induced reactions of atmospheric molecules, especially halogen-containing molecules, might be of far-reaching significance for understanding the important atmospheric processes such as ozone depletion and the formation of the polar ozone hole. These issues will be addressed in detail, based on laboratory measurements presented and discussed in Chapter 4 and atmospheric observations in Chapter 5.

Chapter 3

Conventional Understanding of Ozone Depletion

3.1 The ozone layer and its formation

The ozone (O_3) layer in the earth's stratosphere at 10-50 km above the ground (Fig. 3.1), is an important region both radiatively and chemically, in which many effects such as UV absorption, the ozone hole, global warming/cooling originate. The amount of ozone above a point on the Earth's surface is called as total column ozone and measured in Dobson units (DU). It is typically approximately 260 DU near the tropics and is higher in the temperate zones. There are large seasonal fluctuations in total ozone in the temperate zones with a maximum peak in September/October and a minimum in April/May in the southern hemisphere, and vice versa in the northern hemisphere. Ozone is naturally formed by the solar ultraviolet (UV, $\lambda \leq 240$ nm) photolysis of molecular oxygen, followed by combination with another O_2 molecule to form an O_3 molecule

$$O_2 + hv \ (\lambda \leq 240 \text{ nm}) \rightarrow O + O \qquad (3.1)$$

$$O + O_2 + M \rightarrow O_3 + M \qquad (3.2)$$

where M is a third-body molecule. Near the earth's ground level, ozone is harmful to heath - it is a major component of photochemical smog. In contrast, stratospheric ozone absorbs harmful UV photons from the sun (at wavelengths $\lambda = 240$-320 nm) that can cause harm in forms of DNA damage, skin cancer, cataracts, damage vegetation, *etc*. Therefore, the presence of the ozone layer in the stratosphere is essential to the living creatures on the earth.

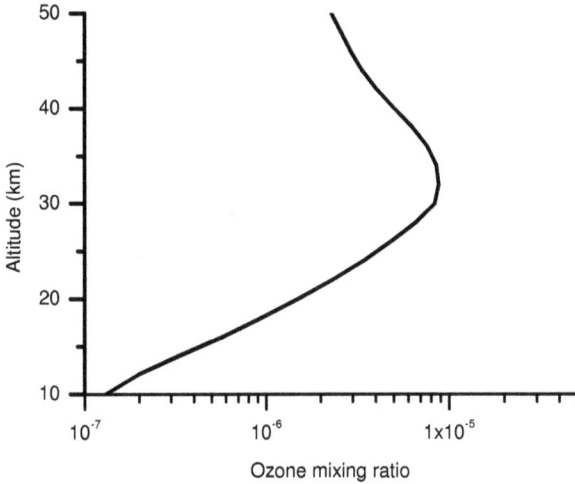

Fig. 3.1. Typical vertical structure of the ozone volume mixing ratio at 30° N in March. Based on data from Brasseur *et al.* [1999].

The solar UV radiation also splits the ozone molecule and the resultant atomic oxygen can also destroy another ozone molecule through the following reactions [Chapman, 1930]

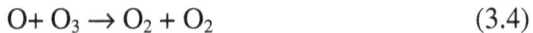

$$O_3 + hv \rightarrow O_2 + O \tag{3.3}$$

$$O + O_3 \rightarrow O_2 + O_2 \tag{3.4}$$

Reactions (3.1)-(3.4) are known as the so-called *Chapman* reactions. Reaction (3.2) becomes slower with increasing altitude, while Reaction (3.3) becomes faster. The concentration of ozone is a balance between these competing production/destruction reactions. In the upper atmosphere, atomic oxygen dominates where UV levels are high. With decreasing altitude, the air density increases, and UV absorption increases. As a result, the O_3 level peaks in the lower stratosphere between 15-25 km. The 'Chapman' reactions provided a satisfactory explanation for the latitudinal and seasonal distribution of atmospheric ozone.

But there was a problem with the Chapman theory. By the 1960s, researchers realized that the loss of ozone given by Reaction (3.4) was too slow. It could not remove enough ozone to reproduce the real values observed in the atmosphere. Other reactions had to occur and control the ozone concentrations in the stratosphere. We'll discuss more about the conventional understanding of ozone loss in Sec. 3.3 of this Chapter.

3.2   The observation of the ozone hole

3.2.1   *The early observation*

It has been argued that the Antarctic ozone hole was first discovered in 1956 and therefore it can't be caused by CFCs. This argument originates from a published paper by Professor G M B Dobson, the scientist who designed the ozone spectrophotometer which has been used for standard ozone measurements since the 1930s and the unit of ozone amount was named after him. In his published paper [Dobson, 1968], Dobson wrote: "One of the more interesting results on atmospheric ozone which came out of the IGY (International Geophysical Year) was the discovery of the peculiar annual variation of ozone at Halley Bay". At that time, the annual variation of ozone at Spitzbergen was fairly well known. Thus, it was expected to see a similar ozone variation but with a seasonal difference of six months. But it turned out that the values for September and October 1956 were about 150 units lower than those expected from the corresponding values at Spitzbergen. And in November, the ozone values suddenly jumped up to those expected from the Spitzbergen results. The observation by Dobson was repeated in the sequent years, as shown in Fig. 3.2; the results are indeed correct not because of instrumental errors. This seasonal variation of polar total ozone has now been well confirmed, which will be discussed again in Chapter 5.

The 1956-observed "hole" of total ozone at Halley is attributed to the formation of the winter vortex over the South Pole, which is maintained late into the spring and then suddenly breaks up in November, and therefore both the ozone values and the stratosphere temperatures suddenly rise. The lower total ozone in the winter and springtime stratosphere over Antarctica than over the Arctic is due to the different atmospheric circulations in the two hemispheres.

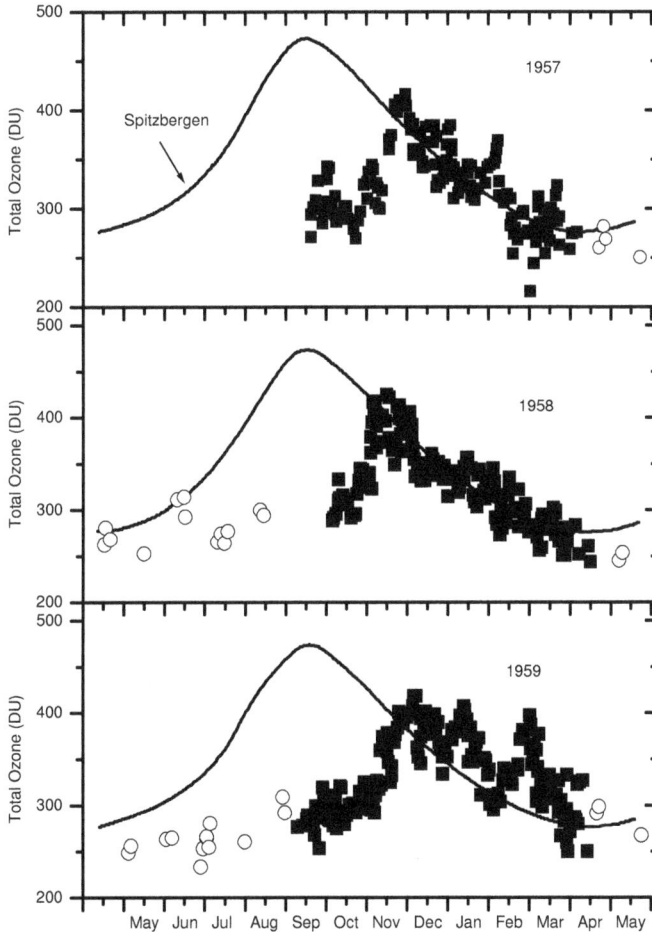

Fig. 3.2. The 1957-1959 observations at Halley Bay, Antarctica. The full curve was for Spitzbergen, shifted by six months. Note the lower values of ozone in the southern winter and spring and the sudden increase in November at the time of the final stratospheric warming. Based on data from Dobson [1968].

It should also be noted that compared with the values observed in the later 1950s, there has been rapid depletion by 50% more in total ozone in the winter and spring stratosphere over Antarctica observed in the past 30 years (Fig. 3.3). The latter is called the modern (Antarctic) ozone hole, which was not discovered until 1985. The origin of the modern ozone hole has generally been well related to anthropogenic emission of chlorofluorocarbons (CFCs) that were once widely used in industries.

It is generally accepted that there was no real ozone hole over Antarctica in the 1950s (before the middle 1970s). However, the observation by Dobson in the 1950s has its own significance in atmospheric science. In this author's opinion, there might be a lack of a proper understanding of its significance in current atmospheric chemistry text. This lack is perhaps the main reason giving room for some sceptics of human major contribution to atmospheric and climate changes to argue against the linking of the well-observed ozone hole to CFCs. This issue will be re-visited in Chapter 5.

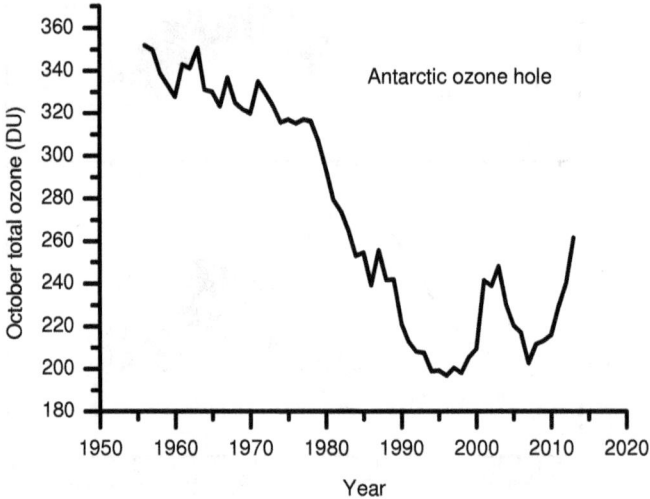

Fig. 3.3. Observed October total ozone at Halley station in the Antarctica during 1956-2013. Only a 3-point smoothing processing was applied to the original observed data by the British Antarctic Survey. Updated from Lu [2010a, 2013].

3.2.2 *The Antarctic Ozone Hole*

Large loss of ozone in the lower stratosphere over Antarctica, the Antarctic Ozone Hole, was discovered by a research group from the British Antarctic Survey (BAS) from data obtained with a Dobson ozone spectrophotometer at Halley Bay station in the 1981-1983 period. They reported their observations of a dramatic October ozone loss in the lower polar stratosphere [Farman, Gardiner and Shanklin, 1985]. Over the period of 1979 to 1985, there was a continuous decline in the springtime ozone abundance from 320 DU to less than 200 DU. Prior to the BAS team's discovery, however, the satellite data by the NASA TOMS (Total Ozone Mapping Spectrometer) team which had been available since 1979 didn't show the dramatic loss of ozone. This is because according to the photochemical models then (see Sec. 3.3), large ozone loss was expected to occur in the middle tropical stratosphere at altitudes of 35-40 km where active chlorine arising from photolysis of CFCs has the highest yield, but *not* to occur in the lower polar stratosphere. Thus, the software processing the raw ozone data from the satellite was programmed to treat very low values of ozone as erroneous readings! When the observations of the BAS team were published, the ozone research community was shocked, as models thought to have advanced sufficiently to reproduce the essentials of stratospheric chemistry and transport/dynamics failed to explain this dramatic ozone loss in the polar stratosphere. Reanalysis of the raw satellite data, after the results from the BAS team were published, confirmed the BAS discovery and showed that the loss was rapid and covered the whole Antarctica continent. The monthly averages of total ozone for the month of October over the Halley Bay station in Antarctica during 1956-2013 are shown in Fig. 3.3. It is clearly seen that the drastic ozone loss occurred after about 1975. By the mid-1990s, the total ozone in October was about 57% its value during the 1970s.

Fig. 3.4 shows one of the biggest ozone holes over Antarctica in October 2006, within which the total ozone has decreased from 292 DU on 15th June to 96 DU on 4th October, by as much as 67%. Note also that the ozone hole covers the whole area of Antarctica, and its size is about 80% of the area of South America. Thus, the large volume and mass of ozone-depleting air after the breakup of the polar vortex may significantly affect the stratospheric ozone levels of the entire hemisphere.

Later atmospheric measurements showed strong evidence that the ozone loss was related to halogen (chlorine)-catalyzed ozone-depleting reactions derived mainly from man-made chlorofluorocarbons (CFCs), which takes place following some special reactions in polar stratospheric clouds (PSCs) existing in the cold dark Antarctic winter. Another ozone-depleting species is nitrogen oxides, which are a by-product of combustion processes, *e.g.*, aircraft emissions.

Although mid-latitude and Arctic depletion has also been observed, the loss is much less and much slower, unlike the sudden and near total loss of ozone over Antarctica.

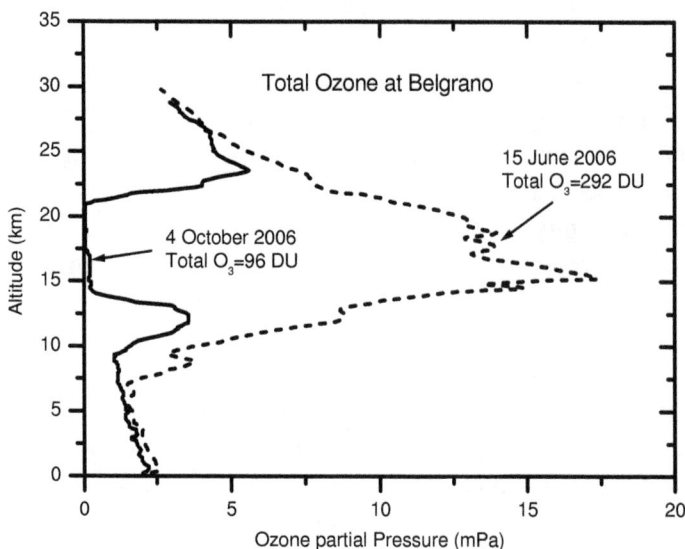

Fig. 3.4. Ozone partial pressure as a function of altitude at Belgrano station (77° S, 35° W), Antarctica in June and October of 2006 (one of the biggest ozone holes).

3.3 Photochemical models of ozone depletion

Modern research on ozone depletion in the stratosphere started in the 1960s. It was realized that the detailed ozone source/sink calculations could not give the balance: the observed levels of ozone were less than would be predicted from the Chapman reactions alone, indicating that some additional sink for ozone must be missing. This gap was then closed by the inclusion of a series of free radical catalytic reaction chains. The stratospheric layer is a stable and somewhat isolated region of the atmosphere; it is difficult for enough reacting species to be transported there to remove ozone continuously and irreversibly. However, there are some trace species that are present in the stratosphere and can destroy ozone catalytically: they are constantly regenerated through reaction cycles, and therefore do not have to be present at concentrations comparable to ozone. The general form of a catalytic cycle is

$$X + O_3 \rightarrow XO + O_2 \qquad (3.5)$$

$$O + XO \rightarrow X + O_2 \qquad (3.6)$$

$$\text{Overall: } O + O_3 \rightarrow 2O_2$$

Radical pairs that can have this catalytic reaction include H/OH, OH/HO_2, NO/NO_2, Cl/ClO, and Br/BrO. To be significant for the sink of stratospheric ozone, the reaction cycle only has to be comparable to the rate of the $(O + O_3)$ reaction (3.4) in the Chapman reactions.

3.3.1 *Catalytic destruction of ozone by HO_x*

The first catalytic cycle to destroy ozone in the atmosphere, involving hydrogen, was proposed by Bates and Nicolet [1950]. The hydroxyl radical, OH, is produced in the stratosphere by the oxidation of H_2O, CH_4, and H_2

$$O(^1D) + H_2O \rightarrow 2OH \qquad (3.7)$$

$$O(^1D) + CH_4 \rightarrow CH_3 + OH \qquad (3.8)$$

$$O(^1D) + H_2 \rightarrow 2H + OH \tag{3.9}$$

The resulting OH (and HO_2) plays an important role in mesospheric and stratospheric ozone chemistry through either direct reaction with odd oxygen or partitioning in the reaction chains of other chemical species.

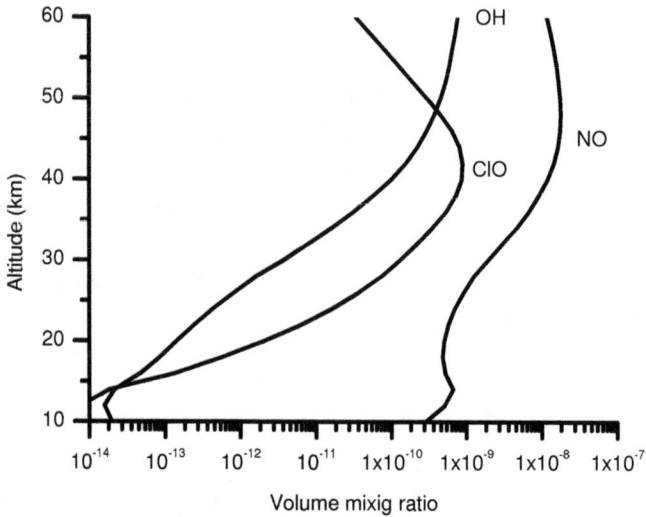

Fig. 3.5. Stratospheric altitude profiles of ozone-depleting HO, NO, and ClO species at 30° N in March. Based on data from Brasseur *et al.* [1999].

At altitudes of 30-40 km, the following hydrogen catalytic cycle that destroys odd oxygen is important

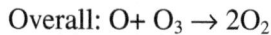

$$OH + O_3 \rightarrow HO_2 + O_2 \tag{3.10}$$

$$O + HO_2 \rightarrow OH + O_2 \tag{3.11}$$

Overall: $O + O_3 \rightarrow 2O_2$

Below about 30 km, where oxygen atoms are rare, the following hydrogen catalytic cycle is dominant

$$OH + O_3 \rightarrow HO_2 + O_2 \tag{3.12}$$

$$O_3 + HO_2 \rightarrow OH + 2O_2 \tag{3.13}$$

Overall: $2O_3 \rightarrow 3O_2$

A typical altitude profile of HO in the stratosphere at 30° N in March is shown in Fig. 3.5.

3.3.2 *Catalytic destruction of ozone by odd nitrogen*

The primary source of active nitrogen in the global stratosphere is from N_2O, which is released from through natural biological processes or anthropogenic activities on the earth's surface. In the stratosphere, a small fraction of N_2O reacts with excited oxygen atoms $O(^1D)$ resulting mainly from UV photolysis of O_3 to produce NO

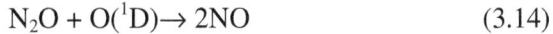

$$N_2O + O(^1D) \rightarrow 2NO \tag{3.14}$$

This reaction occurs predominantly in the middle and upper stratosphere (Fig. 3.5). It was proposed that NO, which is a reactive radical, can destroy ozone via the catalytic cycle [Crutzen, 1970; Johnston, 1971]

$$NO + O_3 \rightarrow NO_2 + O_2 \tag{3.15}$$

$$O + NO_2 \rightarrow NO + O_2 \tag{3.16}$$

Overall: $O + O_3 \rightarrow O_2 + O_2$

Many cycles of this reaction can occur before other processes transform active NO_x into less active forms of odd nitrogen, NO_y.

Note that Reaction (3.16) is the rate-limiting step in the catalytic cycle. It competes with the recombination of the O atom with molecular O_2 to return O_3 (Reaction 3.2) and the photolysis of NO_2 at $\lambda \leq 420$ nm

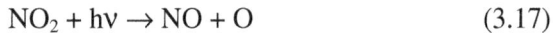

$$NO_2 + h\nu \rightarrow NO + O \tag{3.17}$$

In fact, this reaction, together with Reaction (3.15), constitutes a null cycle, leading to no net change in odd oxygen (O_3). In the lower stratosphere, the rate of Reaction (3.16), *i.e.*, the overall rate of the catalytic cycle, is orders of magnitudes lower than those of the photolysis rate of NO_2 and the null cycle. In the middle and upper stratosphere with the lower air density, however, the level of O atoms, produced mainly by photolysis of O_3, becomes high enough that Reaction (3.16) becomes dominant over Reaction (3.2) (with O_2). Thus, the destruction cycle becomes important (Fig. 3.5).

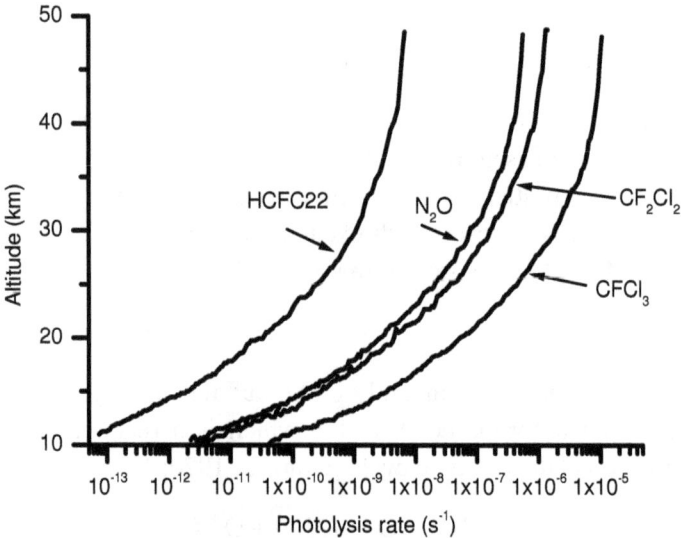

Fig. 3.6. Photolysis rates of CFC-11, CFC-12, HCFC-22 and N_2O at solar zenith angle=0, albedo=0.3. Based on data from Brasseur *et al.* [1999].

3.3.3 *Catalytic destruction of ozone by halogen*

It was proposed in the 1970s that the photolysis of human-made chlorofluorocarbons (CFCs) could provide a significant source of chlorine in the stratosphere, and that chlorine chemistry could lead to catalytic destruction of ozone [Molina and Rowland, 1974; Stolarski and Cicerone, 1974]. CFCs were entirely of anthropogenic origin, and once widely used as important industrial chemicals in refrigeration systems, air conditioners, aerosols, and solvents due to their chemical inertness. However, it is this very chemical property that results in their harmful effects on the ozone layer and global climate. Once emission into the atmosphere from the industrial applications on the Earth's surface, CFCs mix into the free troposphere where there are no known processes for their destruction or removal. This long tropospheric lifetime leads to their effective transport to the stratosphere, where ozone-depletion reactions can take place.

The solar UV *photolysis* of each of CFC molecules in the tropical upper stratosphere leads to the release of one free Cl atom with the formation of a free radical. Taking CF_2Cl_2 as an example, this can be expressed as

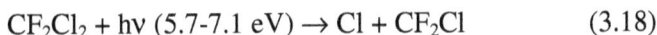

$$CF_2Cl_2 + hv \ (5.7\text{-}7.1 \ eV) \rightarrow Cl + CF_2Cl \qquad (3.18)$$

This stratospheric photolysis mainly occurs in the far UV regime at 175-220 nm. Typical photolysis rates of CFC-11 ($CFCl_3$), CFC-12 (CF_2Cl_2), HCFC-22 and N_2O are shown in Fig. 3.6. The immediate reaction of the CF_2Cl radical with O_2 releases another Cl atom

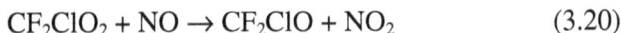

$$CF_2Cl + O_2 + M \rightarrow CF_2ClO_2 + M \qquad (3.19)$$

$$CF_2ClO_2 + NO \rightarrow CF_2ClO + NO_2 \qquad (3.20)$$

$$CF_2ClO + M \rightarrow COF_2 + Cl + M \qquad (3.21)$$

At altitudes of 20-50 km, the Cl atoms released in both the photolysis and the subsequent reaction of the free radical with O_2 destroys ozone via the (Cl, ClO) reaction chain

$$Cl + O_3 \rightarrow ClO + O_2 \qquad (3.22)$$

$$O + ClO \rightarrow Cl + O_2 \qquad (3.23)$$

$$\text{Overall: } O + O_3 \rightarrow O_2 + O_2$$

It was estimated that through this reaction chain, one Cl atom can catalytically destroy up to 100 thousands of O_3 molecules before it is removed from the atmosphere by tropospheric processes.

In the lower stratosphere (below 20 km), the chlorine chain reaction of (3.22) and (3.23) can be diverted through the following reactions

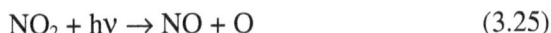

$$ClO + NO \rightarrow Cl + NO_2 \qquad (3.24)$$

$$NO_2 + hv \rightarrow NO + O \qquad (3.25)$$

$$O + O_2 + M \rightarrow O_3 + M \qquad (3.26)$$

The overall result of these reactions plus Reaction (3.22) is no net change in chemical composition (the O_3 concentration) of the atmosphere.

The chain reaction of (3.22) plus (3.23) will be interrupted or terminated by the reaction of Cl and ClO with some stratospheric species, converting to the reservoir species HCl and $ClONO_2$, respectively. The Cl atom reacts with CH_4, H_2, HO_2 and H_2O_2 to form HCl

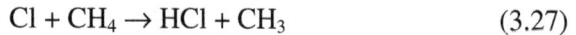

$$Cl + CH_4 \rightarrow HCl + CH_3 \qquad (3.27)$$

$$Cl + H_2 \rightarrow HCl + H \qquad (3.28)$$

$$Cl + HO_2 \rightarrow HCl + O_2 \qquad (3.29)$$

$$Cl + H_2O_2 \rightarrow HCl + HO_2 \qquad (3.30)$$

ClO reacts with NO_2 to generate $ClONO_2$

$$ClO + NO_2 + M \rightarrow ClONO_2 + M \qquad (3.31)$$

The HCl generated in the above reactions is long-lived and is a temporary sink for Cl atoms until it is returned to active Cl through the attack by OH

$$HCl + OH \rightarrow H_2O + Cl \qquad (3.32)$$

$ClONO_2$ is believed to be destroyed mainly by photolysis or via reaction with O atoms, resulting in regeneration of active Cl species

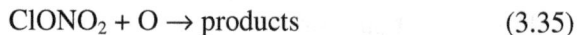

$$ClONO_2 + h\nu \rightarrow Cl + NO_3 \qquad (3.33)$$

$$\rightarrow ClO + NO_2 \qquad (3.34)$$

$$ClONO_2 + O \rightarrow products \qquad (3.35)$$

Similar Br-BrO and I-IO reaction chains can occur in the stratosphere. But the abundances of Br- and I-containing molecules are significantly lower than those of chlorine-containing molecules.

It is interesting to consider the rates of photolysis of CFCs and the relative abundances of the products (Cl, ClO, HCl and $ClONO_2$) in the general stratosphere. The calculations using Eddy diffusion models showed a maximum production rate of Cl atoms from photolysis of CF_2Cl_2 at the altitude of 30-40 km and a maximum removal rate of odd oxygen (O_3) by the Cl-ClO chain at the altitude of 35-40 km [Rowland and Molina, 1975]. These production/removal rates are negligibly small in the lower stratosphere below 20 km (Figs. 3.5 and 3.7).

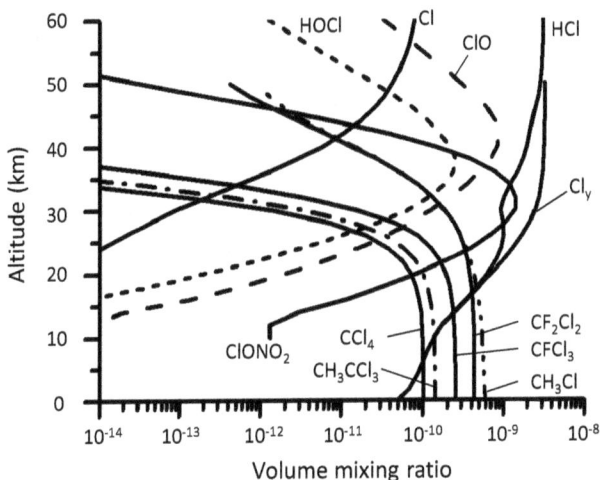

Fig. 3.7. Typical altitude profiles of chlorine species and total Cl (Cl_y) in the stratosphere at 30° N in March. Based on data from Brasseur *et al.* [1999].

The abundances of the chlorine species in the entire stratospheric region have been measured by numerous experimental techniques. Typical altitude profiles of chlorine species in the stratosphere at 30° N in March are shown in Fig. 3.7. The measurements have shown that HCl is the most abundant inorganic species throughout the stratosphere because of its long lifetime, making up over 95% of the Cl loading at altitudes above 45 km. $ClONO_2$ is a significant reservoir species at altitudes below 30 km, at which its formation rate peaks. With increasing altitude, the atmospheric pressure decreases, the formation rate of $ClONO_2$ via Reaction (3.31) drops, and its photolysis rate increases. This leads to a decrease in its mixing rate at higher altitudes and a corresponding increase in the level of ClO.

In the general stratosphere, the ratio of the abundances of active (Cl + ClO) to inactive (HCl and $ClONO_2$) chlorine is low at low altitudes of 15-20 km, while it increases with rising altitudes. The active forms take 2% or less of the total inorganic chlorine below 20 km, while it rises to 35% at 40 km. Note that this rise is mainly due to the above-mentioned decrease in stability of $ClONO_2$ with rising altitude.

3.3.4 *Heterogeneous chemical reactions in the polar stratosphere*

It is obvious that the unexpected discovered ozone hole in the lower Antarctic stratosphere at the altitude of ~18 km in 1985 was neither predicted nor explainable by the photochemical model described in the previous paragraph, which predicted a maximum ozone loss in the middle and upper stratosphere over the tropics where the solar radiation was most intense. The ozone hole must be explained by another mechanism rather than the direct photolysis of CFCs.

The meteorology in the lower stratosphere over Antarctica during winter is very different from that in the general stratosphere. During the winter, sunlight does not reach the lower polar stratosphere. A strong circumpolar wind, known as the *polar vortex*, develops in the middle to lower stratosphere. The strong polar vortex isolates a continent-size body of air, in which polar stratospheric clouds (PSCs) of several km in thickness are formed at very low temperatures (below about -80° C) because there is no sunlight. These PSCs consist of water ice or nitric acid/ice particles with a major composition of H_2O. PSCs play a crucial role in causing drastic ozone loss in the polar stratosphere.

After the Antarctic ozone hole was discovered in 1985, atmospheric chemists proposed a mixed photochemical mechanism, consisting of four major processes: (1) the solar UV photolysis of CFCs produces Cl and ClO that then react with other atmospheric molecules (CH_4 and NO_2) to generate inorganic chlorine species (HCl and $ClONO_2$) in the tropical upper stratosphere at the altitudes of 30-40 km; (2) The reservoirs HCl and $ClONO_2$ are then transported to the lower polar stratosphere via air circulation; (3) *heterogeneous chemical reactions* of these inorganic compounds on the surfaces of PSCs occur to yield photoactive chlorine species in the lower polar stratosphere (15-20 km) in darkness during winter; and (4) upon the return of the sunlight in spring, chlorine atoms are generated to destroy O_3 in the polar stratosphere. The most important heterogeneous reactions believed to occur on the surfaces of PSCs are

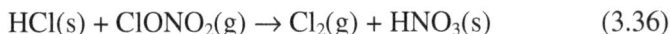

$$HCl(s) + ClONO_2(g) \rightarrow Cl_2(g) + HNO_3(s) \qquad (3.36)$$

$$ClONO_2(g) + H_2O(s) \rightarrow HNO_3(s) + HOCl(g) \qquad (3.37)$$

$$HCl(s) + HOCl(g) \rightarrow H_2O(s) + Cl_2(g) \qquad (3.38)$$

$$HCl(s) + N_2O_5(g) \rightarrow ClONO\ (g) + HNO_3(s) \qquad (3.39)$$

$$H_2O(s) + N_2O_5(g) \rightarrow 2\ HNO_3(s) \qquad (3.40)$$

(s, solid; g, gas). These reactions effective on the surfaces of PSCs convert chlorine from the inactive chlorine reservoir species HCl and $ClONO_2$ (and their bromine counterparts) into more active forms of chlorine such as Cl_2, HOCl and ClONO [Solomon, 1990]. Due to the abundances of HCl and $ClONO_2$ in the lower polar stratosphere (see Fig. 3.7), Reaction (3.36) is believed to be the dominant process. There is also evidence that the heterogeneous reaction mechanism has an ionic pathway, i.e., solvated Cl^- (rather than molecular HCl) on PSC surfaces plays a crucial role in producing the photoactive chlorine species.

In the current context of atmospheric chemistry, upon the formation of PSCs in early winter, heterogeneous reactions (especially Reaction 3.36) take place rapidly and are thought to be the major mechanism for the activation of inert halogenated compounds into photoactive halogens in the dark polar stratosphere, a key step for the subsequent formation of the springtime ozone hole. When sunlight returns in spring, the photoactive chlorine species release chlorine atoms to destroy O_3 in the polar stratosphere

$$Cl_2 + h\nu \rightarrow 2\ Cl \qquad (3.41)$$

$$HOCl + h\nu \rightarrow Cl + OH \qquad (3.42)$$

$$\text{ClONO} + \text{hv} \rightarrow \text{Cl} + \text{NO}_2 \qquad (3.43)$$

In addition, heterogeneous reactions also covert NO and NO_2 into less reactive HNO_3, which remains in PSCs. Hence, the gaseous concentrations of nitrogen oxides are reduced, slowing down the removal rate of ClO via Reaction (3.31) (with NO_2) to form $ClONO_2$. This 'denoxification' also helps to maintain high levels of active chlorine in the polar stratosphere in winter [Crutzen and Arnold, 1986].

The high concentration of ClO in the polar vortex also leads to the formation of the ClO dimer (Cl_2O_2). It is proposed that the subsequent photolysis of Cl_2O_2 regenerates Cl atoms, which then destroy O_3 via the reaction cycle [Molina and Molina, 1987]

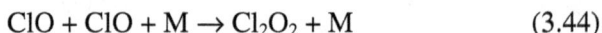

$$\text{ClO} + \text{ClO} + \text{M} \rightarrow \text{Cl}_2\text{O}_2 + \text{M} \qquad (3.44)$$

$$\text{Cl}_2\text{O}_2 + \text{hv} \rightarrow \text{Cl} + \text{ClOO} \qquad (3.45)$$

$$\text{ClOO} + \text{M} \rightarrow \text{Cl} + \text{O}_2 + \text{M} \qquad (3.46)$$

$$2(\text{Cl} + \text{O}_3) \rightarrow 2(\text{ClO} + \text{O}_2) \qquad (3.47)$$

$$\text{Net: } 2\,\text{O}_3 \rightarrow 3\text{O}_2$$

With the disappearance of PSC in the late spring, the active chlorine species are converted back to inactive reservoirs via Reactions (3.27) and (3.31) and the abundances of $ClONO_2$ and HCl are re-established.

3.3.5 *Summary of mechanisms for polar ozone loss in photochemical models*

In summary, several mechanisms are involved in the photochemical models of polar ozone loss. These are:

- The UV sunlight leads to the photolysis of CFCs in the upper stratosphere over the tropic.
- The air circulation transports the photoproducts (inorganic halogen reservoirs) into the lower stratosphere over the polar region.

- Heterogeneous reactions take place and convert the inactive chlorine (bromine) reservoirs to more active forms of chlorine and bromine on the surfaces of PSCs, which are formed in the lower polar stratosphere in cold winter.
- Sunlight returns to the air inside the polar vortex and allows the production of active chlorine (bromine) and initiates the ozone destruction catalytic cycles. No ozone loss occurs without sunlight.

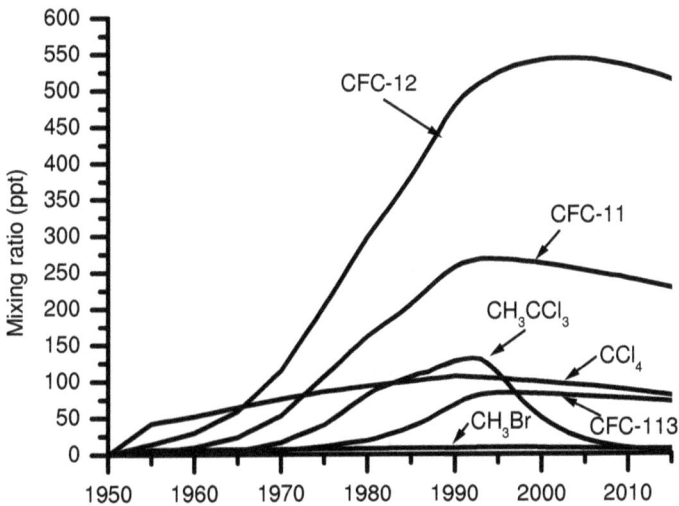

Fig. 3.8. Mixing ratios of the most abundant CFCs: CF_2Cl_2 (CFC-12), $CFCl_3$ (CFC-11), $CF_2ClCFCl_2$ (CFC-113), CCl_4, and CH_3CCl_3, as well as CH_3Br. Based on data from WMO [2010] and IPCC [2013].

3.4 Montreal Protocol

The first international agreement to restrict the production of CFCs came with the signing of the Montreal Protocol in 1987 that ultimately aimed to reduce CFCs by half by the year 2000. Two revisions of this agreement were made, the latest being in 1992. Agreement has been reached on the control of industrial production of many halogenated molecules until the year 2030. The main CFCs have been prohibited to be produced by any of the signatories after the year 1995, except for a limited amount for essential uses such as for medical use. Agreements to phase out the use of other ozone-depleting compounds are also being adopted.

Owing to these regulations on the production of ozone-depleting substances, the atmospheric concentrations of the major man-made substances in the lower atmosphere started to decline in the mid-1990s and reached maximum levels in the stratosphere near the turn of the century, as shown in Fig. 3.8. A corresponding recovery of the Antarctic ozone hole was anticipated [WMO, 1994, 1998].

3.5 Photochemical models versus observations

In the Antarctic vortex, the measured ClO abundance of the order of 1 p.p.b.v. is several hundred times greater than the usual concentration (0.01 p.p.b.v.) in the general stratosphere [Anderson *et al.*, 1991]. This provides strong evidence that the ozone hole is related to chlorine-containing molecules mainly CFCs. Further, observations of the large ozone hole over Antarctica in each spring since the late 1970s appear to support the photochemical mechanisms described above. However, looking closer into the observed data, one can find that there actually exist large discrepancies between photochemical model predictions and observations. Some of them are outlined as follows.

(1) Researchers have pointed out previously that there indeed exit significant gaps in understanding of the partitioning of chlorine in the stratosphere, as demonstrated by some persistent quantitative discrepancies between photochemical models and observed results [Orlando and Schauffler, 1999]. For instance, it has been reviewed that modeled ClO/HCl concentration ratios are consistently larger than those observed at altitudes above about 30 km, whereas they are lower than the values from in situ measurements by a factor of two or more in the lower stratosphere below 20 km.

(2) Orlando and Schauffler [1999] also provided an interesting schematic diagram of chlorine 'photochemical' and dynamical evolution

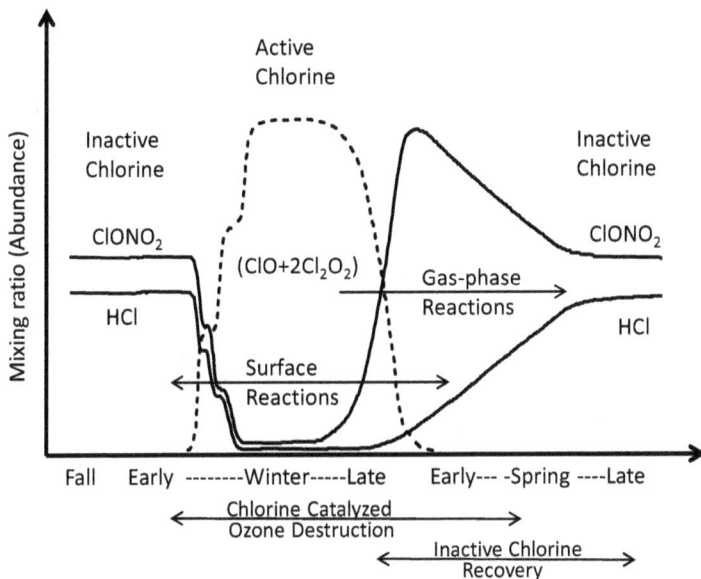

Fig. 3.9. Schematic diagram of active chlorine evolution in polar regions. Based on the observations of Waters *et al.* [1993], Webster *et al.* [1993], Toohey *et al.* [1993] and Roche *et al.* [1994], and adapted from a diagram in WMO [1994] and Orlando and Schauffler [1999]. Note that the conversion of chlorine reservoirs (HCl and ClONO$_2$) into active chlorine is rapidly completed with the presence of PSCs in early winter, and the active ClO species is formed from the very beginning of the dark winter polar stratosphere, so does ozone destruction. Also, the recovery of ClONO$_2$ is rapid in later winter and early spring, and its yield in early spring is much higher than that in late fall.

in polar regions, which was adapted from a diagram originally in the WMO report [1994]. As shown in Figs. 3.9, and 3.10 here, this diagram is instructive and in accord with the observations generally well [Waters *et al.*, 1993; Webster *et al.*, 1993; Toohey *et al.*, 1993; Roche *et al.*, 1994]. It is worth noting that the conversion of chlorine reservoirs (HCl and ClONO$_2$) into active chlorine is quite rapid upon the presence of PSCs in early winter. Indeed, NASA UARS CLAES data have shown that the largest depletion of ClONO$_2$ occurs in June and July, while its recovery is rapid in later winter (August and early September) [Roche *et al.*, 1994], *not* late spring as expected from the photochemical models described in the previous Section. Remarkably, the yield of ClONO$_2$ in early spring is much higher than that in late fall (Fig. 3.9). The latter implies that there are likely other sources of active chlorine than heterogeneous chemical reactions of inorganic chlorine species (HCl and $ClONO_2$) in PSCs.

Moreover, NASA UARS MLS data have also shown that chlorine in the lower stratosphere is almost completely converted to chemically reactive forms in both the northern and southern polar winter vortices [Waters *et al.*, 1993; Webster *et al.*, 1993; Toohey *et al.*, 1993]. The ClO species, the predominant form of chemically-reactive chlorine responsible for stratospheric O_3 destruction, is formed in the polar stratosphere from the very beginning of winter when it is dark (in early June for Antarctica), long before the development of the largest Antarctic ozone hole in September and October. In fact, it has been observed that greatly enhanced ClO is present from early June, with more ClO in August than in September [Waters *et al.*, 1993], though the largest reduction in vortex ozone is usually not observed until September/October. The latter was suggested to be due to diabatic descent of ozone-rich air in the polar vortex, resulting in a net flux of ozone into the lower stratosphere during the earlier winter to counter chemical loss indicated by the enhanced ClO [Waters *et al.*, 1993].

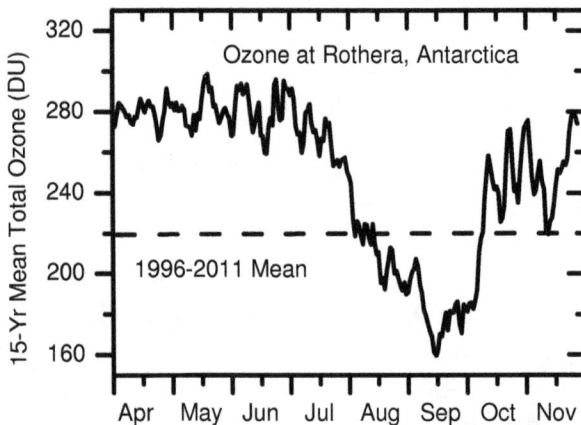

Fig. 3.10. The 15-year mean time series total O_3 data averaged from real-time daily total O_3 data at Rothera in Antarctica over the 16 years (1996-2011) recorded by the British Antarctic Survey (BAS) [Lu, 2013].

However, this attribution seems problematic and the observation must be due to one or more other causes, because the observed data for the period before the mid-1970s when there were no significant stratospheric CFCs showed no considerable changes in total ozone over Antarctica from the end of the fall (April) to the end of winter (September) (see Fig. 3.2 above and Fig. 5.10 in Chapter 5).

(3) The photochemical models would predict no chemical ozone loss in the lower polar stratosphere during the Antarctic winter from May to early August, which is in total darkness. That is, no ozone loss occurs until sunlight returns to the polar vortex in spring and allows the production of active chlorine and initiates the catalytic ozone destruction cycles. In contrast, significant ozone loss over Antarctica and Arctic in the dark winter stratosphere was actually observed [*e.g.*, Becker *et al.*, 1998; Lu, 2013]. Although the 'ozone hole', defined as the areas with total column ozone values below 220 DU in current atmospheric chemistry, is usually observed only after the return of sunlight in early spring, large ozone loss from the normal value above 300 DU to around 220 DU has indeed already occurred in darkness in the winter. This is schematically shown in Fig. 3.9 and actually observed in the data shown in Fig. 3.10 [Lu, 2013], *which is drastically different from the observation by Dobson in the 1950s shown in Fig. 3.2.* More data about this key observation will be further presented and discussed in Chapter 5.

(4) The photochemical models would predict no destruction of CFCs in the lower polar stratosphere in dark winter. In contrast, time-series data from direct satellite measurements have shown significant continuous destruction of CFCs since the end of fall (the very beginning of winter) [Lu, 2010a, 2013], as shown in Fig. 3.11. More observations will also be presented in Chapter 5.

(5) Quantitatively, there exist persistent discrepancies between photochemical models and observed results, as revealed by reviewing the modeled results documented in a series of *WMO Scientific Assessments of Ozone Depletion in 1994-2014.* For instance, it was predicted in the 1994 WMO Report: "Peak total chlorine/bromine loading in the troposphere is expected to occur in 1994, but the stratospheric peak will

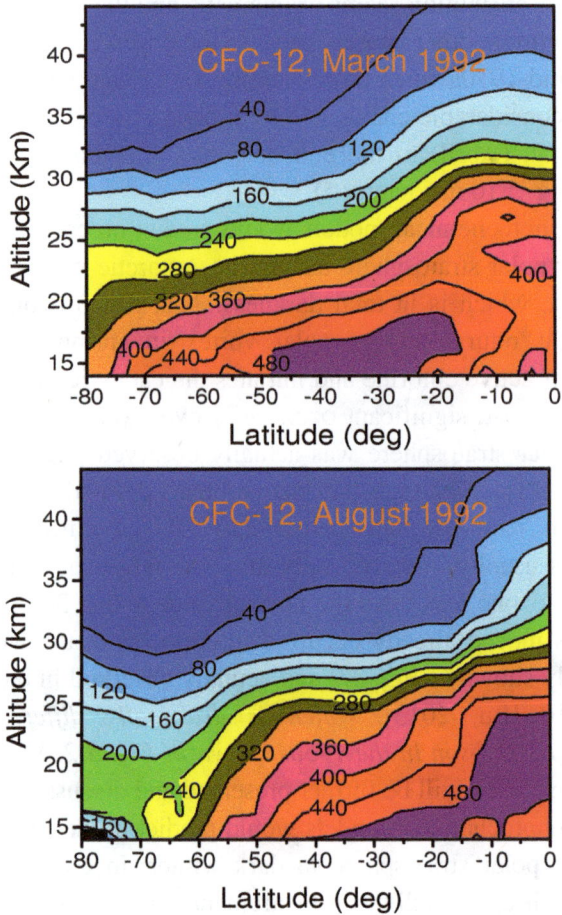

Fig. 3.11. Images of CFC-12 (CF_2Cl_2) levels in ppt in the 27-31 March (fall) and 16-23 August (late winter) in 1992, showing how CFC-12 in the Antarctic stratosphere varies with latitude, altitude and time. Based on data from NASA UARS's CLEAS datasets [Lu, 2013]. Note that a very different behavior is actually seen in the long-lived trace gas CH_4 [see discussion in Chapter 5].

lag by about 3 - 5 years", and therefore "Peak global ozone losses are expected to occur during the next several years". Then the 2010 WMO Report (published in 2011) concluded: "Observed Antarctic springtime column ozone does not yet show a statistically significant increasing trend", in spite of the observed declining of CFCs in the stratosphere due to the Montreal Protocol. Fig. 3.12 shows the projected variations of the Equivalent Effective Stratospheric Chlorine (EESC) (in units of parts per trillion) given in the newest WMO Report [2014], which were calculated from the photochemical models for the midlatitude and polar stratosphere based on global mean tropospheric abundances measured at the surface. It is assumed that, on average, air reaches stratospheric midlatitudes in roughly 3 (±1.5) years and stratospheric polar regions in 5.5 (±2.8) years, close to those (3.0±1.5 and 6.0±3.0 years, respectively) given in the last WMO Report [2010]. The results in Fig. 3.12 show that by 2012, EESC had declined by about 10% in the Antarctic and about 15% in midlatitudes from their peak values of 10–15 years ago [WMO, 2014]. These are due to decreases in atmospheric abundances of CH_3CCl_3, CH_3Br, and CFCs.

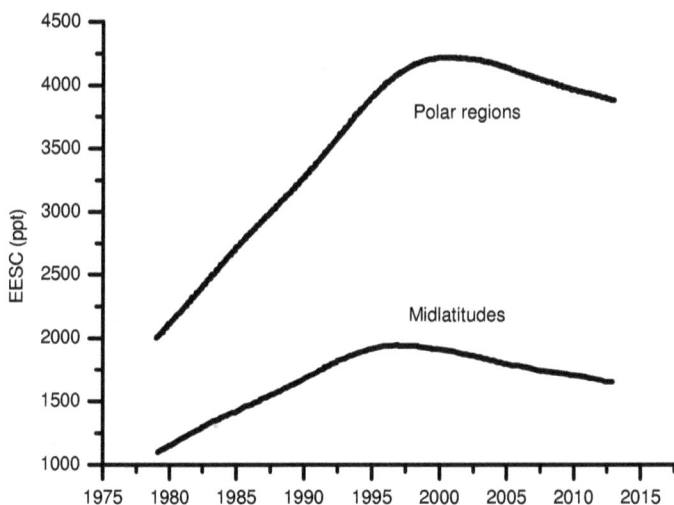

Fig. 3.12. Variations in Equivalent Effective Stratospheric Chlorine (EESC) (in units of parts per trillion) calculated from the photochemical models for the midlatitude and polar stratosphere based on global mean tropospheric abundances measured at the surface. It is assumed that, on average, air reaches stratospheric midlatitudes in roughly 3 (±1.5) years and stratospheric polar regions in 5.5 (±2.8) years. The EESC in the polar regions, where essentially all the ODSs have decomposed to yield active chlorine and bromine that can destroy ozone, is more than a factor of two higher than at midlatitudes. Based on data from WMO [2014].

The Report [WMO, 2014] reviews that total column ozone declined over most of the globe during the 1980s and early 1990s, by about 2.5% in the global mean, but has remained stable since 2000. It also states that there are indications of an increase in global-mean total column ozone over 2000–2012, consistent with (photochemical) model predictions (Fig. 3.13a). The Report also notes: "However, a total column ozone increase that would be attributable to ODS (ozone-depleting substance) decreases has not yet been observed". In the Report, the projected future evolution of tropical total column ozone is strongly dependent on future abundances of CO_2, N_2O, and CH_4 (*e.g.*, as in Representative Concentration Pathways—RCPs), and is particularly sensitive to changes in the tropical upwelling and changes in tropospheric ozone.

The Report [WMO, 2014] also notes that polar ozone depletion continues to occur in the springtime Antarctica and Arctic, and the greater variability in Antarctic springtime polar ozone over the last decade cannot be attributed to recovery from the effect of the ozone-depleting substances according to the photochemical models (Fig. 3.13b).

The WMO Report [2014] certainly gives very conscientious and painstaking reviews of the current status of ozone research and is instrumental to policy makers. On the other hand, we might need to be somewhat cautious about the observations reviewed there. The observed data of total column ozone at midlatitudes come actually from interrupted multiple ground- and space-based measurements shown in Fig. 3.13a, which are different from the NASA satellite measurements for polar ozone changes shown in Fig. 3.13b. The former shows a continuous increase in midlatitude ozone since the mid-1990s, which appears to be consistent with what is expected from the photochemical model (see Fig. 3.12). However, as long as the data from the same source as for polar total ozone (NASA satellites) are used, the midlatitude ozone has actually shown a continuous decreasing trend since 1979, that is, *no recovery in midlatitude ozone has been observed* [Lu, 2013]. This is clearly shown in Fig. 3.14a, which shows a striking difference from Fig. 3.13a given in the WMO Report [2014].

Another important issue is that although the midlatitude total ozone data shown in the Report [WMO, 2014] appear to agree fairly well with

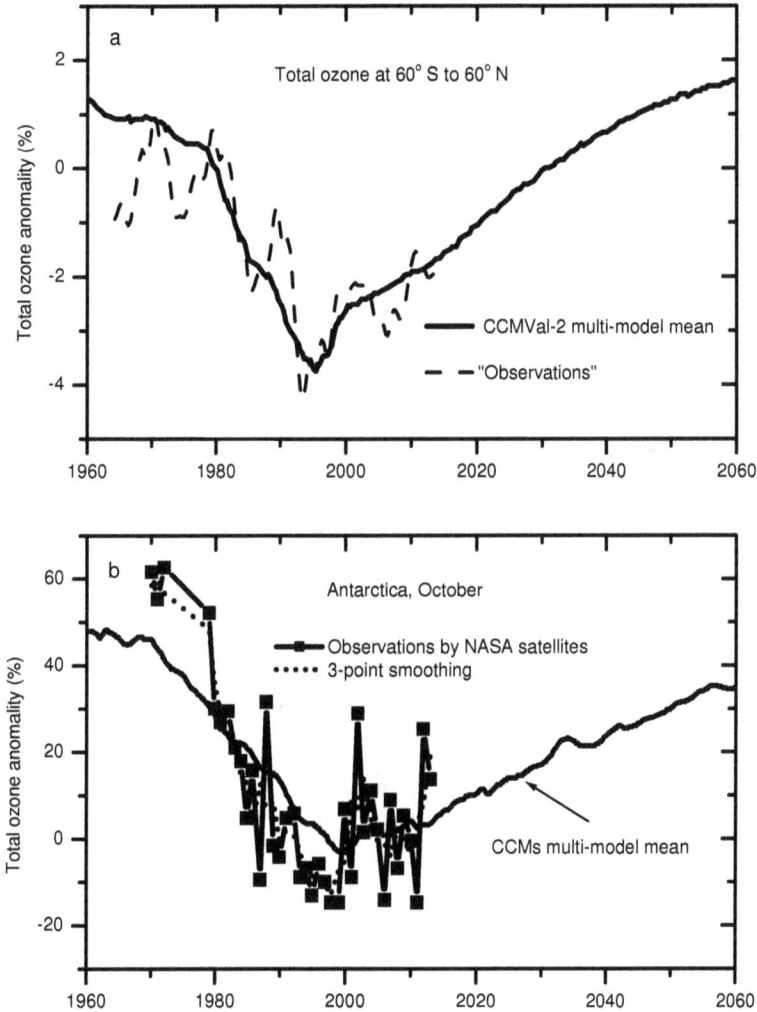

Fig. 3.13. Top panel (a): Variation in average total column ozone at midlatitudes (60° S to 60° N) between 1960 and 2060, from the multiple-model mean (MMM) of CCMVal-2 simulations, compared with the 'observed' column ozone changes between 1965 and 2013 (dash line) from mixed ground, balloon and satellite observations. Bottom panel (b): Total column ozone changes for the Antarctica (60°-90° S) in October, from the CCM multi-model mean (MMM) Antarctic total ozone anomalies in percent relative to a 1998 to 2008 base period, compared with the observations by NASA satellites. Based on data from WMO (2014).

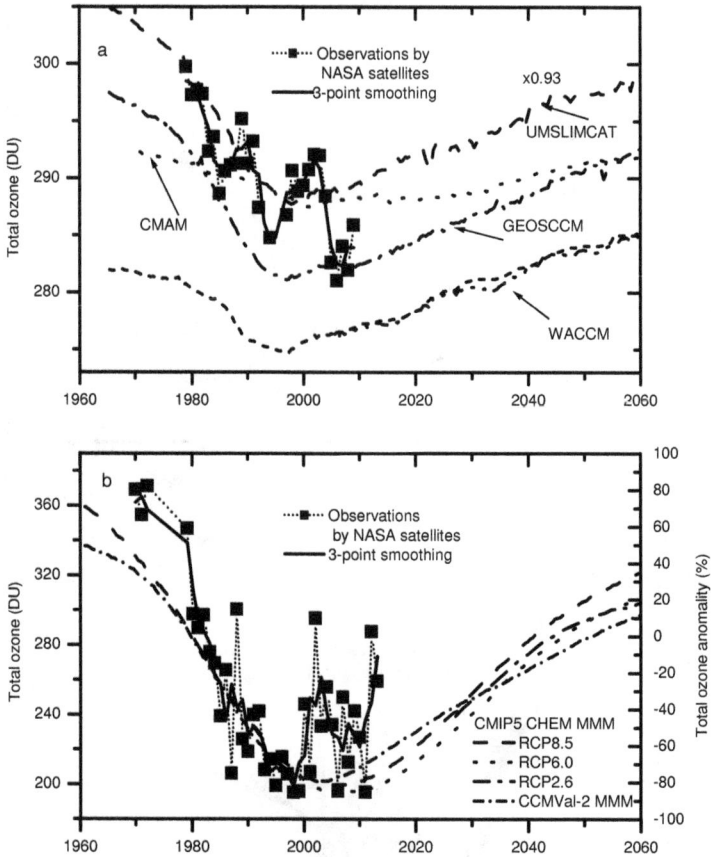

Fig. 3.14. Top panel (a): Comparison of annual means total ozone column changes at mid-latitudes (60°S to 60°N) observed by NASA satellites and simulated by various chemistry-climate models: UMSLIMCAT (Unified Model Single-Layer Isentropic Model of Chemistry and Transport), CMAM, GEOSCCM, and WACCM. The simulations used the Representative Concentration Pathway (RCP) 6.0 scenario of greenhouse gases (GHGs), except for UMSLIMCAT, which uses the RCP4.5 scenario. WACCM is coupled to an interactive ocean; the other simulations use prescribed sea surface temperatures. Bottom panel (b): Total column ozone changes for the Antarctica (60°-90° S) in October, from the multi-model means (MMMs) of CCMVal-2 and CMIP5 CHEM models in percent relative to the 1980 Antarctic total ozone anomaly, compared with the observations by NASA satellites. Based on data from Lu [2013] for mid-latitudes (60°S to 60°N) and WMO [2014] for the October Antarctica (60°-90° S) with the same NASA satellites. Besides the original observed data, a 3-point smoothing is applied (the solid line penetrating the symbols).

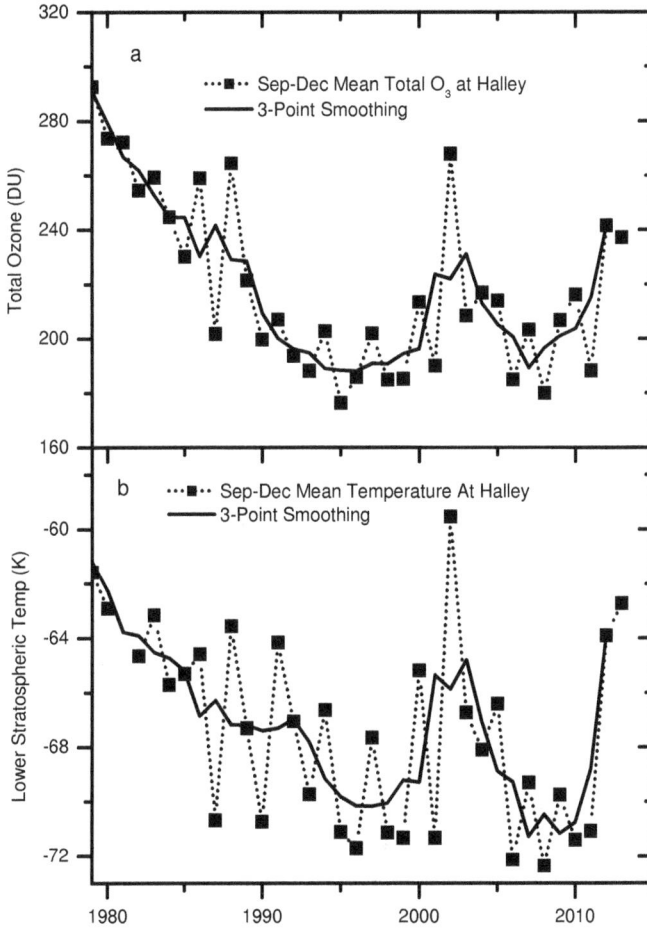

Fig. 3.15. Observed time-series 4-month mean total ozone and lower stratospheric temperatures at Halley station in the Antarctic-ozone-hole months during 1979-2013. The symbols are averages of original observed data, while the solid lines are the 3-point smoothed observed data. Based on data from Lu [2013, 2014].

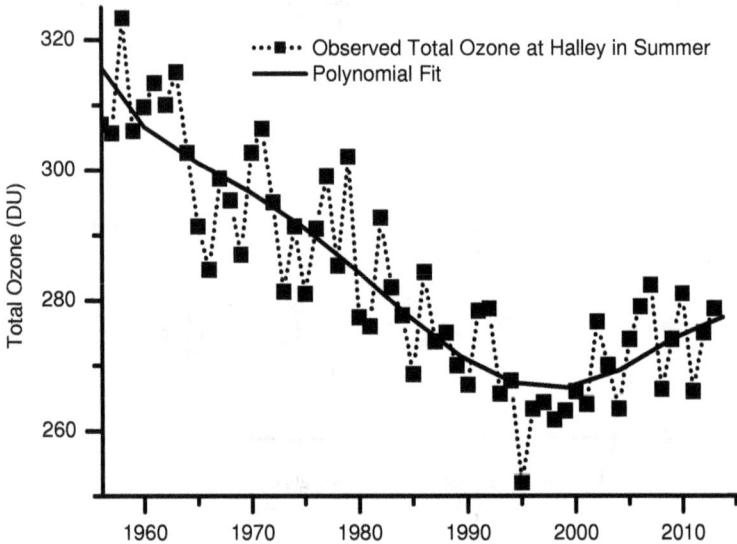

Fig. 3.16. Observed time-series 3-month mean summer-time total ozone at Halley, Antarctica in January, February and March over the period of 1956-2013. A polynomial fit to the observed data is also shown. It is clearly shown that ozone at Halley has exhibits a solid recovery since around 1995. Updated from Lu [2010a].

the multiple-model mean (MMM) of CCMVal-2 simulations, there exist very large differences between the simulated results, from model to model, as also shown in Fig. 3.14a.

(6) For polar total ozone changes, the state-of-the art photochemical models cannot reproduce the well-observed pronounced modulations of ozone loss in the springtime Antarctic ozone hole and of resultant stratospheric cooling in 11-year cycles [Lu, 2010a, 2013, 2014c], which is shown in Figs. 3.14b and 3.15.

(7) Furthermore, polar ozone data available from the longest-recorded BAS since 1956 have shown a clear and steady recovery in the summertime total ozone over the Antarctic station, Halley, since around 1995, giving an instantaneous response to the declining of ozone-depleting substances (mainly CFCs). This was reported previously [Lu, 2010a], and is clearly shown in Fig. 3.16. This observation shows a

distinct difference from the predictions of photochemical models (see Fig. 3.12).

(8) Chemistry transport models (CTMs) might partially reproduce the observed ozone, but their simulations require the use of *observed* temperatures and winds and thus do not have the capability to predict future changes of the ozone hole. The ability of current photochemical models including CTMs to predict future ozone hole trends is very limited; improving their predictive capabilities for the ozone hole is one of the greatest challenges in the ozone research community [*e.g.*, Manney *et al.*, 2011].

3.6 Concluding remarks

In summary, there exist persistent quantitative discrepancies between photochemical models and observations, even though the models include a large number of parameters. There are still large gaps in our understanding of the activation of chlorine in the stratosphere. This conclusion is similar to that made by some researchers a decade ago [*e.g.*, Orlando and Schauffler, 1999].

According to the photochemical models, neither a clear recovery (increase) in total column ozone at midlatitudes nor a recovery in springtime polar ozone attributable to the declining of the ozone-depleting substances regulated by the Montreal Protocol has been observed [WMO, 2014]. The state-of-the-art photochemical models have included the effects of non-halogen greenhouse gases (GHGs), CO_2, N_2O, and CH_4. It is also predicted that the evolution of the ozone layer in the late 21st century would largely depend on the atmospheric abundances of these non-halogen GHGs. In particular, increases of CO_2, and to a lesser extent N_2O and CH_4, would cool the stratosphere radiatively and thus elevate global ozone. However, these models give results with very large uncertainties, varying from model to model.

The photochemical models appear to be able to roughly explain the springtime ozone hole. Upon comparing the models with the observed data in detail, however, one can find that there indeed exist significant discrepancies between photochemical models and observations. Notably, current photochemistry-climate models cannot reproduce the observed 11-year cyclic variations of polar ozone loss, nor can they capture the essential features of polar stratospheric cooling [Lu, 2010a, 2013]. An ozone increase (recovery) that would be attributable to the observed decreases in ozone-depleting substances has not yet been observed according to the photochemical models [WMO, 2014]. In contrast to the predictions of photochemical models, *a solid and clear recovery in the summer total ozone over Antarctica has indeed been observed (see Fig. 3.16).* These observations indicate that some important processes are missing in current models of ozone depletion chemistry. There is still a need to place the Montreal Protocol on a firmer and more precise scientific basis.

It is important to point out that there exist physical processes rather than photochemical reactions, especially dissociative electron attachment (DEA) of molecules, which can give a similar distribution of halogen species in the stratosphere, given the observed increasing electron density with rising altitude in the general stratosphere (similar to the distribution of solar UV photons, see Fig. 1.9). In particular, ozone-depleting halogenated molecules both organic and inorganic, such as CFCs, HCl and $ClONO_2$, are extremely effective for dissociative electron transfer (DET) reactions, especially on the surfaces of PSC ice particles. The DEA and DET processes and relevant publications in the literature should not have been long ignored and excluded in our understanding of ozone depletion and the ozone hole in the WMO Reports [2006, 2010, 2014]. This issue will be presented and discussed in Chapters 4 and 5.

Chapter 4

The Cosmic-Ray-Driven Theory of the Ozone Hole: Laboratory Observations

4.1 Introduction

In Chapter 3, we discuss the photochemical models of ozone depletion and the ozone hole. Here, we present a fairly different physical mechanism in this chapter, which will be centered on electron-driven reactions of atmospheric molecules. This will focus more on the contributions of this author and co-workers to this topic since the late 1990s, though others' relevant work will also be discussed.

4.2 Dissociative electron attachment to halogenated gases

There is a long history of experimental and theoretical studies of dissociative electron attachment (DEA) of halogenated molecules including chlorofluorocarbons (CFCs) to low-energy free electrons since the 1950s [Hickam and Berg, 1958; Curran, 1961; Bansal and Fessenden, 1972; Christophorou, 1976; Illenberger *et al.*, 1979; Peyerimhoff and Buenker, 1979; Christophorou *et al.*, 1984; Oster *et al.*, 1989; Chu and Burrow, 1990]. By the late 1970s, it had been well observed that DEA of gaseous CFC molecules to low-energy free electrons near zero eV is an extremely efficient process, with measured DEA cross sections approximately 10000 times the photodissociation cross sections of CFCs peaking near 7.6 eV [Bansal and Fessenden, 1972; Christophorou, 1976; Illenberger *et al.*, 1979; Oster *et al.*, 1989].

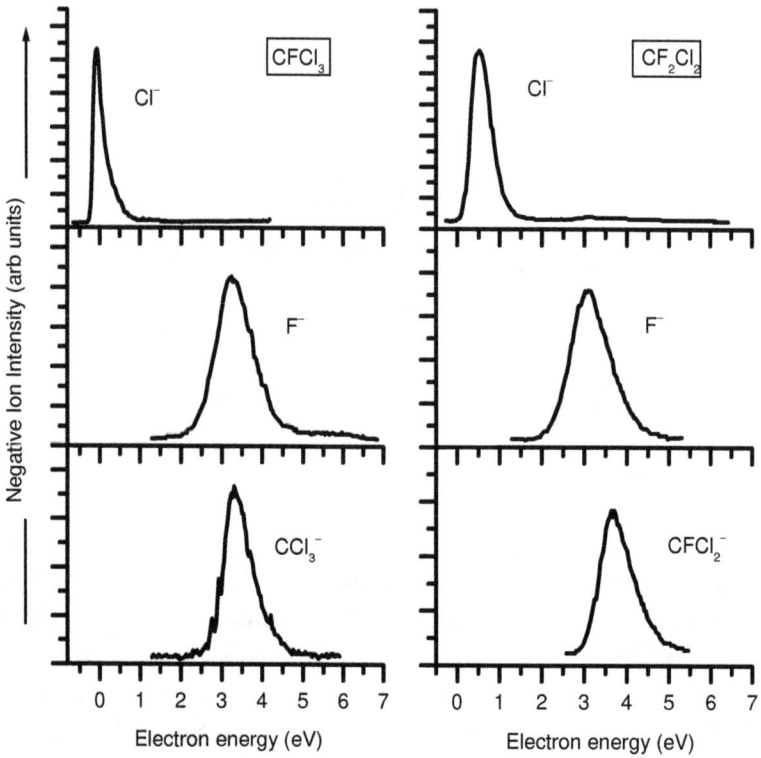

Fig. 4.1. Negative ion yield from dissociative electron attachment to $CFCl_3$ (CFC-11) and CF_2Cl_2 (CFC-12). Based on data from Illenberger *et al.* [1979].

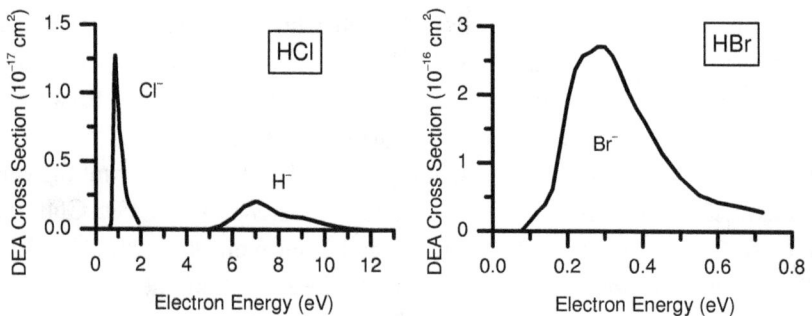

Fig. 4.2. Dissociative electron attachment cross sections of HCl and HBr. Based on data of HCl from Petrovic *et al.* [1988] and of HBr from Christophotou *et al.* [1968].

In particular, Illenberger *et al.* [1979] reported measurements with a good electron energy resolution that DEA of CFC molecules such as CF_2Cl_2 (CFC-12) and $CFCl_3$ (CFC-11) to low-energy free electrons near zero eV is an extremely efficient process (Fig. 4.1)

$$e^- (\sim 0eV) + CF_2Cl_2 \rightarrow CF_2Cl_2 *^- \rightarrow Cl^- + CF_2Cl \qquad (4.1)$$

$$e^- (\sim 0eV) + CFCl_3 \rightarrow CFCl_3 *^- \rightarrow Cl^- + CFCl_2. \qquad (4.2)$$

The electron attachment rate constants of gaseous CF_2Cl_2 and $CFCl_3$ are $\sim 2.0 \times 10^{-9}$ and $\sim 2.0 \times 10^{-7}$ cm^3 s^{-1}, respectively [Christophorou *et al.*, 1984], and their corresponding DEA cross sections at near 0 eV are $\sim 1.1 \times 10^{-16}$ cm^2 and 9.5×10^{-15} cm^2, respectively [Illenberger *et al.*, 1979]. These DEA cross sections are extremely large, $\sim 10^4$ times the photodissociation cross sections of CFCs in the order of $10^{-20} - 10^{-19}$ cm^2.

DEAs to inorganic halogenated molecules such as HCl and HBr were also studied [Petrovic *et al.*, 1988; Christophotou *et al.*, 1968]. As shown in Fig. 4.2, these molecules also have DEA resonances at electron energies near zero eV. Their DEA cross sections are also large, in the order of 10^{-17} to 10^{-16} cm^2.

Table 4.1. Electron attachment rate constants for halogenated molecules at 300 K.

Molecules	Rate constant k (cm^3 s^{-1})	Molecules	Rate constant k (cm^3 s^{-1})
CH_3Cl	$<5\times10^{-15,\,a}$	CF_2Cl_2	$1.7\times10^{-9,\,f}$; $1.9\times10^{-9,\,e}$
CH_3Br	$6.0\times10^{-12,\,b}$	$CFCl_3$	$2.4\times10^{-7,\,e}$
CH_3I	$7\times10^{-8,\,c}$	CCl_4	$2.8\times10^{-7,\,c}$
CH_2Cl_2	$4.8\times10^{-12,\,d}$	CHF_2Cl	$<3.3\times10^{-13,\,g}$
$CHCl_3$	$4.7\times10^{-9,\,e}$	$CHFCl_2$	$1.5\times10^{-12,\,g}$
CF_3Cl	$4.2\times10^{-13,\,e}$	$ClONO_2$	$1.1\times10^{-7,\,h}$; $2.9\times10^{-7,\,h}$

a. Christodoulides *et al.* [1975]; b. Alge *at al.* [1984]; c. Christophorou [1976]; d, Schultes *et al.* [1975]; e. Burns *et al.* [1996]; f. Wang *et al.* [1998]; g. Christodoulides *et al.* [1978]; h. Van Doren *et al.* [1996].

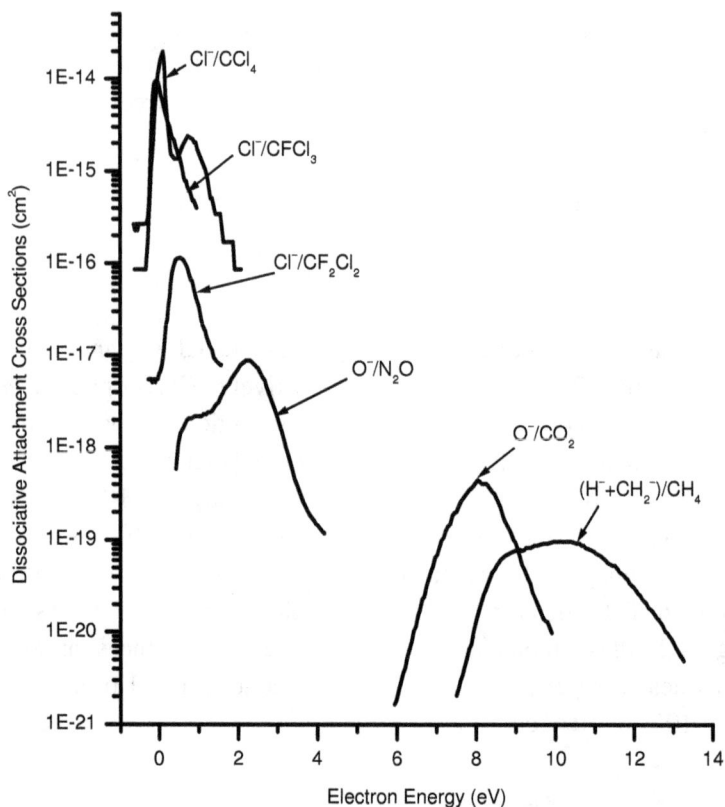

Fig. 4.3. Dissociative electron attachment cross sections versus electron energy for important atmospheric molecules (CFCs, N_2O, CO_2 and CH_4). Based on data from Illenberger *et al.* [1979], Oster *et al.* [1989], and Christophorou *et al.* [1984].

Electron attachment rate constants for various halogenated gases at room temperature (300 K) are summarized in Table 4.1, while DEA cross sections as a function of electron energy of some important atmospheric molecules are shown in Fig. 4.3. The measured electron attachment rate constants (DEA cross sections) of CFCs are extremely large, up to $\sim 3 \times 10^{-7}$ cm^3 s^{-1} ($\sim 2 \times 10^{-14}$ cm^2).

The DEA reactions of CFCs (Eqs. 4.1 and 4.2) are exothermic. This can be deduced from thermodynamical data. The energetics of DEA to a molecule AB to form a negative ion B$^-$ can be expressed as (see Fig. 2.1)

$$E_e = D(A\text{-}B) - E_A(B) + E_{ex}, \qquad (4.3)$$

where E_e is the electron energy, $D(A\text{-}B)$ the bond dissociation energy, which can be calculated by $D(A\text{-}B) = \Delta H_f^0(A) + \Delta H_f^0(B) - \Delta H_f^0(AB)$ with ΔH_f^0 being the heat of formation for each neutral species, $E_A(B)$ the electron affinity of the neutral B, and E_{ex} is the excess energy carried by the fragments. The lowest electron energy threshold E_e^{th} for the DEA is

$$E_e^{th} = D(A\text{-}B) - E_A(B) . \qquad (4.4)$$

With $E_A(Cl) = 3.61$ eV, $D(CF_2Cl\text{-}Cl) = 3.58$ eV and $D(CFCl_2\text{-}Cl) = 3.21$ eV [Dispert and Lacmann, 1978], Eq. 4.4 gives $E_e^{th} = -0.03$ eV for CF_2Cl_2 and $E_e^{th} = -0.4$ eV for $CFCl_3$. Thus, dissociation of CFCs can occur not only by attachment of free electrons but for transfer of weakly-bound electrons. The latter is called the dissociative electron transfer (DET) process, which, for example, was observed in collisions of highly excited Rydberg atoms or neutral alkali atoms with gaseous $CFCl_3$ and CCl_4 [Foltz *et al.*, 1977; Dispert and Lacmann, 1978].

Immediately after the experimental measurements, Peyerimhoff and co-workers [Peyerimhoff and Buenker, 1979; Lewerenz *et al.*, 1985] made the first theoretical studies of the DEAs of CFCs and pointed out that the DEA process, effectively reducing the amount of Cl released from the photolysis of CFCs, must be seen in competition to the photodissociation process and must be considered as a factor in evaluating stratospheric ozone depletion.

As discussed in Chapter 1, the major source producing electrons in the stratosphere is simply the atmospheric ionization by cosmic rays (CRs). Striking the atmosphere, the ionization of molecules by CRs generates an enormous number of low energy secondary electrons. However, the free electron concentration drops sharply with altitude: it is $\sim 10^3$ electrons cm^{-3} at ~ 85 km, and $\sim 10^1$ electrons cm^{-3} at ~ 50 km (see Fig. 1.9). Below this height, the detected density of *free* electrons is very low. This is because most of the generated free electrons are rapidly captured by stratospheric molecules (mainly O_2) to produce the primary negative ion O_2^- by termolecular electron attachment to O_2 due to the large abundance of the latter (Reaction 2.22). DET from O_2^- to CFCs such as CF_2Cl_2 and $CFCl_3$ can take place:

$$O_2^- + CF_2Cl_2 \rightarrow O_2 + CF_2Cl_2^{*-} \rightarrow O_2 + Cl^- + CF_2Cl, \qquad (4.5)$$

$$O_2^- + CFCl_3 \rightarrow O_2 + CFCl_3^{*-} \rightarrow O_2 + Cl^- + CFCl_2. \qquad (4.6)$$

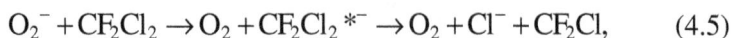

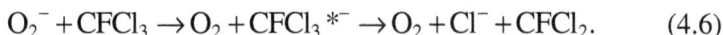

But the rate constants of these DET reactions in the gas phase were measured to be low, with $k=2.1\times10^{-10}$ cm^{3}s^{-1} and $k=7.6\times10^{-10}$ cm^{3}s^{-1} for Reaction 4.5 and 4.6, respectively, and no DET reactions from the hydrates of negative ions O_2^-, O_3^-, CO_3^-, NO_2^- and NO_3^- were found [Fehsenfeld *et al.*, 1979]. The gaseous DEA/DET process was therefore thought to be an insignificant sink for CFCs in the lower atmosphere, though it was once identified as the most interesting prospect (and also the most controversial) in the 1970s [Fehsenfeld *et al.*, 1979; Smith and Adams, 1980; Torr, 1985]. Despite the general agreement that this understanding of negative-ion chemistry in the stratosphere was rather speculative [Smith and Adams, 1980; Torr, 1985], the DEA/DET process has been completely excluded in current atmospheric chemistry context. However, this complete neglect of electron-induced reactions of halogenated molecules as an efficient process for the destruction of the ozone layer is perhaps a big mistake [Lu and Madey, 1999b; Lu and Sanche, 2001a; Lu, 2009, 2010a, 2013].

4.3 Discovery of extremely effective dissociative electron transfer (DET) of halogenated molecules on ice

4.3.1 *DET vs DEA*

The DET process is similar to DEA to electrons, but there are some essential differences. As discussed in Chapter 2 (Sec. 2.3.1), DEA occurs when a low-energy (0-20 eV) *free, unbound* electron resonantly attaches to a molecule to form a transient negative ion (TNI) state, which then dissociates into a neutral and an anionic fragment: $e^- + AB \rightarrow AB^{*-} \rightarrow A + B^-$. In contrast, DET occurs by rapid electron transfer of *a weakly-bound electron* localized at an atom/molecule or in a polar medium to a foreign molecule, forming a TNI that then dissociates. As reviewed recently [Lu, 2010a], DET reactions of molecules such as CFCs which have a strong DEA resonance with free electrons at near zero eV in the gas phase can effectively occur when these molecules are adsorbed on polar ice surfaces and dissolved in polar liquids. This is because the potential energy curve of the TNI AB*$^-$ is lowered by the polarization potential E_p of 1–2 eV to lie below that of the neutral AB in the Franck-Condon (electron-transition) region. To evaluate the effect of halogenated molecules on atmospheric ozone depletion, the present review and discussion will mainly be focused on DET reactions of halogenated molecules adsorbed on polar ice surfaces or present in polar liquids. Taking CF_2Cl_2 adsorbed on the surface of H_2O ice ($E_p \approx 1.3$ eV) [Lu *et al.*, 2002] as an example, the DEA and DET processes are illustrated in Fig. 4.4. In contrast to the DEA process, the DET process has two main characteristics: (1) the lifetime of a weakly-bound trapped electron in polar media is orders of magnitudes longer than that of a free electron in the gas phase or a quasi-free electron in nonpolar media; and (2) the autodetachment of the TNI AB*$^-$ once formed cannot occur in DET. These properties can greatly enhance the capture probability of the electron and the dissociation probability of the molecule in a DET reaction [Lu, 2010a].

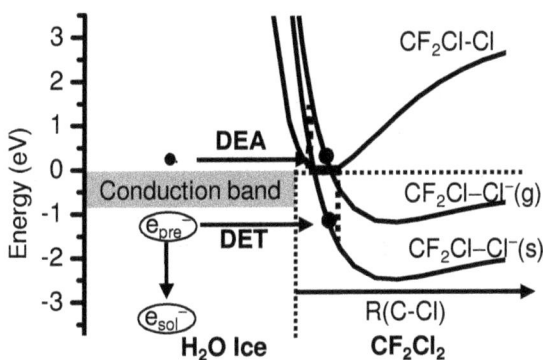

Fig. 4.4. Potential energy curves (PECs) for dissociative electron attachment (DEA) of CF_2Cl_2 to a ~0 eV free electron in the gas phase (g) and for dissociative electron transfer (DET) of a weakly-bound pre-solvated electron (e_{pre}^-) to CF_2Cl_2 adsorbed on H_2O ice surface (s). Without being captured by a CFC, the e_{pre}^- would proceed to a solvated state (e_{sol}^-). The PECs of gas-phase CF_2Cl_2 and CF_2Cl_2*$^-$ are constructed from gaseous thermodynamic data, while the PEC of CF_2Cl_2*$^-$ adsorbed on H_2O ice surface (s) is obtained with the polarization energy of ~1.3 eV. Adapted from [Lu, 2010a].

4.3.2 *Electron-stimulated desorption (ESD) experiments of CFCs adsorbed on surfaces*

From Eq. 4.3, one can see that the kinetic energies of dissociation fragments arising from DEAs/DETs of molecules to near 0 eV electrons are very low. This causes no serious problems for experiments in the gas phase. However, it is extremely difficult to detect the desorbing yield of those halogen negative ions (*e.g.*, Cl^-) on a solid surface. This is because most of the resultant anions are trapped at the surface by the image potential (typically ≥1.0 eV), and therefore the probability of negative ions desorbing from the surface is extremely small. According to the studies by Polanyi and co-workers [Dixon-Warren *et al.*, 1991, 1993], only a very small fraction (a maximum of ~10^{-7}) of the Cl^- ions formed by DEAs of nearly zero eV (weakly-bound hot) electrons to halomethanes adsorbed on a metal surface is able to overcome the image-potential barrier and desorb into the vacuum. Thus, the detection of desorbing Cl^- ions is an extremely technically challenging task. In the late 1990s, few laboratories in the world possessed this capability; those of Professor John Polanyi at the University of Toronto and Professor Theodore Madey at Rutgers—The State University of New Jersey were two really exceptional. For detection of negative ions with an extremely low yield, a pulsed gating method must be applied to veto the negative ion detector upon arrival of a large number of secondary electrons generated by primary electrons/photons, i.e., to open the detector just before the arrival of negative ions to be detected. This can avoid over-warming (pre-saturation) at the detector especially when a high voltage is applied to achieve the highest detection efficiency. The relatively unique facility at Rutgers allowed observing true DEA/DET reactions of intact CFCs with low electron doses and avoiding the artificial effects of their reaction products and sample damage.

Researching on a US NSF-funded, renewed project aimed to study elastic and inelastic processes during transmission of low-energy (≤10 eV) ions through nanoscale ultrathin surface overlayers, this author had to generate low-energy anions from the metal substrate in the first step and then studied the physical properties (changes in kinetic energy, angular distribution, and the yield) of ions in transport through the ultrathin overlayers. Prior to this author's joining the Laboratory of Surface Modification at Rutgers as a postdoctoral fellow in early 1997, the group led by Madey had been productive on the project over the previous three year period (1994-1997), and they just published a comprehensive review article in *Surface Science Reports* on the topic [Akbulut, Sack and Madey, 1997]. Obviously Lu and Madey faced the challenge to make new breakthroughs from this fairly well-studied but still interesting NSF-funded project. By accident, this author found a gas cylinder in the Laboratory, introduced the gas into an ultrahigh vacuum (UHV) chamber, and analyzed it by a UTI-100C quadrupole mass spectrometer (QMS) that had been connected to the UHV chamber. It turned out that the gas was CF_2Cl_2 (CFC-12), one of the most important ozone-depleting gases.

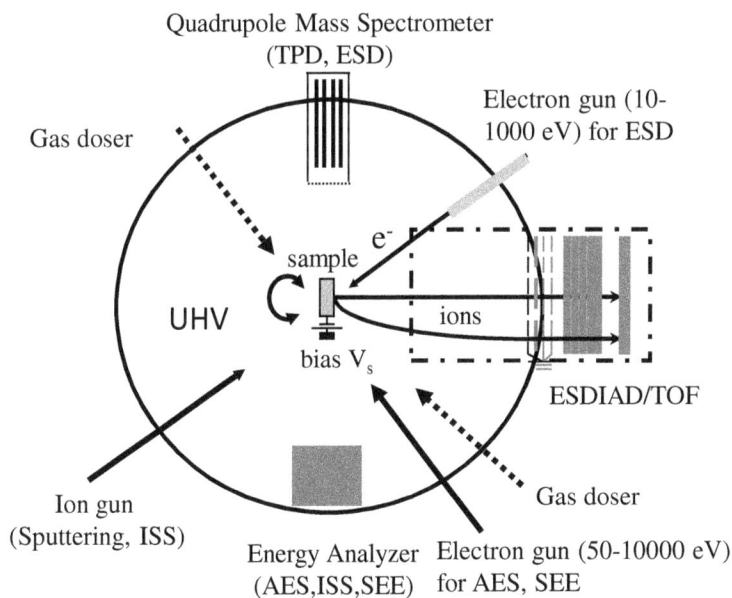

Fig. 4.5. Schematic diagram of an ultrahigh vacuum (UHV) surface analyzed system equipped with AES, LEED, TDS, ESD/ESDIAD and ISS.

The Lu-Madey experiments were conducted in an excellent UHV chamber that constantly reached to a base pressure $\sim 4\times10^{-11}$ torr. As schematically shown in Fig. 4.5, the chamber was equipped with apparatus for Auger electron spectroscopy (AES), low-energy electron diffraction (LEED), thermal desorption spectroscopy (TDS), ion scattering spectroscopy (ISS), as well as an electron stimulated desorption ion angular distribution (ESDIAD) detector with time-of-flight (TOF) capability for mass- and angle-resolved ion detection. The ESDIAD was developed as a surface science technique by Theodore Madey and John Yets in the 1970s [Madey and Yets, 1971]. The ESDIAD/TOF detector in Madey's laboratory at Rugters consisted of four grids, five microchannel plates (MCPs), and a position-sensitive resistive-anode encode (RAE) (Fig. 4.6). The latter was connected to a position-analyzing computer to obtain a direct acquisition of two-dimensional digital data. This permitted a direct measurement of the yield and angular distribution of a specific cation or anion species. By pulsing the primary electron beam and gating a retarding potential at the

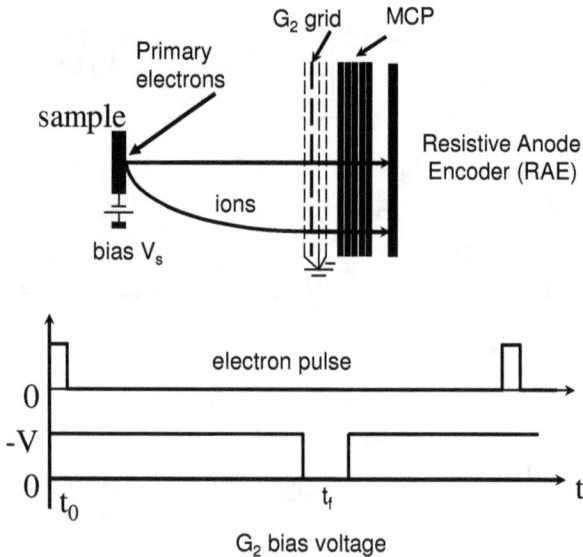

Fig. 4.6. ESDIAD detector with a time of flight (TOF) capability. The gate for the detector is only open just before desorbing negative ions arrive at the detector. Adapted from Akbulut, Sack and Madey [1997].

entrance grid G to quench the secondary electron signal received by the MCPs, one could obtain an extremely high efficiency for detecting desorbing anions. The latter is a crucial step for detection of anions arising from DEAs/DETs of halogenated molecules adsorbed on a surface.

In the Lu-Madey experiments, a Ru(0001) crystal was cooled to 25 K with a closed-cycle helium refrigerator and heated to 1600 K by electron bombardment. The surface was cleaned by sputtering using 1 keV Ar^+ and annealing in oxygen; its cleanliness was checked by AES and work-function measurements. The purity of CF_2Cl_2 and coadsorbate (H_2O, NH_3, CH_3OH, $(CH_3)_2CO$, CH_4, Xe, Kr, etc.) gases was checked by mass spectra obtained with a QMS as each of the gases was introduced into the chamber. CF_2Cl_2 and coadsorbate gases were dosed normally onto the surface in sequence at 25 K with two separate directional dosers, and their relative coverages were determined using TDS. In these studies, one monolayer (ML) was defined as the coverage corresponding to the saturation of the monolayer peak(s) in thermal desorption spectra, i.e., the onset of the multilayer peak [Lu *et al.*, 1998]. In measurements of negative/positive ions, a bias voltage of −/+100 eV was applied to the sample for increasing the detection efficiency. The electron current was adjustable between 0.01 nA and 20 nA with a beam spot ~1mm^2 and the collection time for each data point was 5 seconds, to avoid detector saturation and to minimize beam damage. Secondary electron energy spectra were recorded with a concentric hemispherical electrostatic analyzer (50 mm radius), and a bias voltage of −10 V was applied to the sample in order to measure the low-energy threshold.

Lu and Madey first studied anion formation in electron-stimulated desorption (ESD) of CF_2Cl_2 adsorbed on a clean Ru(0001) surface with an incident electron beam at hundreds of eV (200-300 eV). They measured the yields of Cl^- and F^- as a function of CF_2Cl_2 coverage on the surface, and found that the anions were mainly generated by DEAs to CF_2Cl_2 of low-energy (0–3 eV) secondary free electrons emitted from the metal substrate [Lu *et al.*, 1998].

Perhaps the most striking and interesting observation made by Lu and Madey in October 1998 was that the yields of anions were enhanced *by up to ~3×10^4 times* when CF_2Cl_2 was coadsorbed with ~ 1 monolayer of polar molecular ice (H_2O, NH_3) on the metal substrate exposed to an electron beam of 250 eV [Lu and Madey, 1999a, 1999b]. For observations of the giant Cl^- enhancements in the ESDIAD measurements, reduced incident electron currents at 0.01-1.0 nA had to be used. The coverages of CF_2Cl_2 and polar molecules were carefully determined from TDS spectra, where one monolayer (ML) of H_2O referred to a bilayer with a density of $\sim 1.0 \times 10^{15}$ molecules/cm^2 and one ML of CF_2Cl_2 was defined as the coverage corresponding to the saturation of the monolayer peak in TDS spectra, i.e., the onset of the multilayer peak.

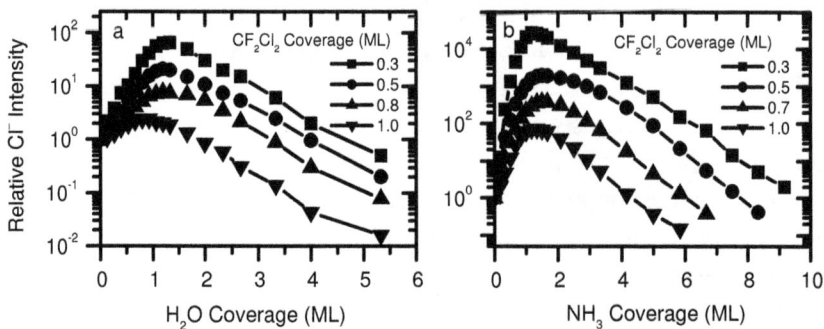

Fig. 4.7. Relative Cl^- yields desorbing from 250 eV primary electrons incident onto various amounts of CF_2Cl_2 covered Ru(0001) at ~ 25 K as a function of (a) H_2O or (b) NH_3 coverage, where Cl^- yields are normalized to the initial value at zero H_2O/NH_3 coverage. Based on data from Lu and Madey [1999b].

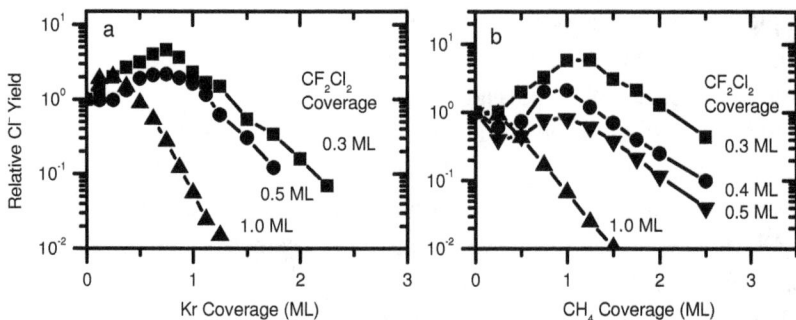

Fig. 4.8. Same as Fig. 4.7, but for coadsorption with (a) krypton (Kr) and (b) methane (CH_4). Based on data from Lu and Madey [2000, 2001].

The variations of the Cl⁻ yield versus H_2O and NH_3 film thickness for various CF_2Cl_2 pre-coverages are re-plotted here as Fig. 4.7. In the case of H_2O coadsorption, it is seen that for the lowest CF_2Cl_2 coverages, the Cl⁻ yield increases greatly with the initial coadsorption, exhibits a maximum enhancement by nearly two orders of magnitude at about one ML of H_2O, and finally decreases to zero intensity at 4.0-5.0 ML H_2O overlayer. With higher H_2O thickness, the Cl⁻ yield decreases, which is expected due to elastic and inelastic scattering as the desorbed ions pass through the H_2O film and due to the limited penetration depth of low-energy secondary electrons in H_2O. Similar results were observed for the coadsorption of NH_3, but the enhancement is about two orders of magnitude larger than for H_2O coadsorption: the maximum Cl⁻ enhancement for 0.3 ML CF_2Cl_2 is a factor of ~3×10^4. Also, F⁻ enhancements were observed for either H_2O or NH_3 coadsorption, but the magnitude of enhancement for F⁻ is much smaller than for Cl⁻ with an identical CF_2Cl_2 coverage [Lu and Madey, 1999a, 1999b].

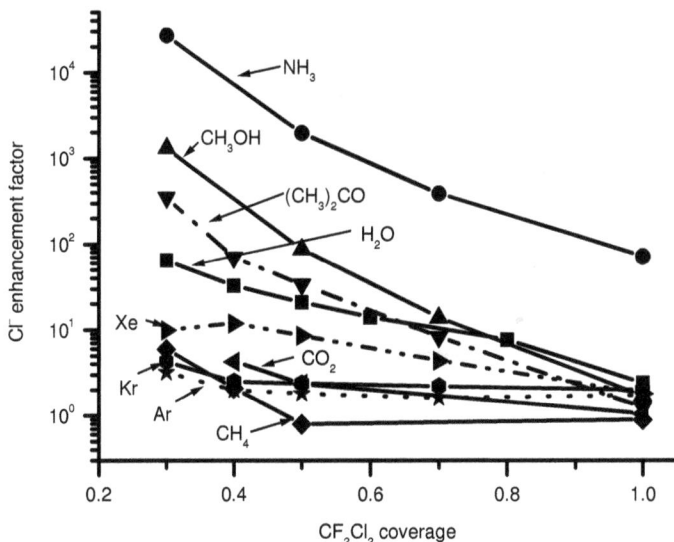

Fig. 4.9. Maximum magnitudes of Cl⁻ enhancements versus CF_2Cl_2 pre-coverage on Ru(0001) for coadsorption with polar H_2O, CH_3OH, $(CH_3)_2CO$, NH_3 and nonpolar Xe, Kr, Ar, CH_4 and CO_2 at 25 K. Based on data from Lu and Madey [1999a, 1999b, 2000, 2001].

Lu and Madey also studied the anion enhancements for coadsorption of pre-deposited CF_2Cl_2 on Ru(0001) with various polar and non-polar gases, including polar H_2O, NH_3, CH_3OH, $(CH_3)_2CO$, and nonpolar Xe, Kr, Ar, CH_4 and CO_2. As shown in Figs. 4.8 and 4.9, the results show that the large enhancements by more than one order of magnitude occur for coadsorption with polar molecules only [Lu and Madey, 2000, 2001].

Similar anion enhancements were observed for ESD of 0.3 ML CF_2Cl_2 adsorbed on top of H_2O-precovered Ru surfaces with various H_2O thicknesses [Lu and Madey, 1999b]. The variation of the Cl^- desorption yield is re-plotted as Fig. 4.10. At the H_2O coverage of ~1 ML, the Cl^- yield exhibits a maximum of two orders of magnitude higher than that without the presence of H_2O. With larger H_2O spacer thickness, the Cl^- yield decreases. This is due to the finite tunneling depth (2~4 ML) of low-energy secondary electrons from the substrate through the ice film [Gilton *et al.*, 1989; Chakarov and Kasemo, 1998]. The higher

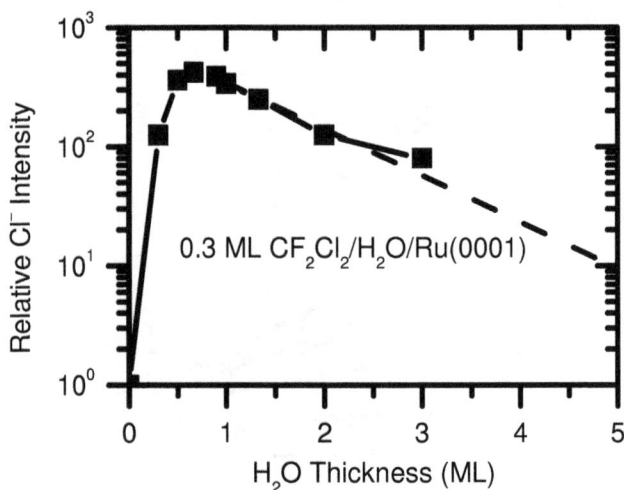

Fig. 4.10. Relative Cl^- yields desorbing from 250 eV primary electrons incident onto 0.3 ML CF_2Cl_2 adsorption on top of H_2O-precovered Ru(0001) surfaces at ~25 K as a function of the H_2O spacer thickness. The dash-dot line is to extrapolate the Cl^- yield to the H_2O spacer thickness of 5 ML, which is at least one order of magnitude less than the maximum Cl^- yield at ~1 ML H_2O. Based on data from Lu and Madey [1999b].

desorbed Cl⁻ yield for CF_2Cl_2 adsorbed on top of a water-ice surface (Fig. 4.10) than for its coadsorption with water on a metal substrate (Fig. 4.7) is due to a lower ion scattering loss probability and a lower image-potential attraction (a higher desorption probability) on the water-ice surface than on the metal substrate.

4.3.3 *Proposal of a dissociative electron transfer (DET) mechanism*

To find a correct mechanism to explain the ESD observations described above, Lu and Madey had to first consider several possible scenarios that had already existed in the literature by the late 1990s. First, the possible effects of the metal substrate on the observed anion enhancements must be considered. It was observed that the F⁻ yield enhancements were much smaller than those of the Cl⁻ yield, and both F⁻ and Cl⁻ enhancements were much smaller (less than one order of magnitude) when CF_2Cl_2 was coadsorbed with nonpolar atoms such as rare gases (Xe, Kr and Ar) or nonpolar molecules (CH_4 or CO_2) on Ru [Figs. 4.8 and 4.9]. It was therefore concluded that the metal substrate could only play a minor role (such as through secondary-electron yield change, work-function variation and image-potential effects) in enhancing the anion yields by the presence of polar H_2O/NH_3 ice [Lu and Madey, 2001].

Second, DEA cross sections of molecules in the condensed-phase medium might be orders of magnitude larger than those in the gas phase [Sanche *et al.*, 1995; Fabrikant *et al.*, 1997]. This enhancement had been described by a R-matrix calculation: the polarization interaction between the TNI and the medium lowers the potential curve of the TNI and thus increases the survival probability of the resonance against autodetachment [Fabrikant *et al.*, 1997]. However, this condensed-matter effect is expected to be most effective for molecules that have DEA resonances at electron energies *far above zero eV* and adsorb on rare-gas films, but ineffective for molecules (*e.g.*, CCl_4) having DEA resonances near zero eV [Bass *et al.*, 1996] or adsorbed on polar molecular films (*e.g.*, H_2O film) [Huels *et al.*, 1994].

Third, negative ion yields in ESD may be strongly affected by the change in the surrounding environment in the condensed phase. For instance, it was observed by Akbulut, Sack and Madey [1997] that the F^- yield from ESD of ~1 monolayer (ML) of PF_3 adsorbed on a Ru(0001) is enhanced by 1.5~4 when a ~1 ML rare-gas (Xe, Kr) or water film is deposited on the PF_3 monolayer. This enhancement was attributed to a dielectric screening effect: the dielectric layer on the surface induces a potential barrier that increases the survival probability of desorbing ions [Akbulut, Sack and Madey, 1997].

Furthermore, in ESD of fractional-monolayer molecules (ABs) adsorbed on top of thick rare gas (RG) films (tens of monolayers), negative-ion enhancements due to formation of anionic excitons (RG^{*-}) at 7-10 eV were observed by Rowntree *et al.* [1993]. And they proposed a mechanism involving resonant coupling between a *core-excited* negative-ion state (AB^{*-}) of the molecule and an anionic exciton (RG^{*-}): $RG^{*-} + AB \rightarrow AB^{*-} + RG$, followed by dissociation of AB^{*-}. Moreover, much larger cross sections for charge trapping in molecules adsorbed on a glassy n-hexane (nHg) film than adsorbed on a Kr film were observed [Nagesha and Sanche, 1998]. This was attributed to electron trapping in image states of the substrate; the nHg film had a negative electron affinity (*i.e.*, the vacuum level is below the conduction-band minimum), which enhanced the lifetime of low-energy free electrons trapped in image states (caused a slow decay into the substrate).

Cowin and co-workers [Marsh *et al.*, 1988] also made an interesting observation of photoelectron-induced dissociation of CH_3Cl on Ni(111) under UV irradiation, though it was not clear whether the fragment of CH_3 resulted from the DEA of a low-energy photoexcited *free* electron or from the DET of a weakly-bound subvacuum photoexcited electron from the metal. Particularly they also observed an enhancement by ~50 times of the yield of the neutral CH_3 fragment in photoreduction of CH_3Cl adsorbed on Ni(111) surface with the presence of ~1 ML H_2O spacer layer [Gilton *et al.*, 1989]. The CH_3-yield enhancement then quickly decreased with increasing H_2O thickness and was about one order of magnitude lower at 5 ML H_2O due to the limited tunneling depth (2~4 ML) of low-energy electrons in ice. The observed ESD Cl^--yield enhancements shown in Fig. 4.10 quite resembles this observation, though the CH_3Cl coverage was not given in the paper of Gilton *et al.* Gilton *et al.* [1989] attributed the CH_3-yield enhancement to a strong inelastic scattering interaction between the photoexcited electrons and the H_2O, compared with the monotonic and slow decrease of the CH_3 signal with increasing Xe spacer thickness. This interpretation without involving pre-solvated electrons trapped in H_2O and associated DET, however, cannot explain the observed much larger anion enhancements in electron-induced dissociation of CF_2Cl_2 by the presence of NH_3 than by H_2O, since inelastic electron scattering by NH_3 (with a dipole moment of 1.47 D) is weaker than by H_2O (1.84 D).

Lu and Madey also studied the modifications by coadsorbates of the secondary-electron spectrum, the work function and the adsorption sites of the CF_2Cl_2-precovered surface, and their relative contributions to the anion enhancement. They found that these factors did not play a major role in the large anionic enhancements [Lu and Madey, 2001].

None of the mechanisms mentioned above could explain the observed giant enhancements of F^- and Cl^- yields in ESD of a fractional monolayer of CF_2Cl_2 adsorbed on Ru(0001) by coadsorption with some polar molecules, such as H_2O and NH_3. Therefore, Lu and Madey had to extend their search scope of the clues underlying the anion enhancements to the observations beyond condensed-phase and surface science experiments.

Fortunately, Lu and Madey noticed a basic fact that an excess electron can become self-trapped (solvated) in a polar medium. The solvated electron was first observed in liquid NH_3 by Weyl [1864] and in liquid H_2O by Boag and Hart [1963] about 100 years later. The advent of femtosecond ($1fs=10^{-15}$ s) time-resolved laser spectroscopy in 1987 provided an unprecedented understanding of the dynamics of electron solvation in water, as first studied by Migus *et al.* [1987]. By the end of 1990s, it became clear that prior to the formation of the equilibrium-state solvated electron (e_{sol}^-), the excess electron in bulk water is located at precursor states with finite lifetimes less than 1 picosecond ($1ps=10^{-12}$ s), the so-called prehydrated electron (e_{pre}^-) [Rossky and Schnitker, 1988; Long *et al.*, 1990; Silva *et al.*, 1998]. And negatively charged water clusters $e^-(H_2O)_n$ (n=2-69) were also first observed by Haberland and

Bowen's groups in 1980s-1990s [Armbruster *et al.*, 1981; Coe *et al.*, 1990].

Furthermore, Lu and Madey also noticed a critical observation that gaseous NH_3 had indeed long been employed as a reagent gas to enhance the detection efficiency of organic and inorganic halogenated molecules in ammonia-enhanced anion mass spectrometry though the role of gaseous NH_3 was unknown there [Currie and Kallio, 1993; Chaler *et al.*, 1998]. Most interestingly, the solvation of electrons generated by adding electron donors (alkali metals) into polar media such as liquid ammonia had already been adopted as an effective method for dehalogenation of environmentally hazardous halogenated materials including CFCs. In the latter, halogen atoms were reduced to halogen ions and dechlorination of CFCs was also observed to be much more efficient than defluorination [Oku *et al.*, 1988; Mackenzie *et al.*, 1996]. These observations implied a common reaction mechanism that should be operative in all the gas, liquid and solid phases.

After the notice of the above key observations and a painstaking thought, Lu and Madey [1999a, 1999b, 2000, 2001] proposed a *dissociative electron transfer* mechanism to explain the observed anion enhancements in ESD of CFCs adsorbed on polar ice surfaces. They proposed that secondary electrons with energies of nearly 0 eV, produced by bombardment of the metal substrate with high-energy (200-300 eV) electrons, are injected and trapped in the polar molecular (H_2O/NH_3) layer; the giant anionic enhancements are due to transfer of trapped electrons (e_t^-) in polar ice to CFCs that then dissociate into Cl^- and a neutral fragment. The DET process, e.g. for CF_2Cl_2, can be expressed as

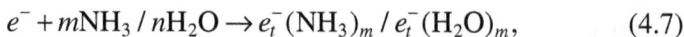

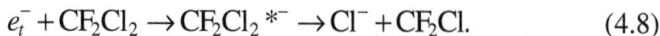

$$e^- + mNH_3 / nH_2O \rightarrow e_t^- (NH_3)_m / e_t^- (H_2O)_m, \qquad (4.7)$$

$$e_t^- + CF_2Cl_2 \rightarrow CF_2Cl_2 {*}^- \rightarrow Cl^- + CF_2Cl. \qquad (4.8)$$

In contrast to e_t^- in polar media, low-energy secondary electrons in nonpolar media (*e.g.*, rare-gas films) remain quasi-free and thus have extremely short residence times (lifetimes) in femtoseconds (10^{-15} s), decaying quickly into the metallic substrate without transfer to CFCs. Therefore the anionic enhancements via the DET mechanism in nonpolar media is very limited, in spite of the fact that coadsorption of rare-gas atoms leads to a larger yield of secondary electrons from the metal substrate [Lu and Madey, 2001]. Moreover, according to the DET mechanism, the anion enhancement factor *EF* can be expressed as:

$$EF=(Y_{DA}+Y_{DET})/Y_{DA}=1+ Y_{DET}/Y_{DA}, \qquad (4.9)$$

where Y_{DA} is the amount of CFC molecules dissociated by DEA of low-energy free electrons from the metal and Y_{DET} the amount of CFC molecules dissociated by DET of trapped electrons in ice. Y_{DA} is usually proportional to the CFC coverage unless all secondary electrons are depleted (not the case under normal experimental conditions). However, it should be noted that only a small percent of low-energy free electrons become trapped electrons in the ice layer of ~1.0 ML and they have a much longer lifetime to react with CFC molecules. Thus, there is an upper limit for the Y_{DET} value: the maximum value, $(Y_{DET})_{max}$, is equal to the total number of trapped electrons (Y_e), which depends on specific experimental conditions. This leads to a result that at very low CFC coverages, the Y_{DET} is proportional to the CFC coverage; at high CFC coverages, $Y_{DET}=(Y_{DET})_{max}=Y_e$. Consequently, the anion enhancement factor *EF* should reduce with rising CFC coverages at low to intermediate coverages, while $EF \approx 1$ (no enhancement) at high CFC coverages. This is in good agreement with the observed results.

Most of the Cl⁻ ions arising from the DET reaction expressed as reactions (4.7) and (4.8) are trapped at the surface by the image potential, as the desorption probability of Cl⁻ ions resulting from near 0 eV electrons is extremely low (10^{-7}–10^{-6}) [Dixon-Warren *et al.*, 1991, 1993; Lu and Madey, 1999b]. Note that based on the results shown in Fig. 4.10, the DET cross sections of CF_2Cl_2 adsorbed on H_2O ice, after removing the possible effects of the metal substrate, was measured to be ~1.0×10^{-14} cm², which is *six orders of magnitude* higher than the photodissociation cross section (10^{-20} cm²) of CF_2Cl_2 [Lu and Madey, 1999b]. Thus, a very low electron dose must be used in order to make real measurements of the DET reaction.

After the finding of Lu and Madey [1999a, 1999b], Langer *et al.* [2000] reported no anion enhancements for electron-induced reactions of

CF_2Cl_2 with NH_3. However, the experimental conditions of Langer *et al.* were very different from those of Lu and Madey. In those gas-phase or cluster experiments, two problems could exist. First, Langer *et al.* used a commercial QMS detection system, which did not have the capability to detect desorbing Cl^- ions resulting from DEA/DET resonances at near 0 eV on a metal surface even for thick multilayers of adsorbed CF_2Cl_2. And they used a large incident electron current of 40-50 nA. Langer *et al.* [2000] thus concluded that "desorption at very low electron energies is not operative". Second, detailed TDS spectra of CF_2Cl_2 adsorbed the surface were not recorded in those experiments, which are often required to determine adsorbate coverages reliably. The latter is also important since the anion enhancement is strongly dependent on CFC coverage (see Eq. 4.9).

The negative results of Langer *et al.* [2000] stimulated Madey and co-workers [Solovev *et al.*, 2004] to revisit the ESD experiments using both ESDIAD and higher-sensitivity QMS systems. The latter was modified to have the TOF capability by using a pulsed electron beam and a pulsed detector gate. The main results of Solovev *et al.* is reproduced as Fig. 4.11, which have substantially confirmed the giant Cl^- enhancements by $>10^3$ times for submonolayers of CF_2Cl_2 co-adsorbed with NH_3 on Ru(0001), originally observed by Lu and Madey. Solovev *et al.* also observed some differences: the Cl^- enhancements observed by the QMS detector, are significantly smaller those with the ESDIAD detector, e.g., the Cl^- enhancement factor at 1 ML CF_2Cl_2 by 1.5 ML NH_3 coadsorption decreased from 150~200 to less than 50. These differences were attributed to technical origins such as mass resolution, background subtractions and different collection angles for QMS and ESDIAD detectors. Unfortunately, a key experimental difference was not mentioned. The ESDIAD detector at Rutgers [Fig. 4.6] has an extremely high sensitivity, orders of magnitude higher than the QMS detector with a single channel electron multiplier at Rutgers and probably in most commercial QMS systems. Thus, the required electron dose with the ESDIAD detector is much lower than that required by a QMS. High electron doses can cause significant damage to the sample, in particular when DET cross sections of CF_2Cl_2 are greatly enhanced up to 10^{-14} – 10^{-12} cm^2 by H_2O and NH_3 [Lu and Madey, 1999b]. For those experiments, low electron doses $\leq 1 \times 10^{12}$ cm^2 are required to achieve reliable measurements. Lu and Madey were aware of this critical condition, and hence low electron currents (≤ 1 nA) and a short data correction time of only 5 s were used for detection of the maximum Cl$^-$ enhancements by factors of up to 10^4 in their ESDIAD experiments [Lu and Madey, 1999a, 1999b, 2000, 2001]. Notably, Solovev *et al.* [2004] used the same facility with similar electron currents but a data collection time of 60 s, one order higher than that used by Lu and Madey. This can reasonably explain the slight differences in the ESDIAD results: at high CF$_2$Cl$_2$ coverage (1 ML), the maximum Cl$^-$ enhancement by ~1 ML NH$_3$ measured by Solovev *et al.* (Fig. 4.11b) was no less than that observed by Lu and Madey (Fig. 4.7b), but at the lowest CF$_2$Cl$_2$ coverage (0.3 ML) a smaller maximum Cl$^-$ enhancement was measured by Solovev *et al.* The most likely origin was that for 0.3 ML CF$_2$Cl$_2$, an electron current much larger than that for 1ML CF$_2$Cl$_2$ was required and a significant decomposition of the CF$_2$Cl$_2$ coadsorbed with ~1 ML NH$_3$ could not be avoided even within a single measurement (60 s) if the electron beam current was not reduced. Overall, it is not surprising to observe smaller anion enhancements with a QMS than with an ESDIAD detector, since the former required a much larger electron dose and tended to cause damage to the sample.

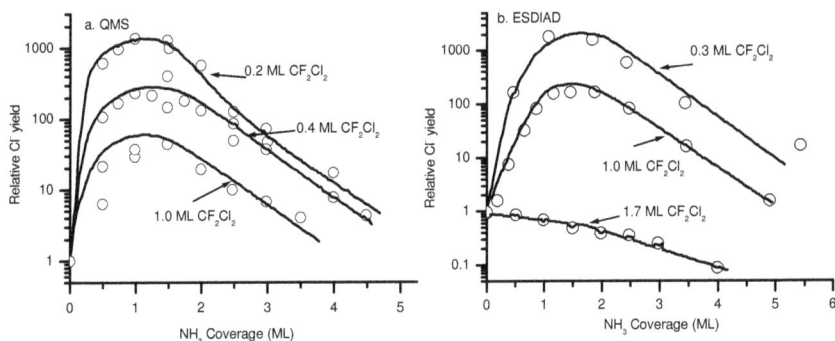

Fig. 4.11. Relative Cl$^-$ yield for various CF$_2$Cl$_2$ precoverages on Ru(0001) as a function of NH$_3$ coverage at 25 K. (a) Detected by QMS and (b) Detected by ESDIAD. Plotted with data from Solovev *et al.* [2004].

Faradzhev *et al.* [2004] also reported their measurements of the DET cross sections of CFCs adsorbed on top of 5 ML H_2O using temperature programmed desorption (TPD). Unfortunately, however, their results were essentially misinterpreted. With energy-undefined, yield-unknown secondary electrons generated by the X-ray source or a defocused 180 eV electron source from a QMS filament, these authors claimed obtained "direct" measurements of the absolute DET cross sections of CCl_4 or CF_2Cl_2 adsorbed on ice. Their measured dissociation rates of 1ML CF_2Cl_2 and 1 ML CCl_4 increased in an H_2O (D_2O) environment by ~2–3 times, which is *not* smaller than that reported by Lu and Madey [1999a, 1999b], as seen in Fig. 4.7a. However, the maximum absolute cross sections for decomposition of 0.25 ML CFCs adsorbed on top of 5 ML H_2O, using 180 eV incident electrons from the QMS filament, were measured to be $1.0\pm0.2 \times 10^{-15}$ cm^2 for CF_2Cl_2 and $2.5\pm0.2 \times 10^{-15}$ cm^2 for CCl_4. The latter is even less than the reported DEA cross sections of CCl_4 at 0 eV in the gas phase (1.3×10^{-14} cm^2) [Illenberger, 1982] and adsorbed on Kr film (5×10^{-15} cm^2) [Bass *et al.*, 1996]. Faradzhev *et al.* [2004] attempted to compare their measured dissociation cross sections of CCl_4 or CF_2Cl_2 adsorbed on ice directly with those obtained by Lu and Madey [1999b] in ESDIAD and by Lu and Sanche [2001b] in electron trapping experiments. There was a serious problem, which must be pointed out: Faradzhev *et al.* did not consider the large loss of low-energy secondary electrons in transmission from the metal substrate through 5 ML H_2O to the outmost CFC layer, as robustly seen in the previous experiments by Gilton *et al.* [1989], Chakarov and Kasemo [1998] and Lu and Madey [1999b, see Fig. 4.10]. According to these previous studies, the tunneling depth of low-energy secondary electrons in H_2O ice is 3-4 ML [Gilton *et al.*, 1989; Chakarov and Kasemo, 1998] and the secondary-electron induced dissociation cross section of CF_2Cl_2 (CH_3Cl) adsorbed on top of 5 ML H_2O is about one order of magnitude lower than on 1 ML H_2O [Gilton *et al.*, 1989; Lu and Madey, 1999b], as shown in Fig. 4.10. Thus, the TPD experiment by Faradzhev *et al.* [2004] would have given a maximum dissociation cross section of 1.0×10^{-14} cm^2 for CF_2Cl_2 on 1 ML H_2O ice if the attenuation of secondary electrons in ice were taken into account. This value would then be identical to the cross section of ~1.0×10^{-14} cm^2 measured by ESDIAD of Cl$^-$ from

CF$_2$Cl$_2$ on 1 ML H$_2$O ice with 250 eV incident electrons [Lu and Madey, 1999b] and of 1.3x10^{-14} cm^2 measured by electron trapping with ~ 0 eV electrons incident onto CF$_2$Cl$_2$ adsorbed on 5 ML H$_2$O ice [Lu and Sanche, 2001b]. Thus, the discrepancy could be removed if the secondary electron flux reacting with CFCs is carefully calibrated in the TPD experiments.

Remarkably, the advantage of the ESD experiments reviewed above is the capability of directly measuring the dissociation products (anions) from the DET reactions. On the other hand, it requires an anion detector of an extremely high sensitivity, and cautions must be taken to remove the effects of the metal substrate and the potential artificial effects mentioned above in order for reliable determination of the absolute DET cross sections [Lu and Madey, 2001].

4.3.4 *Observations of DET reactions from electron trapping measurements*

In a series of experiments, Lu and Sanche [2001a, 2001b, 2001c, 2003, 2004] directly used a low-energy (0-10 eV) electron beam reaching zero eV to examine the DET mechanism for large enhancements in dissociation of halogenated molecules (CFCs, HCFCs and HCl, *etc*) adsorbed on ice films. As shown in Fig. 4.12, a Kr spacer film of 10 ML was used to isolate any possible effects of the Pt substrate and to facilitate the growth of a uniform H$_2$O/NH$_3$ film. In the experiments, electron trapping in a dielectric film was measured by the low energy electron transmission (LEET) method developed by Sanche and co-workers [Marsolais *et al.*, 1989]. A magnetically collimated electron beam (0-10 eV) having an energy resolution of 40 meV is produced by a trochoidal monochromator. A LEET spectrum records the electron current transmitted through a dielectric film as a function of incident electron energy E, which has a sharp onset at the vacuum level defined as zero eV. If electrons are trapped in the film with a lifetime longer than the detection limit of ms, the onset curve will shift to a higher energy by ΔV. An electron trapping coefficient A$_s$(E) is defined as d(ΔV)/dt at t=0. For charging by a submonolayer of CF$_2$Cl$_2$ on the H$_2$O or NH$_3$ film deposited on the Kr spacer film, the electron-trapping cross section $\sigma(E)$ is obtained by [Lu and Sanche, 2001b, 2001c, 2003, 2004]

$$As(E) = \frac{\sigma(E)\rho_o J_o}{\varepsilon_o} (\frac{L_1}{\varepsilon_1} + \frac{L_2}{\varepsilon_2}). \tag{4.10}$$

Here, ρ_0 is the surface density of electron-trapping molecules, J_0 the incident electron density and ε_0 the vacuum permittivity. ε_1 (1.91) and ε_2 (3.3 and 3.4, respectively, for H_2O and NH_3) are the dielectric constants of the Kr and polar molecular films. L_1=32.6 Å is the film thickness for 10 ML Kr and L_2=13.0 and 11.7 Å for 5 ML H_2O and 5 ML NH_3, respectively. In the experiments, the spectra of $A_s(E)$ were measured for 0.1-0.2 ML halogenated molecules (ρ_0 is estimated to be $\sim 6\times10^{13}$, 5.8×10^{13}, 7.4×10^{13}, 5.0×10^{13} and 1.2×10^{14} /cm^2 for CF_2Cl_2, $CFCl_3$, CHF_2Cl, CH_3CF_2Cl and HCl, respectively) condensed on a 10 ML Kr

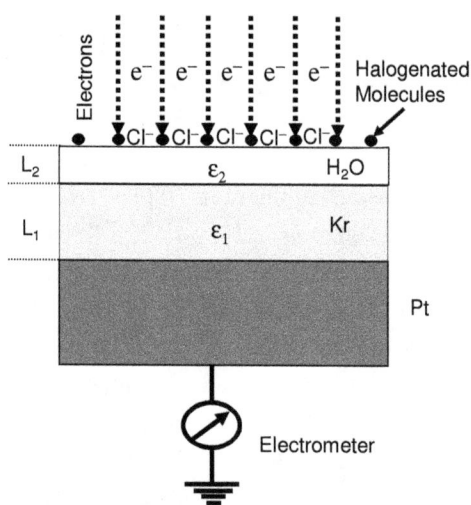

Fig. 4.12. Schematic diagram of the electron trapping setup. A 10 monolayer (ML) Kr film as a spacer was first deposited onto the Pt substrate at ~20 K; 5 ML H_2O was subsequently deposited on the Kr film. Afterward, a submonolayer of electron attaching molecules is adsorbed on top of the H_2O film. The variation of the electrostatic potential at the film surface, caused by electron trapping, is monitored by measuring the shift in the energy onset of electron transmission through the film. Adapted from Lu and Sanche [2004].

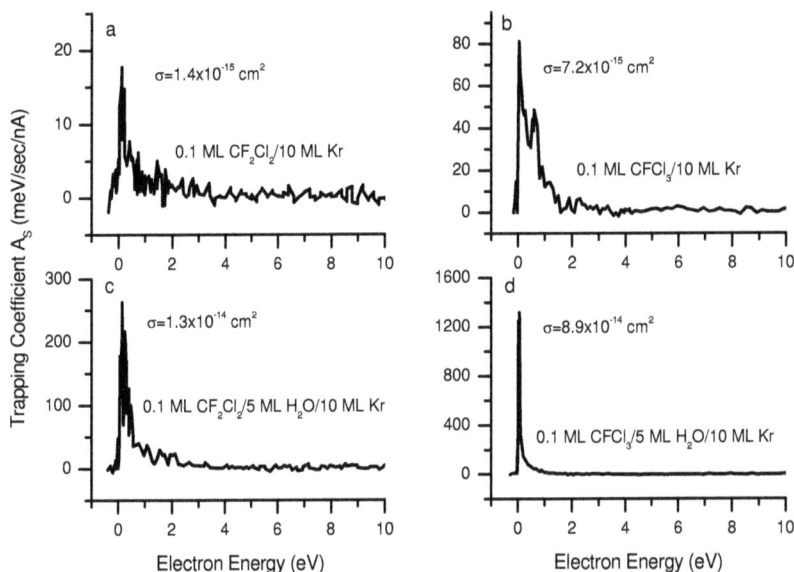

Fig. 4.13. Electron trapping coefficient A_s versus electron energy for 0.1 ML CF_2Cl_2 and 0.1 ML $CFCl_3$: (a) and (b) on 10 ML Kr; (c) and (d) on 5 ML H_2O on 10 ML Kr at ~20 K. Measured peak DA cross sections are indicated. Based on data from Lu and Sanche [2001b, 2003, 2004].

film with and without the presence of a 5 ML H_2O/NH_3 layer. The latter has properties sufficiently close to the bulk ice, and minimizes the experimental error in measured trapping cross section $\sigma(E)$ due to the uncertainty in H_2O/NH_3 thickness (L_2) since $L_1/\varepsilon_1 \gg L_2/\varepsilon_2$ in Eq. 4.10.

The measured $A_s(E)$ for 0.1 ML CF_2Cl_2 adsorbed on a 10 ML Kr surface is shown in Fig. 4.13a, from which a peak trapping cross section $\sigma=1.4\times10^{-15}$ cm^2 at nearly 0 eV was derived, which is about 14 times the DEA cross section in the gas phase. As shown in Fig. 4.13b, in contrast, a DEA resonance peaking at 0 eV with a cross section of $\sigma=7.2\times10^{-15}$ cm^2 was observed for $CFCl_3$, which is slightly smaller than the gas-phase DEA cross section (~1×10^{-14} cm^2), along with a much weaker peak appearing around 6.0 eV [Lu and Sanche, 2003]. The larger DEA cross section for CF_2Cl_2 adsorbed on the Kr surface than the gas-phase cross section has been well explained by the condensed phase effects on the lifetime, decay channels and energy of anion resonances [Lu and Sanche, 2001b], well described in the R-matrix model by Fabrikant [1994, 2007]. In contrast, the smaller DEA cross section for $CFCl_3$ adsorbed on the Kr surface than its gas-phase DEA cross section at zero eV has been attributed to the reduction in nuclear wave function overlap between the neutral AB and the anion AB^{*-} states in the Franck-Condon region [Lu and Sanche, 2003].

Of particular interest are the results of $A_s(E)$ for 0.1 ML CF_2Cl_2 and $CFCl_3$ condensed on 5ML H_2O pre-deposited on the 10 ML Kr surface, which are shown in Figs. 4.13c and d, respectively. Evidently, the presence of the polar molecular films leads to a complete quenching of DEA resonances at electron energies larger than ~1.0 eV but increases the electron trapping cross section near 0 eV sharply. However, the trapping coefficient in Figs. 4.13c and d may include contribution from long-lived electrons trapped in the pure H_2O film [Lu and Sanche, 2001b]. After subtracting the latter, a trapping cross section $\sigma=1.3\times10^{-14}$ cm^2 at ~ 0 eV for 0.1 ML CF_2Cl_2 on 5 ML H_2O was obtained, which is about one order of magnitude larger than that on Kr or two orders higher than the gaseous cross section [Lu and Sanche, 2001b]. This is quite close to the DET cross section of ~1.0×10^{-14} cm^2 measured by Lu and Madey [1999b]. Similarly, a trapping cross section of ~8.9×10^{-14} at ~ 0 eV for 0.1 ML $CFCl_3$ on 5 ML H_2O was measured, which is nearly one order higher than the gaseous cross section [Lu and Sanche, 2004].

The results of $A_s(E)$ for 0.1 ML CHF_2Cl (HCFC-22) and 0.1 ML CH_3CF_2Cl (HCFC-142b) adsorbed onto the 10 ML Kr surface, with the data from Lu and Sanche [2003], are reproduced as Figs. 4.14a and b, respectively. From these $A_s(E)$ values, the DEA cross sections for HCFC-22 and -142b were measured to be ~4.2×10^{-16} cm^2 broadly peaking at 0.78 eV and ~7.8×10^{-16} cm^2 at 0.89 eV, respectively, which are 2 to 3 orders of magnitude larger than their gaseous DEA cross sections. For HCFC-22, there are also two small peaks at 4.7 eV and 8.0 eV. The results of $A_s(E)$ obtained for 0.1 ML CHF_2Cl and 0.1 ML CH_3CF_2Cl adsorbed on 5ML H_2O predosed onto the 10 ML Kr surface, observed by Lu and Sanche [2004], are reproduced in Figs. 4.14c and d, respectively. The measured trapping cross sections at ~ 0 eV are ~5.1×10^{-15} and ~4.9×10^{-15} cm^2 for CHF_2Cl and CH_3CF_2Cl on H_2O ice, respectively, which are approximately 3-4 orders of magnitudes larger than their respective DEA cross sections in the gas phase.

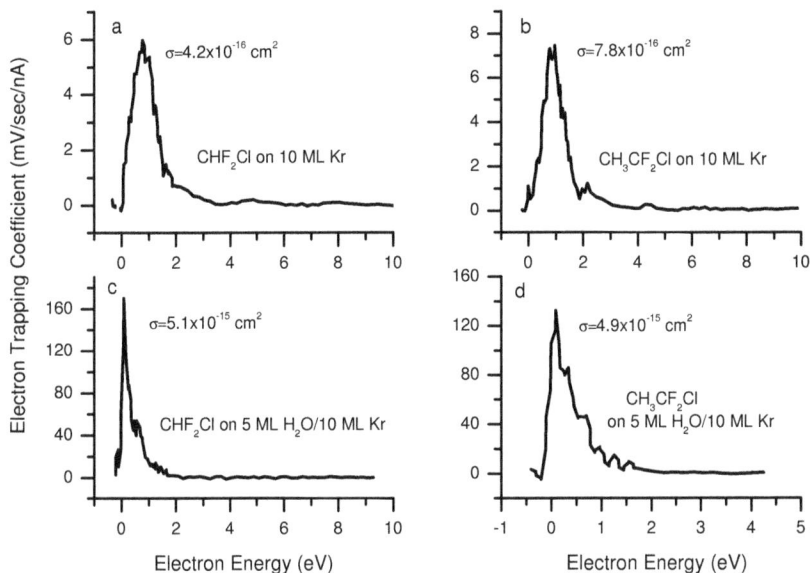

Fig. 4.14. Electron trapping coefficient A_s versus electron energy for 0.1 ML CHF_2Cl and 0.1 ML CH_3CF_2Cl: (a) and (b) on 10 ML Kr; (c) and (d) on 5 ML H_2O on 10 ML Kr at ~20 K. Measured peak DA cross sections are indicated. Based on data from Lu and Sanche [2003, 2004].

Lu and Sanche [2001c] also measured the $A_s(E)$ for HCl adsorbed on 10 ML Kr and on 5ML H_2O predosed onto 10 ML Kr, from which electron trapping cross sections $\sigma=(1.1\pm0.3)\times10^{-16}$ and $(4.0\pm1.2)\times10^{-15}$ cm^2 at ~0 eV were derived respectively (Fig. 4.15). Evidently, the presence of H_2O greatly increases electron trapping at ~0 eV by about two orders of magnitude, compared with the DEA cross section of gaseous HCl.

It is also worthwhile noting the following facts. (1) The condensed-phase effects lead to a *decrease* of the DEA cross section at zero eV of $CFCl_3$ (CCl_4) adsorbed on a non-polar Kr film. (2) For halogenated molecules (CFCs, HCFCs, HCl, *etc.*), except electron-trapping cross sections at near 0 eV, the DEA resonances at electron energies ≥1 eV observed in the gas phase and on the Kr surface are almost completely suppressed when they are adsorbed on the ice surface. More evidently for HCFCs, the higher-energy sides of the DEA resonances at electron energies ≥ 1 eV are significantly quenched when these molecules are adsorbed on H_2O ice, as clearly seen in Figs. 4.14c and d. This is also evident for CF_4, which has a DEA resonance peak around 6.0 eV observed for CF_4 adsorbed on the Kr surface by Bass *et al.* [1995], but the resonance completely disappears on the H_2O ice surface [Lu and Sanche, 2004]. These observed results demonstrate that the H_2O ice causes a quenching effect on the DEA resonances above 1.0 eV. In striking contrast, the electron trapping cross sections of these halogenated molecules at 0 eV are greatly enhanced by the presence of the H_2O ice films. It is evident that such enhancements cannot be explained by the above-mentioned condensed phase effects observed for molecules adsorbed on non-polar Kr films. Instead, these results have confirmed the DET mechanism expressed in Reactions (4.7) and (4.8). The excess electrons near 0 eV are rapidly thermalized and trapped in the polar H_2O ice to become weakly-bound pre-solvated/prehydrated electrons (e_{pre}^-). During its lifetime, e_{pre}^- is effectively resonantly transferred to a halogenated molecule that has an anion resonance in the energy range of e_{pre}^- ($-1.5 \sim -1.0$ eV) (Fig. 4.4). Note that the fully-solvated electrons with a binding energy of ~ 3.5 eV in the H_2O bulk can hardly contribute to the DETs of halogenated molecules, as well demonstrated in the experiments in liquid water [Wang *et al.*, 2006, 2008a, 2008b; Wang and Lu, 2007].

Strictly speaking, the above electron trapping measurements could not distinguish whether the electron is trapped as the Cl^- fragment or a stabilized molecular ion (*e.g.*, a $CF_2Cl_2^-$). However, it has been pointed out that even for gaseous CFCs with DEA resonances near 0 eV, the dissociation probability of the AB^{*-} transient state lies near unity once an electron is attached [Illenberger *et al.*, 1979]. This is most likely to be true for AB^{*-} states resulting from DETs of halogenated molecules with e_{pre}^- in H_2O, since their autodetachment cannot occur (see Fig. 4.4) [Wang *et al.*, 2008a; Lu, 2010a]. And desorption of Cl^- from the surface is negligible. Thus, the measured charges are approximately equal to electrons trapped as Cl^- ions, *i.e.*, σ approximately corresponds to the

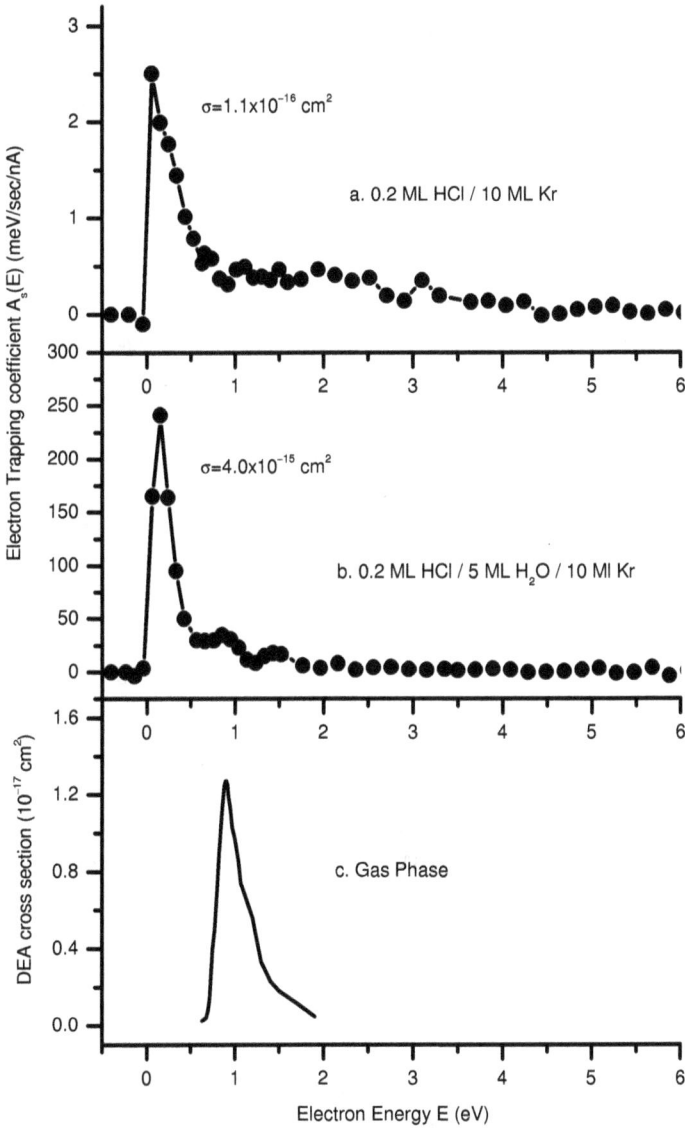

Fig. 4.15. a and b. Electron trapping coefficient A_s versus electron energy for 0.2 ML HCl on 10 ML Kr and on 5 ML H_2O deposited on 10 ML Kr at ~20 K; measured peak DA cross sections are indicated. c. DEA cross section of gas-phase HCl versus electron energy. a and b are based on data from Lu and Sanche [2001c], whereas c is based on data from Petrovic *et al.* [1988].

DEA/DET cross section. The DET cross section of CF_2Cl_2 on H_2O ice measured in electron trapping experiments [Lu and Sanche, 2001b] is nearly identical to the value estimated from the ESD measurements [Lu and Madey, 199b]. Thus, the low-electron trapping experiments at 0-10 eV with an excellent energy resolution (40 meV) provided an ideal complement to the ESD measurements, well confirming the DET mechanism.

4.3.5 *Observations of DET from femtosecond time-resolved laser spectroscopic measurements*

Since the late 1980s, researches on electron solvation dynamics have continued in water solution [Migus *et al.*, 1987; Rossky and Schnitker, 1988; Long *et al.*, 1990; Silva *et al.*, 1998; Laenen *et al.*, 2000; Wang *et al.*, 2008b], and on ice surface [Gahl *et al.*, 2002; Baletto *et al.*, 2005; Onda *et al.*, 2005; Bovensiepen *et al.*, 2009]. Prior to 2008, many experimental and theoretical studies gave very diverse lifetimes and physical natures of the prehydrated electron e_{pre}^- states in liquid water. But Wang *et al.* [2008b] then resolved the long-standing controversies, and showed that e_{pre}^- states are electronically excited states and have lifetimes of ~200 and 500 fs after the identification and removal of a coherent spike effect hidden in pump-probe kinetic traces. The e_{pre}^- lifetime of ~500 fs is consistent with the theoretical prediction by Rossky and Schnitker [1988] and recovers the earlier result observed by Long *et al.* [1990]. The coherent spike effect was also later found in other (many) pump-probe spectroscopic measurements [Luo *et al.*, 2009].

Earlier femtosecond time-resolved spectroscopic studies by Gahl *et al.* [2002] and Onda *et al.* [2005] on thin amorphous ice films (a few monolayers) on a metal or insulator substrate showed lifetimes of less than 1 ps for the presolvated electrons. By first-principles molecular dynamics simulations of the ice surface at the temperatures close to those found in PSCs (150-200 K), Baletto *et al.* [2005] found interesting results that there could exist very stable surface-bound states for trapping electron at the ice surface due to the structural rearrangement induced by an excess electron. They proposed that the surface molecular rearrangement leads to an increase of the number of dangling OH bonds pointing towards the vacuum and to the formation of an electrostatic barrier preventing the decay of the electron into the bulk solvated state. Indeed, very interesting results have subsequently been observed by Bovensiepen *et al.* [2009], who observed long-lived trapped electrons with a lifetime up to minutes at the crystalline ice surface. Fig. 4.16 shows the energy and intensity/yield of the observed trapped electrons in timescales over 17 orders of magnitude from 10^{-14} to 10^3 s. They also performed first-principle calculations, which lead to a conclusion that the observed long-lived trapped electrons is due to electron trapping at pre-existing structural defects on the surface of the crystalline ice [Bovensiepen *et al.*, 2009].

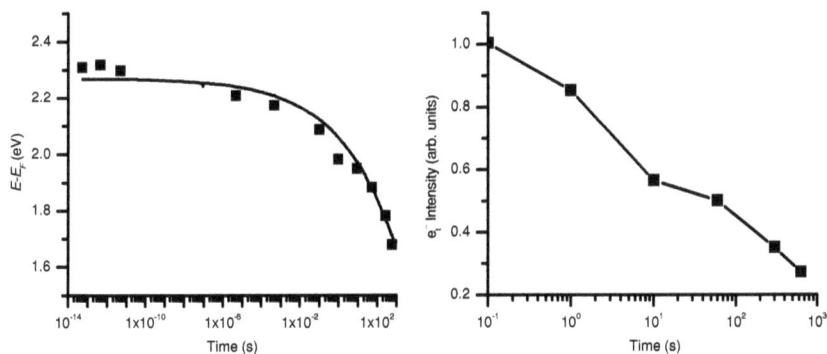

Fig. 4.16. (Left) The temporal evolution of e_t^- binding energies as a function of time over 17 orders of magnitude (10^{-14} to 10^3 s). (Right) The e_t^- intensity integrated over the peak as a function of time up to 630 s. Based on data from Bovensiepen *et al.* [2009].

The above-mentioned results have provided a strong foundation for the cosmic-ray-driven DET reaction mechanism of halogenated molecules adsorbed on ice surfaces in PSCs. As mentioned above, DEAs of many halogen(Cl, Br and I)-containing molecules (such as CFCs) are exothermic, and therefore DEAs of these molecules can effectively occur at zero eV electrons in the gas phase. The exothermic energies of the DEA reactions on H_2O ice or in liquid water will be enhanced by 1−2 eV due to the effect of the polarization potential, as shown in Fig. 4.4. This leads to strong resonances of anion states of Cl-, Br- and I-containing molecules with e_{pre}^- that is weakly-bound at -1.5 to -1.0 eV [Wang *et al.*, 2008b]. Thus, quite effective resonant DET can occur for organic and inorganic (ozone-depleting) halogenated molecules (CFCs, HCl, $ClONO_2$, etc) on H_2O ice, as demonstrated in the experiments reviewed above and below.

Real-time observation of the reaction transition state is the most direct method for observation of molecular reactions and bond breaks, according to the seminal contributions of Nobel laureates Drs. Polanyi and Zewail [1995]. This came true with the advent of femtosecond (fs) time-resolved laser spectroscopy (fs-TRLS) in the late 1980s, as first demonstrated by the pioneering work of Zewail [2000]. Following this

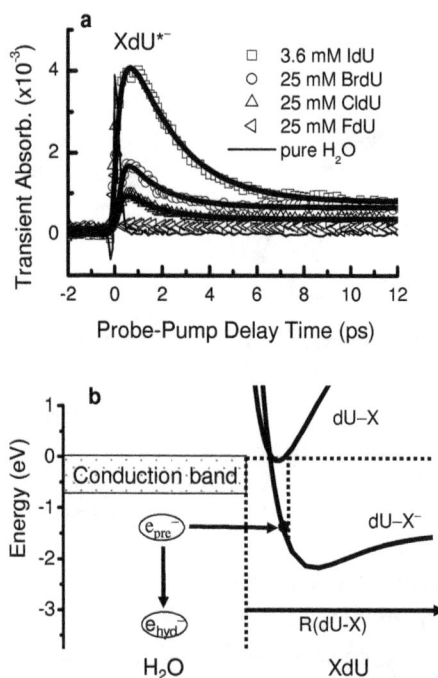

Fig. 4.17. (a) Femtosecond transient absorption kinetic traces of the TNI XdUs*$^-$ of DET reactions of XdUs (X=F, Cl, Br and I) with a pre-solvated electron (e_{pre}^-) at room temperature. The sharp peak at time zero for the pure water trace (fine solid line) is the coherent spike. For XdUs, the symbols are the experimental data, while the solid lines are the best fits to the measured data. (b) Schematic diagram for the DET reaction of e_{pre}^- with a XdU. Adapted from Wang and Lu [2007] and Wang *et al.* [2008b].

strategy, Lu and co-workers [Wang *et al.*, 2006; Wang and Lu, 2007; Wang *et al.* 2008a, 2008b] have obtained direct, real-time observations of the TNI states of the DET reactions of Cl-, Br- and I-containing molecules, such as halopyrimidines (XdUs, X=Cl, Br or I) and CCl_4, with e_{pre}^- in liquid water/methanol solution:

$$e_{pre}^- + XdUs \rightarrow XdUs^{*-} \rightarrow X^- + dU^\bullet \qquad (4.11)$$

$$e_{pre}^- + CCl_4 \rightarrow CCl_4^{*-} \rightarrow Cl^- + CCl_3^\bullet. \qquad (4.12)$$

As shown in Fig. 4.17, the formation time of $XdUs^{*-}$ or CCl_4^{*-} just corresponds to the lifetimes of e_{pre}^-, which are within 1.0 ps after the electron is generated by a fs laser pulse. These results clearly show that the DET reactions occur mainly with the pre-solvated electron, rather than with the fully solvated electron. These observations therefore disprove a notion held in radiation chemistry and biology since the 1970s.

With a less direct but still very interesting approach monitoring the dynamics of trapped electrons using fs time-resolved two-photon photoemission (2PPE) spectroscopy, Ryu *et al.* [2006] and Bertin *et al.* [2009] have also observed the DET reactions of CFCs adsorbed on H_2O ice on a metal substrate. Ryu *et al.* [2006] showed that the lifetime of an electron photoexcited from a metal substrate and trapped in the ice film is significantly decreased by co-adsorption of submonolayer $CFCl_3$, which was explained by DET of the "solvated" electron from the ice to $CFCl_3$. It should be noted that the transient electron species with a lifetime of 120 fs observed by Ryu *et al.* is in a weakly-bound state at 1.2 eV below the vacuum level and is a pre-solvated electron rather than a fully solvated electron that has a binding energy of ~3.5 eV [Coe *et al.*, 1990]. As shown in Fig. 4.18, similar results were subsequently observed by Bertin *et al.* [2009] for $CFCl_3$ adsorbed on the crystalline ice surface, but they observed long-lived trapped electrons with a lifetime up to minutes, trapping at pre-existing structural defects on the surface of the crystalline ice. The results of Bertin *et al.* indicate that DET of the trapped electrons from the ice to $CFCl_3$ is highly efficient, since an extremely low $CFCl_3$ coverage of only ~0.004 monolayer can completely deplete all the trapped electrons generated in the ice under their experimental conditions (Fig. 4.18). The observations by Bertin *et al.* [2009] are certainly of great interest, but such long-lived trapped electrons may not be unique to the crystalline ice surface. Indeed, long-lived trapped electrons with a lifetime ≥ms were also observed at the amorphous ice film deposited on Kr at 20 K in the electron trapping experiments [Lu and Sanche, 2001b]. The 2 PPE experimental results by Ryu *et al.* [2006] and Bertin *et al.* [2009] also agree with the observations of Lu and Madey [1999a, 1999b, 2000, 2001] that the anionic-yield enhancement factor due to the DET process is strongly dependent on the coverage of CFCs. Bertin *et al.* [2009] also explore the implication of their results for photo-enhanced DET reactions of CFCs in the lower polar stratosphere with the returning of sunlight in spring. However, it should be noted that most of the CFC molecules (albeit high levels in the start of winter) have been depleted in the lower polar stratosphere during winter, so that the concentrations CFCs are low in spring (see Fig. 3.11, and Chapter 5).

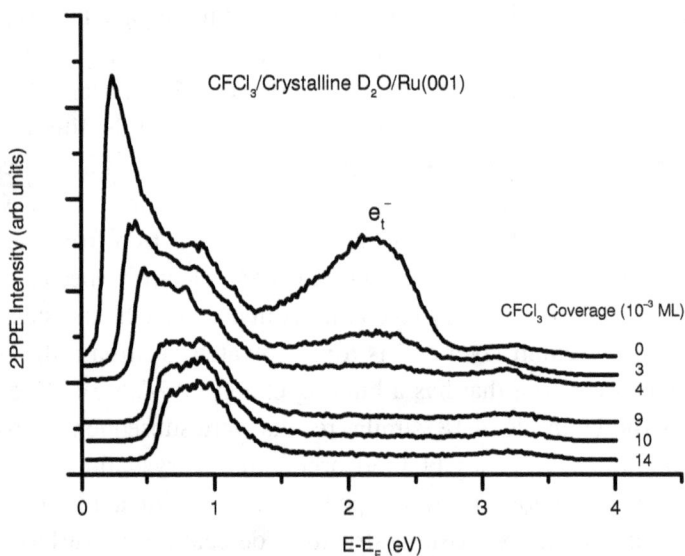

Fig. 4.18. 2PPE intensity as a function of $E–E_F$ for different $CFCl_3$ coverages on crystalline D_2O (coverage ~4.5 BL) deposited on Ru(001) at ~30 K. The trapped electrons (e_t^-) appear at 2.2 eV above the Fermi level on ice surface and disappear as the $CFCl_3$ coverage is more than 0.004 ML. Based on data from Bertin *et al.* [2009].

Most recently, a very large DET cross section up to 4×10^{-12} cm^2 for CFCl$_3$ on D$_2$O ice was measured by Stähler *et al.* [2012]. This value is larger than those of $\sim 8.9 \times 10^{-14}$ for 0.1 ML CFCl$_3$ on H$_2$O [Lu and Sanche, 2004] and $\sim 1 \times 10^{-14}$ for CF$_2$Cl$_2$ adsorbed on H$_2$O ice [Lu and Madey, 1999b; Lu and Sanche, 2001a] while it is comparable to $\sim 6 \times 10^{-12}$ cm^2 for CF$_2$Cl$_2$ adsorbed on NH$_3$ ice [Lu and Madey, 1999b].

4.3.6 *DEA/DET reactions of ClONO₂*

Using a flowing–afterglow Langmuir probe apparatus with mass spectral analysis, Viggiano and co-workers [Van Doren *et al.*, 1996] made a very interesting investigation of the DEA reaction of thermal electrons with chlorine nitrate (ClONO$_2$) in the gas phase at 300 K. This reaction is of particular interest because of the relevance of ClONO$_2$ to the formation of the polar ozone hole and of the large number of possible reaction pathways energetically accessible even with low energy electrons near zero eV

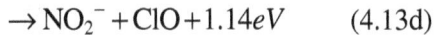

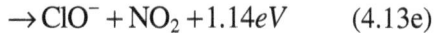

$$e^- (\sim 0eV) + ClONO_2 \rightarrow ClONO_2^- + > 2.1eV \qquad (4.13a)$$

$$\rightarrow NO_3^- + Cl + 2.21eV \qquad (4.13b)$$

$$\rightarrow Cl^- + NO_3 + 1.89eV \qquad (4.13c)$$

$$\rightarrow NO_2^- + ClO + 1.14eV \qquad (4.13d)$$

$$\rightarrow ClO^- + NO_2 + 1.14eV \qquad (4.13e)$$

These DEA reactions are highly exothermic with large exothermic energies indicated [Van Doren *et al.*, 1996].

The measured electron attachment rate constant of ClONO$_2$ is $1.1(\pm 50\%) \times 10^{-7}$ cm^3 s^{-1} and proceeds principally through the dissociative channels (4.13d), (4.13b) and 4.13e), leading to the major product anions NO$_2^-$ ($\sim 50\%$), NO$_3^-$ ($\sim 30\%$), and ClO$^-$ ($\sim 20\%$). The parent ion $ClONO_2^-$ and Cl^- are also observed but are the minor products, <2% and <6%, respectively [Van Doren *et al.*, 1996].

The maximum cross section σ_{max} for *s*-wave electron attachment to a point molecule is $\sigma_{max}=\pi\lambda^2$, where $\lambda=\lambda_e/2\pi$ and λ_e is the electron wavelength [Klots, 1976; Christophorou *et al.*, 1984]. The corresponding maximum rate constant k_{max} at 300 K is 5×10^{-7} cm^3 s^{-1}. Klots [1976] has given an expression for the maximum cross section σ_{max} which takes into account the polarization attraction between the electron and molecule. Van Doren *et al.* [1996] averaged the Klots expression over a Maxwell velocity distribution and obtained an electron capture rate constant at 300 K of 2.9×10^{-7} cm^3 s^{-1} for $ClONO_2$. This value seems very reasonable, in view of the fact that $ClONO_2$ has an electron affinity of 2.9 eV (2.2–3.1) eV [Van Doren *et al.*, 1996; Seeley *et al.*, 1996], which is even larger than that (2.1 eV) of CCl_4. The latter has the well-known largest measured electron attachment rate constant (DEA cross section) of $(2.5–4.1)\times10^{-7}$ cm^3 s^{-1} ($\sim2\times10^{-14}$ cm^2) among molecules in the gas phase [Christophorou *et al.*, 1984; Oster *et al.*, 1989].

The mass spectral data of Van Doren *et al.* [1996] suggest that three or four of the five exothermic DEA channels (Reactions 4.13a-4.13e) are operative at 300 K. These DEA resonances are not resolved in electron energy in their study, but occur typically at very low electron energies near zero eV due to the reaction exothermicities. As described in Sec. 2.3.1 of Chapter 2, different DEA channels are associated with different states of the TNI resonances formed upon attachment of an electron to the molecule. The authors noted that the substantially larger branching fraction observed for formation of the NO_3^- + Cl pair than the Cl^- + NO_3 pair may reflect the larger electron affinity of NO_3 (3.937±0.014) eV [Miller, 2003] as compared with Cl ~3.613 [Miller, 2003]. However, the larger branching fraction of NO_2^-+ClO as compared with ClO^-+NO_2 arises likely from the different dissociating TNI states involved, since the electron affinities of NO_2 and ClO are nearly the same [Miller, 2003].

Viggiano and co-workers [Seeley *et al.*, 1996] also made *ab initio* calculations of the structures of $ClONO_2$ and $ClONO_2^-$ and the electron affinity of $ClONO_2$. Their calculations show that the optimized geometry of neutral $ClONO_2$ is planar. The addition of an electron causes substantial changes to the structure, in particular, the Cl–O bond that is much longer than that in neutral $ClONO_2$ (2.3 vs 1.7 Å). The structure and charge distribution indicate that $ClONO_2^-$ is planar and can be thought of as $Cl \cdot NO_3^-$, with the negative charge mainly accumulated at the NO_3 group. They also calculated that $ClONO_2$ has an adiabatic electron affinity of 2.9 eV and $ClONO_2^-$ has a vertical detachment energy (VDE) of 4.5 eV. These data can be used to construct the potential energy curves of $ClONO_2$ and $ClONO_2^-$, as shown in Fig. 4.19, illustrating the DEA reactions observed in experiments and the likely DET reactions on ice surfaces. It is shown that the potential energy surface for gaseous $ClONO_2^-$ crosses that of the neutral $ClONO_2$ in the

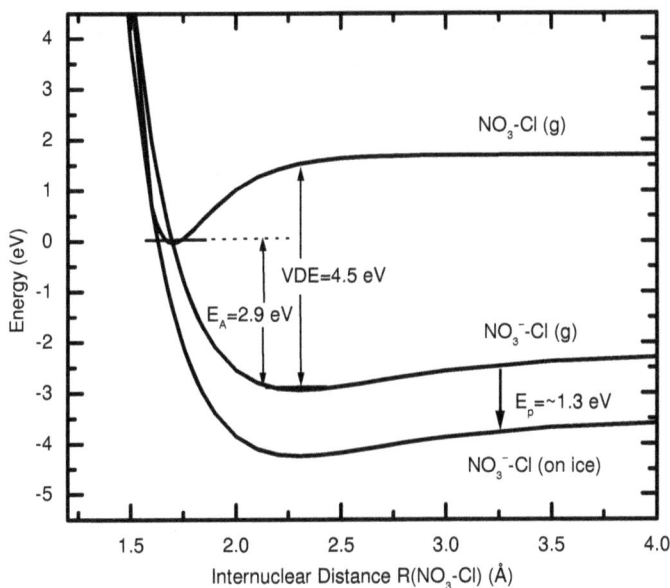

Fig. 4.19. Approximate potential energy curves of neutral $ClONO_2$ and $ClONO_2^-$ anion in the gas phase (g) and on the surface of ice. Dissociative electron attachment (DEA) to $ClONO_2$ of a low-energy (~0 eV) free electron is highly effective in the gas phase, while dissociative electron transfer (DET) to $ClONO_2$ of a weakly-bound (prehydrated) electron (e_{pre}^-) or a partially hydrated electron bound at $\geq$−2.5 eV is highly effective on the surface (s) of H_2O ice. The potential energy curves for the gas phase are constructed, based on thermodynamic data obtained from *ab initio* calculations by Seeley *et al.* [1996], while the curve of $ClONO_2^-$ adsorbed on H_2O ice surface (s) is obtained with the polarization energy of ~1.3 eV [Lu *et al.*, 2002].

vicinity of the equilibrium geometry of $ClONO_2$. If the TNI $ClONO_2^-$ formed in low energy electron attachment, dissociation of the anion will most likely lead to NO_3^- + Cl, which should have a dissociation probability close to unity. The latter is similar to the DEAs of CCl_4, $CFCl_3$, and CF_2Cl_2. It was also suggested that formation of the dissociative attachment products NO_2^- + ClO may arise from excited states of the anion (electronic or rovibrational) whose geometry more closely resembles that of the neutral [Van Doren *et al.*, 1996].

No experiments have been reported for the DET reactions of $ClONO_2$ with the pre-hydrated (trapped) electrons (e_{pre}^-) trapped at ice surfaces, probably due to the commercial unavailability of this molecule, which was only synthesized in individual laboratories. As also shown in Fig. 4.19, however, there is every reason to expect that these DET channels will be extremely effective for $ClONO_2$, according to the potential energy curve of $ClONO_2^-$ on the surface of ice, which highly resembles those of $CFCl_3^-$ and CCl_4^- arising from the DET reactions with e_{pre}^- [Lu and Sanche, 2004; Wang *et al.*, 2008a]. The corresponding DET reactions of $ClONO_2$ can be expressed as

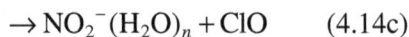

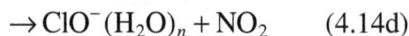

$$e_{pre}^- (H_2O)_n + ClONO_2 \rightarrow NO_3^- (H_2O)_n + Cl \qquad (4.14a)$$

$$\rightarrow Cl^- (H_2O)_n + NO_3 \qquad (4.14b)$$

$$\rightarrow NO_2^- (H_2O)_n + ClO \qquad (4.14c)$$

$$\rightarrow ClO^- (H_2O)_n + NO_2 \qquad (4.14d)$$

Further experimental studies of the DET reaction of $ClONO_2$ adsorbed on the H_2O ice surface will be of significant interest.

4.3.7 *Temperature dependent DEA/DET cross section*

As reviewed in Secs. 4.3.2-4.3.5, electron-induced dissociations of halogen-containing molecules are enhanced by orders of magnitude when a submonolayer of them is coadsorbed with or adsorbed on water ice at 20-30 K. It is known that the presence of polar stratospheric clouds consisting of water ice plays a key role in forming the Antarctic/Arctic ozone hole. Since the Antarctic stratosphere temperature in winter is 180-200 (~190) K, much higher than achievable ice deposition temperatures under ultrahigh vacuum (UHV) conditions, it is required to understand the temperature effect of the electron-induced dissociation of halogen-containing molecules such as CFCs.

The temperature dependences of the Cl$^-$ yield resulting from electron-induced dissociation of a submonolayer of CF_2Cl_2 adsorbed on amorphous and crystalline H_2O ice films grown at 25 K and 150 K respectively, were studied in the UHV chamber under the pressure of 10^{-11} torr (Fig. 4.5) [Lu *et al.*, 2001]. For adsorption on amorphous H_2O ice, the Cl$^-$ yield exhibits a sharp peak between 50 and 70 K, which, together with a sharp drop in electron trapping coefficient of ice near 50 K, is associated with a phase transition in ice. The sharp peak does not appear for CF_2Cl_2 on crystalline ice; instead, an exponential-like increase of the Cl$^-$ yield with temperature up to about 107 K is observed. The Cl$^-$ yield then sharply decreases to nearly zero for T≥125 K due to the desorption of CF_2Cl_2 from the surface under the UHV conditions. Indeed the temperature programmed desorption spectrum of CF_2Cl_2 on crystalline ice exhibits a desorption peak at 120 K. This made it impossible to investigate the temperature dependences of the electron-induced dissociation up to the winter polar stratospheric temperature of 180-200 K.

Fortunately, Le Garrec *et al.* [1997] measured electron attachment rate coefficients for the molecules SF_6, CF_3Br, and CF_2Cl_2 at temperatures between 48 and 170 K. Their results clearly demonstrate the strong effect of internal vibrational energy of the molecules on the attachment/dissociation process of CF_3Br and CF_2Cl_2.

An early review [Smith *et al.*, 1984] showed that DEA of CF_2Cl_2 has a strong temperature dependence at temperature 300-600 K and this has been ascribed to the existence of an activation barrier, estimated to be 0.15 eV. The results of Le Garrec *et al.* [1997] merge smoothly with those of Smith *et al.* The data of electron attachment rate constants exhibit an excellent Arrhenius (exponential) dependence on temperature at T ≥ 120 K. But Le Garrec *et al.* did not show data at temperatures of

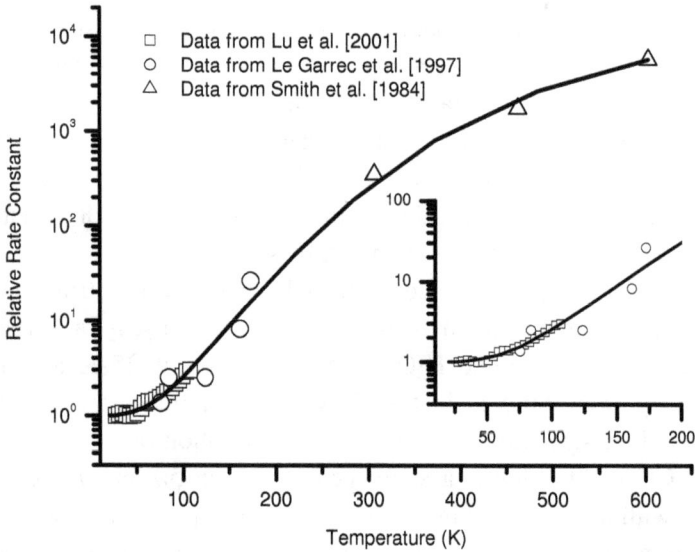

Fig. 4.20. Relative rate constant for dissociative electron attachment to CF_2Cl_2 as a function of temperature. Based on data from measurements of Lu *et al.* [2001], Le Garrec *et al.* [1997] and Smith *et al.* [1984].

48-75 K due to their experimental uncertainty. Interestingly, when the data of Lu *et al.* [2001] for temperatures of 25-107 K and those of Le Garrec *et al.* for 75-170 K and Smith *et al.* for 300-600 K are normalized and plotted together, a smooth exponential increase of the attachment/dissociation rate constant with temperature is obtained, as shown in Fig. 4.20. From this curve, we can obtain that the dissociation cross section of CF_2Cl_2 at T=~200 K (the stratospheric temperature) is about 30 times that at 25 K. Although a smooth curve is obtained in Fig. 4.20, it should be noted that this cross section increase by 30 times should be taken as the lower limit as the DET efficiency is supposed to depend not only on the electron attachment cross section that is affected by the internal vibrational energy of the molecule but also on the electron-transfer rate from ice to the molecule. Both effects should lead to an increase in DET cross section with rising temperature.

Table 4.2. Electron-induced dissociation cross sections of halogenated molecules in the gas phase (G) at 300K, condensed phase (on Kr) and on polar ice surfaces at 20-30 K.

Reaction	Products	Cross section σ (cm^2)
$e^- + CF_2Cl_2$ (CFC-12)$\rightarrow$	$Cl^- + CF_2Cl^\bullet$	1.1×10^{-16} (G)[a]; 1.4×10^{-15} (Kr)[b,c]; $(1.0-1.3)\times10^{-14}$ (H$_2$O ice)[b,c]; 6.0×10^{-12} (NH$_3$ ice)[b]
$e^- + CFCl_3$ (CFC-11)$\rightarrow$	$Cl^- + CFCl_2^\bullet$	9.5×10^{-15} (G)[a]; 1.0×10^{-14} (G)[e]; 7.2×10^{-15} (Kr)[f]; 8.9×10^{-14} (H$_2$O ice)[g]; $(0.33-4)\times10^{-12}$ (D$_2$O ice)[h]
$e^- + CHF_2Cl$ (HCFC-22) $\rightarrow$	$Cl^- + CHF_2^\bullet$	4.8×10^{-19} (G)[i]; 2.0×10^{-18} (G)[j]; 4.2×10^{-16} (Kr)[f]; 5.1×10^{-15} (H$_2$O ice)[g]
$e^- + CH_3CF_2Cl$ (HCFC-142b) $\rightarrow$	$Cl^- + CH_3CF_2^\bullet$	7.8×10^{-16} (Kr)[f]; 4.9×10^{-15} (H$_2$O ice)[g]
$e^- + HCl\rightarrow$	$Cl^- + H^\bullet$	$\sim1.3\times10^{-17}$ (G)[k]; 1.1×10^{-16} (Kr)[l]; 4.0×10^{-15} (H$_2$O ice)[l]
	$H^- + Cl^\bullet$	$\sim7.0\times10^{-19}$ (G)[k]
$e^- + ClONO_2\rightarrow$	$NO_2^- + ClO^\bullet$	$\sim2\times10^{-14}$ (G)[m]

a. Illenberger *et al.* [1979]; b. Lu and Madey [1999b]; c. Lu and Sanche [2001b]; e. Klar *et al.* [2001]; f. Lu and Sanche [2003]; g. Lu and Sanche [2004]; h. Stähler *et al.* [2012]; i. Jarvis *et al.* [1997]; j. Brüning *et al.* [2000]; k. Fedor *et al.* [2008]; l. Lu and Sanche [2001c]; m. estimated from Van Doren *et al.*[1996], in analogous to CCl$_4$.

In summary, there is every reason to conclude that the DET mechanism for halogenated molecules adsorbed on polar molecular ice, originally proposed by Lu and Madey, has been well observed in various laboratory measurements with various methods. This mechanism has also been confirmed by several theoretical studies [Tachikawa and Abe, 2007; Tachikawa, 2008; Bhattacharya *et al.*, 2010; Fabrikant *et al.*, 2012]. As reviewed recently [Lu, 2010a], it is now very well established that electron-induced dissociations of organic and inorganic chlorine, bromine and iodine containing molecules such as CFCs, HCl and ClONO$_2$ can be greatly enhanced via the DET mechanism by the presence of polar media in various phases. The measured DEA cross sections in the gas phase at 300K and the DET cross sections of ozone-depleting halogenated molecules on solid surfaces at 20-30K are summarized in Table 4.2. The DET cross sections at polar stratospheric temperature of ~190 K can approximately be obtained via amplifying the values at 20-30 K by 30 times.

4.4 Possible charging mechanism of polar stratospheric clouds via $N_2^-(^2\Pi_g)$ resonance

Nitrogen (N_2) is the major component (78%) of the Earth's atmosphere. Moreover, there exist ice particles in the winter lower polar stratosphere, where the electron production rate from cosmic rays exhibits a maximum. An investigation of electron trapping in water ice via a N_2^- anion resonance is of great interest, which likely has important implications for ozone-depleting processes leading to the ozone hole.

As discussed in Sec. 2.3 of Chapter 2, resonances in electron-molecule collisions are highly efficient energy transfer processes, in which the incident electron is captured by a target molecule to form a TNI resonance [Schultz, 1973b]. The subsequent decay of the resonance may lead to DEA, vibrational and rotational excitation, or electronic excitation of the molecule. In gaseous N_2, the electron scattering cross sections exhibit a *shape* resonance of symmetry $^2\Pi_g$ around 2.3 eV, which decays by autoionization into vibrational states of N_2: $e^- + N_2 \rightarrow N_2^-(^2\Pi_g) \rightarrow N_2$ (v=0,1,2,...) $+ e^-$ (see Fig. 2.2 in Chapter 2); DEA does not occur [Schultz, 1973b]. In the condensed phase, for N_2 molecules adsorbed on a metal or dielectric surface, the $N_2^-(^2\Pi_g)$ resonance is also observed in electron scattering experiments, but its energy is lowered to ~1.0 eV due to its polarization of the medium [Demuth *et al.*, 1981; Sanche and Michaud, 1990]. The vibrational excitation cross sections on a metal substrate were found to be smaller than in the gas phase; this feature was first attributed to a decrease of the resonance lifetime for adsorbed molecules [Demuth *et al.*, 1981]. Moreover, for N_2 coadsorption with a submonolayer of H_2O, the $N_2^-(^2\Pi_g)$ resonance was found to be almost completely "quenched" in electron scattering experiments [Jacobi *et al.*, 1990]. However, the resonance may still be efficient for N_2 molecules physisorbed on a surface, as predicted theoretically by Djamo *et al.* [1993]. These authors showed that the autoionization of $N_2^-(^2\Pi_g)$ on a surface mostly leads to electrons finally going into the substrate that cannot be observed in reflected electron scattering experiments. Finally, Lu *et al.* [2002], using the electron trapping apparatus the same as for DEA cross section measurements

[Fig. 4.12], presented a method to detect this process by measuring the electron trapping in a H_2O layer on which N_2 molecules are physisorbed.

Fig. 4.21 shows the energy dependencies of electron-trapping coefficient $A_s(E)$ for (a) 0.2 ML N_2 adsorbed on top of a 10 ML Kr film; (b) 0.2 ML N_2 on a 3 ML H_2O film grown on the Kr film, where the $A_s(E)$ for the pure H_2O film on Kr is also shown. Fig. 4.21a shows that no electron trapping occurs in the Kr film; electrons emitted from the autoionization of the $N_2^-(^2\Pi_g)$ resonance transmit quickly through Kr into the metallic substrate. In contrast, trapping of near 0 eV incident electrons is readily observed for H_2O ice. It is seen clearly in Fig. 4.21b that, in addition to the trapping peak near 0 eV, the presence of N_2 on the polar molecular films produces a new structure peaking at an energy of about 1.0 eV. This peak is similar to the envelop shape of the $N_2^-(^2\Pi_g)$ resonance observed in the electron scattering spectra for N_2 adsorbed on a rare gas film or on a metal, where the lowest energy peak is shifted from the gas-phase energy of 2.3 eV to ~1.0 eV due to the polarization energy of the condensed film or the image potential of the metal [Demuth et al., 1981; Sanche and Michaud, 1990].

From the difference spectrum (see the insets in Fig. 4.21b) between the $A_s(E)$ spectra with and without the adsorption of 0.2 ML N_2, Lu et al. [2002] obtained the absolute cross section σ for electron trapping in ice caused by adsorbed N_2 molecules with the use of Eq. 4.10. Fig. 4.21c shows the obtained σ as a function of the H_2O film thickness. The measured σ increases from the lowest H_2O thickness and then reaches a nearly saturation value of $\sigma \approx 5.5 \times 10^{-16}$ cm^2 for film thicknesses ≥ 4 ML. This value is very similar to the sum (~5 $\times 10^{-16}$ cm^2) of the vibrational cross sections for excitations of the $v=1$ to 5 levels of N_2 via resonant scattering at ~2.3 eV in the gas phase [Schultz, 1973b]. The trapping peak at ~1.0 eV is therefore evident of electron transfer from the $N_2^-(^2\Pi_g)$ resonance. Indeed, the results show that the $N_2^-(^2\Pi_g)$ resonance *does not* quench for N_2 adsorbed on a H_2O ice surface, but remains a very efficient process with most of the electrons finally being trapped in the polar molecular film. In other words, the electron transfer process from the $N_2^-(^2\Pi_g)$ to the ice is so efficient that nearly all electrons from the shape resonance are trapped in the ice film. This observation is in good agreement with the theoretical prediction for the case of N_2

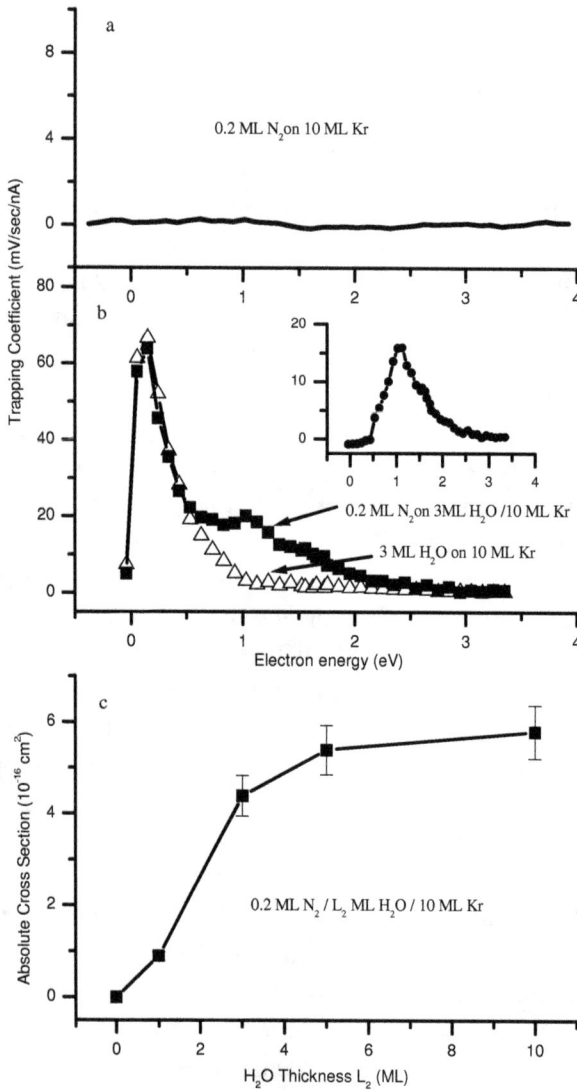

Fig. 4.21. Trapping coefficient A_s as a function of incident electron energy for 0.2 ML N_2 on (a) 10 ML Kr (a) and (b) 3 ML H_2O isolated from the Pt substrate by a 10 ML Kr film. The inset in (b) is the A_s difference between the spectra for the H_2O film with and without the presence of 0.2 ML N_2. (c) The solid squares are for the measured absolute cross section σ for electron trapping in H_2O ice via the $N_2^-(^2\Pi_g)$ resonance at ~1.0 eV (the maximum) versus the H_2O film thickness. Based on data from Lu *et al.* [2002].

adsorbed on a metal surface: most electrons via the resonance are scattered into the metal and thus unobservable in scattering experiments [Djamo *et al.*, 1993]. According to the theoretical analysis, the total vibrational excitation cross section is close to that in the gas phase due to two competing effects: the decrease in the resonance lifetime and the lowering in the resonance energy.

Lu *et al.* [2002] also studied the mechanism as to how electrons are transferred from $N_2^-(^2\Pi_g)$ to the solvated state in the polar ice film. Two possible mechanisms could cause the final trapping of electrons in H_2O ice: (1) $N_2^-(^2\Pi_g)$ autoionizes to produce a low energy electron, which is then thermalized and trapped in ice via inelastic scattering; (2) the electron in $N_2^-(^2\Pi_g)$ is transferred directly into a pre-existing trap in the ice. Mechanism (1) was ruled out by the observation that the trapping peak around 1.0 eV is absent when N_2 is isolated from the H_2O film by a Kr layer. In the latter case, the electron from the autoionization of $N_2^-(^2\Pi_g)$ would transmit through the top Kr film and reach the H_2O layer, while no electron trapping in the H_2O layer was observed. Thus, the observation is clearly inconsistent with the mechanism (1).

Based on the above observations, it was proposed that the mechanism (2), *i.e.*, direct electron transfer from the $N_2^-(^2\Pi_g)$ shape resonance into a trap in the polar film is responsible for the electron trapping peak near 1.0 eV. The extra electron in $N_2^-(^2\Pi_g)$ can make a superinelastic transition to a preexisting trap below the vacuum level in the polar film, leaving N_2 with excess vibrational energy. The electron in the preexisting trap can then become more deeply trapped via relaxation of the medium, and be detected in electron-trapping measurements. Superinelastic electron transfer should depend on the availability of the preexisting traps in the H_2O film, which increases from the lowest H_2O coverages and saturates as the film changes to bulk ice at high coverages (≥ 4 bilayers). Thus, the electron trapping cross section via the $N_2^-(^2\Pi_g)$ resonance shows a corresponding behavior (Fig. 4.21c).

In addition, charging of ice surfaces via the $O_2^-(^2\Pi_g)$ resonance is also possible. This charging phenomenon via negative ion resonances is likely to be significant for charging (electron trapping) of PSC ice particles present in the lower polar stratosphere in the winter season due to the abundances of N_2, O_2 and H_2O clusters as well as numerous electrons produced by the ionization of cosmic rays.

4.5   Charge-induced adsorption of molecules on polar stratospheric clouds

Atmospheric chemists have puzzles about how CFCs and other halogenated molecules could be adsorbed on ice particles in PSCs at temperatures of 180-200 K. They have thought that halogenated molecules such as CFCs and $ClONO_2$ are non-sticky to ice surface and have used such an argument to make strong criticisms of the cosmic-ray-driven electron-induced reaction mechanism of the ozone hole [Harris *et al.*, 2002; Grooß and Müller, 2011]. However, it is fairly well-known to surface scientists that although CFCs are not sticky to the *pure and static* H_2O ice surface at the stratospheric temperature under the vacuum condition, dynamical and other physical properties of ice surfaces under polar stratospheric conditions can make drastic differences, causing adsorption, diffusion and trapping of molecules into ice [Girardet and Toubin, 2001]. Experimental studies [Lu *et al.*, 2001; Souda, 2009] showed that the interactions of CF_2Cl_2 molecules with crystalline and amorphous water films are rather complicated. Souda [2009] concluded that the interactions of CF_2Cl_2 with ice surfaces are significantly different from that with an inert (static) graphite substrate; CF_2Cl_2 molecules do not simply physisorb on the surface of ice and might be incorporated into the bulk of the amorphous ice film at temperatures ≥ 57 K.

In this Section, we will discuss another important mechanism, that is, charge-induced adsorption of molecules on surfaces, which is well known to surface physicists but alien to atmospheric researchers.

4.5.1   *A universal electrostatic coupling mechanism for electron-induced adsorption*

First, electric-field-induced chemisorption is a long known effect in the field of surface science [*e.g.*, Kreuzer and Wang, 1994]. Second, it was also long observed [Caplan *et al.*, 1982; Bloch *et al.*, 1996] that surface charging can occur at thin (1–10 nm) films of the SiO_2/Si system, one of the most important technological systems, in the presence of ambient O_2. Since oxygen was observed to be most effective in gas-assisted surface charging over a 10^{-2}–760 torr pressure range, it was proposed that oxygen possibly forms O_2^- on the surface [Caplan *et al.*, 1982; Levine and Bernstein, 1987; Bloch *et al.*, 1996].

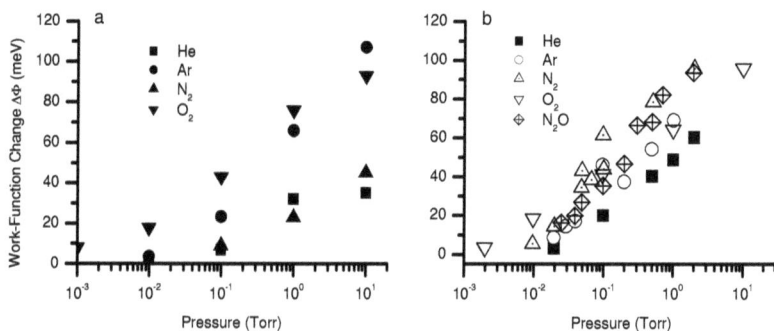

Fig. 4.22. Work-function change $(\Delta\Phi)$ measurements, reflecting adsorption of the various gases on the SiO_2/Si surface versus ambient gas pressure. (a) Measured by a Kelvin proble; (b) Measured by the photoelectron emission. Based on data from Shamir *et al.* [1999] and Shamir and van Driel [2000].

Furthermore, a very interesting study was reported by van Driel and co-workers [Shamir *et al.*, 1999; Shamir and van Driel, 2000]. Using contact potential difference (CPD) and multiphoton-photoemission techniques, both highly sensitive to surface adsorption, they showed that a variety of gases (He, Ar, H_2, N_2, O_2, CO, and N_2O), including most un-sticky rare gases, can be adsorbed nondissociatively on the SiO_2/Si surface for gas pressure in the range 10^{-3}–10^2 torr *even at room temperature (300 K)*. The adsorption can be enhanced through surface charging via photoemitted electrons from the Si substrate, leading to mutual electron-gas transient trapping. Their results, shown in Fig. 4.22 here, can be explained well by a universal electrostatic coupling mechanism, in which an electrostatic field generated by the presence of negative charge on the surface polarizes the adsorbed gas particles leading to binding energies in the range of up to 400 meV [Shamir *et al.*, 1999]. The negative charge (electrons) can either naturally exist or be photo-induced at the surface by laser pulses. This electrostatic coupling mechanism is in good agreement with the observed data, and should be universally applicable to charge-induced adsorption of gases on various surface systems [Shamir *et al.*, 1999; Shamir and van Driel, 2000].

4.5.2   *Charge-induced adsorption of halogenated molecules on H₂O ice surfaces*

Due to the larger polarizability for halogen-containing molecules than the common gases (He, Ar, H_2, N_2, O_2, CO), the effect of charge-induced adsorption is expected to be even more significant for halogenated gases. To investigate the universal electrostatic coupling mechanism for adsorption of halogenated molecules on H_2O ice surfaces and the associated implications for polar stratospheric processes, we recently used CCl_4 absorption on ice with trapped electrons as a model study system [Liu, 2012; Lu and Liu, unpublished].

First, a charged ice surface was needed in order to compare with the uncharged surface in this study. The charged surface can readily be prepared by doping alkali-metal atoms on ice surface deposited on a metal substrate. Both experiments and quantum chemistry calculations have shown that alkali-metal atoms (Na, K, Cs) autoionize on ice surfaces, forming metastable surface trapped electrons transiently stabilized at *the ice-vacuum interface* with a lifetime of several hundreds of seconds depending on the temperature [Vondrak *et al.*, 2006a, 2006b, 2009; Horowitz and Asscher, 2012]. During the lifetime of these metastable surface electrons, they are loosely associated with their parent alkali-metal ions and not completely solvated by H_2O. The autoionization of alkali atoms can serve as an electron donor to make the ice surface charged easily.

Second, we needed an alkali atom (potassium, K) source, a gas introduction for CCl_4 exposure and a Kelvin probe for surface work function measurements installed in our UHV chamber with a base pressure at low 10^{-10} Torr. It was necessary to avoid contamination of the Kelvin probe detector, which must be installed at a position different from the alkali-metal atom source. Therefore, it required at least minutes to complete one measurement of charge-induced adsorption in our experiments. Therefore, we could not detect the transiently trapped electrons at the surface of crystalline H_2O ice held at a temperature higher than 145 K. For the latter case, the trapped electrons quickly tunnel through the ice film into the metal (Cu) substrate beyond the lifetimes of the trapped electrons in seconds (see Fig. 4.16). Nevertheless, we could easily demonstrate the CCl_4 adsorption induced by electrons trapped at ice surfaces at temperatures below 120 K. A similar mechanism should occur for adsorption of molecules on PSC ice surfaces in the winter polar stratosphere at higher temperatures of 180-200K. There should be no doubt that ice particles in PSCs can be effectively charged (electrons can be effectively trapped in PSC ice particles without the presence of a metal substrate) under continuous cosmic ray radiation.

In this experiment, the adsorption on freshly-prepared (uncharged) and K-deposited ice surfaces and associated work function changes were monitored by the surface-sensitive Kelvin probe in a UHV chamber. The samples were prepared as the following. First, the Cu substrate was cleaned by annealing to 900 K and cooled back to 95 K by use of liquid N_2. Then an ice film (about 15 ML) was grown on the Cu(111) substrate at 100 K at a pressure of 5×10^{-9} torr. After the potassium was deposited onto the ice surface from a getter source at 6 A for 1-2 minutes, the work function of the ice film was largely lowered by about 2.8 eV. This significant work function decrease can be attributed to the metastable trapped surface electrons donated by K atoms undergoing auto-ionization on the ice surface.

The freshly prepared (uncharged) ice surface and K-adsorbed ice surface at ~100 K were then respectively exposed to CCl_4 by background exposure at 1×10^{-8} torr in the UHV chamber for 10 minutes while the work function changes of the ice surfaces were monitored. As shown in Fig. 4.23, for the uncharged ice surface exposed to CCl_4, there was only a negligible decrease in the surface work function observed for exposures up to 6 Langmuir (L, $1L \equiv 10^{-6}$ torr·s). This indicates that only a small amount of CCl_4 molecules were adsorbed on the uncharged ice surface at 100 K for such a background exposure.

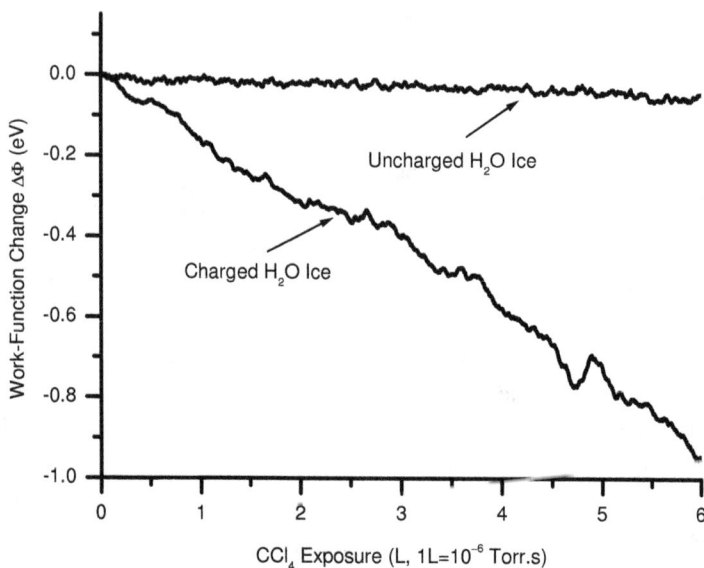

Fig. 4.23. Work function changes of a 15 ML H_2O ice film (uncharged) and a potassium-deposited ice film (charged) during CCl_4 exposure at 1×10^{-8} torr for 10 minutes at ~100K.

A contrast difference was observed for the charged ice surface, on which the work function change was dramatically different from what observed on the uncharged ice surface. Exposure of CCl_4 resulted in a large decrease by about 1.0 eV in work function change, indicating a significant adsorption of CCl_4 on the charged ice surface. These results have clearly demonstrated that the charged ice surface exhibits very different properties from the uncharged ice surface. The charged surface creates an electric field which polarizes gas molecules and then causes them effectively adsorbed on the surface by electrostatic binding. Once gas molecules are adsorbed on the surface, the work function change of the ice surface can be detected.

The results shown in Fig. 4.23 have clearly shown that the universal electrostatic coupling mechanism for electron-induced surface adsorption operates at lower temperatures (100 K), the same at room temperature (300 K) as demonstrated by van Driel and co-workers [Shamir *et al.*,

1999; Shamir and van Driel, 2000]. Such a mechanism is expected to cause effective adsorption of halogen-containing molecules on surfaces of PSC ice present in the lower polar stratosphere under the known strongest cosmic ray ionization.

As reviewed in the previous Sections, the observed results indicate sufficiently that to understand atmospheric processes properly, it is required to include multidisciplinary research and to keep an open mind. Knowledge of different areas in physics and chemistry is vital to unravel real science underlying the formation of the ozone hole.

4.6 The cosmic-ray-driven electron-induced reaction (CRE) mechanism of the ozone hole

The discovery of large enhancements in electron-induced dissociations of CFCs by the presence of polar molecular ices has been reviewed in Sec. 4.3.2 above. Since then, the study of the implications of electron-induced reactions of halogenated molecules for stratospheric ozone depletion has revived. Lu and Madey [1999b] first noticed that the existence of polar stratosphere clouds (PSCs) consisting of ice particles in the winter polar stratosphere can make it different from the general stratosphere: the electron-induced physics and chemistry of molecules in the winter polar stratosphere can be drastically different from that in the general stratosphere without the presence of PSCs. Thus, the observed giant enhancements in DET of CFCs adsorbed on ice surfaces can have great implications for ozone depletion in the polar stratosphere [Lu and Madey, 1999b].

As mentioned in Sec. 1.3.2 in Chapter 1, the primary source of electrons in the stratosphere is the ionization of the atmospheric constituents by continuous cosmic rays. It can be expected that polar stratospheric clouds or polar molecular clusters are filled with self-trapped electrons. It should be interesting to estimate the dissociation rates induced by cosmic rays for CFCs adsorbed on H_2O ice in PSCs. The rate, k_{CR}, for the production of electrons by CR ionization is ~45 cm^{-3} s^{-1} in air at an altitude of ~15 km in the general stratosphere [Cole and Pierce, 1965]. In the winter polar stratosphere, an estimate gives an ionization rate k_{CR} on the surface of ice in PSCs: $k_{CR}=45\times\rho_{ice}/\rho_{air}\approx2.3\times10^5$ cm^{-3} s^{-1}, where ρ_{ice} and ρ_{air} are the densities of water ice and the air at the altitude of 15 km, respectively [Lu and Sanche, 2001a]. Even if a mean lifetime of 10 ms is assumed for prehydrated electrons in PSC ice in the polar stratosphere (see Fig. 4.16), the prehydrated electron density $n(e_{pre}^-)$ will be 2.3×10^3 cm^{-3}, which is comparable to the observed ion density in the stratosphere.

The absolute DET cross sections (σ) for CF_2Cl_2 and $CFCl_3$ adsorbed on the surface of water ice at 20-25 K have been measured to be $\sim1.3\times10^{-14}$ cm^2 [Lu and Madey, 1999b; Lu and Sanche, 2001b] and $\sim8.9\times10^{-14}$ cm^2 [Lu and Sanche, 2004], respectively. Extrapolating to the winter lower polar stratospheric temperature of 180-200 K, *i.e.*, increasing the DET cross sections by 30 times (see Sec. 4.3.7), gives the corresponding DEA cross sections of $\sim3.9\times10^{-13}$ cm^2 and $\sim2.7\times10^{-12}$ cm^2. These values are 3.5×10^3 and 2.8×10^2 times their corresponding DEA cross sections in the gas phase (see Table 4.2), respectively. Thus, the DET rate constants of CF_2Cl_2 and $CFCl_3$ adsorbed on PSC ice surfaces will be approximately 6.7×10^{-6} and 6.7×10^{-5} cm^3s^{-1}, respectively. As a result, the electron-induced dissociation rate, R, of a CF_2Cl_2 molecule on the surface of ice particles in a PSC is estimated to be $\sim6.7\times10^{-6}$ $cm^3s^{-1} \times 2.3\times10^3$ $cm^{-3} =\sim1.5\times10^{-2}$ s^{-1}; the corresponding R for $CFCl_3$ is $\sim1.5\times10^{-1}$ s^{-1}. Nevertheless, it should be noted that the electron density $n(e_{pre}^-)$ available for the DET reactions of CFCs and other halogenated molecules is expected to be less than the total e_{pre}^- density (2.3×10^3 cm^{-3}), due to the competing electron attaching reactions of other atmospheric molecules co-adsorbed on the PSC ice surface. Therefore, the limiting factor for the DET reactions of halogenated molecules on PSCs will be the available density of e_{pre}^-.

The above estimates indicate that in the winter lower stratosphere, electron-induced dissociation of CFCs once *adsorbed on the surface of ice particles* in PSCs by cosmic rays is very fast, with lifetimes of CFCs likely in the order of minutes to hours. Note that this is very different from CFCs present *in the gas phase* in the polar stratosphere.

Moreover the experiments reviewed in Sec. 4.5 have indicated that PSC ice surfaces can become charged due to the cosmic ray ionization. Since the trapped electron density is linearly proportional to the CR intensity, it is reasonably expected that the amount of halogenated molecules such as CFCs and $ClONO_2$ adsorbed on PSC ice surfaces will also be linearly proportional to the CR density and the gaseous concentration of halogenated molecules.

A strong enhancement in DET reactions of CFCs with trapped electrons in PSCs can play a crucial role in creation of the Antarctic ozone hole. It is known that the following reactions of the resultant Cl^- ions are fast [Sander and Crutzen, 1996; Oum *et al.*, 1998]

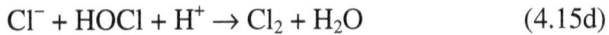

$$OH + Cl^- \rightarrow HOCl^- \qquad (4.15a)$$

$$HOCl^- + H^+ \rightarrow Cl + H_2O \qquad (4.15b)$$

$$Cl^- + NO_3 \rightarrow Cl + NO_3^- \qquad (4.15c)$$

$$Cl^- + HOCl + H^+ \rightarrow Cl_2 + H_2O \qquad (4.15d)$$

The rate constants of reactions (4.15a)-(4.15d) are not known in the vapor phase, but their values in the aqueous phase are 4.3×10^9 $M^{-1}s^{-1}$, 2.1×10^{10} $M^{-1}s^{-1}$, 1.0×10^7 $M^{-1}s^{-1}$, and 1.8×10^4 $M^{-2}s^{-1}$, respectively [Sander and Crutzen, 1996]. Due to the abundances of OH, H^+, NO_3 and other radicals in the stratosphere, it is expected that Cl^- can be rapidly converted to Cl atoms [Lu and Madey, 1999b]. In addition, the reaction of Cl^- with O_2^+ can result in OClO, another abundant species found in the Antarctic vortex [Solomon, 1990; Toon and Turco, 1991].

Another important process that is well known in radiation chemistry and biology is the reaction of the neutral radical formed by a DEA/DET reaction of halogenated molecules, which can make significant biological damage [Lu, 2010c]. Halogenated molecules such as CFCs are very stable in chemistry, in which the C-Cl bond dissociation energies are very large, *e.g.*, $D(CF_2Cl-Cl)=3.58$ eV and $D(CFCl_2-Cl)=3.21$ eV [Dispert and Lacmann, 1978; Illenberger *et al.*, 1979]. This means that the neutral fragments such as $CF_2Cl^\bullet$ and $CFCl_2^\bullet$ resulting from the DEA/DET of CFCs are highly reactive radicals, which can effectively react with stratospheric molecules (O_2 and O_3). Take the DEA/DET reactions of CF_2Cl_2 and $CFCl_3$ as examples, the immediate reactions of the resultant $CF_2Cl^\bullet$ and $CFCl_2^\bullet$ free radicals with O_2 release Cl atoms

$$CF_2Cl^{\bullet} + O_2 + M \rightarrow CF_2ClO_2 + M \qquad (4.16a)$$

$$CF_2ClO_2 + NO \rightarrow CF_2ClO + NO_2 \qquad (4.16b)$$

$$CF_2ClO + M \rightarrow COF_2 + Cl + M \qquad (4.16c)$$

and

$$CFCl_2^{\bullet} + O_2 + M \rightarrow CFCl_2O_2 + M \qquad (4.17a)$$

$$CFCl_2O_2 + NO \rightarrow CFCl_2O + NO_2 \qquad (4.17b)$$

$$CFCl_2O + M \rightarrow COFCl + Cl + M \qquad (4.17c)$$

$$COFCl + h\nu \rightarrow FCO + Cl \qquad (4.17d)$$

These reactions have similarly been proposed for the reactions of the identical radicals resulting from the photolysis of the CFCs [Orlando and Schauffler, 1999], as mentioned in Chapter 3.

However, the following radical reactions should also be very interesting and significant, leading to direct destruction of ozone and formation of Cl atoms that can further destroy O_3

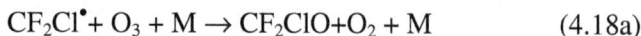

$$CF_2Cl^{\bullet} + O_3 + M \rightarrow CF_2ClO + O_2 + M \qquad (4.18a)$$

$$CF_2ClO + M \rightarrow COF_2 + Cl + M \qquad (4.18b)$$

and

$$CFCl_2^{\bullet} + O_3 + M \rightarrow CFCl_2O + O_2 + M \qquad (4.19a)$$

$$CFCl_2O + M \rightarrow COFCl + Cl + M \qquad (4.19b)$$

It is worth noting that these reactions should be highly effective, since the oxidative potentials (>3.0 eV) of $CF_2Cl^{\bullet}$ and $CFCl_2^{\bullet}$ radicals are much larger than the bond dissociation energy $D(O-O_2)=1.05\pm0.02$ eV in O_3 [Gole and Zare, 1972]. Thus, the direct O_3-destroying reactions (4.18) and (4.19) should be more effective than the radical-O_2 reactions (4.16)-(4.17).

Furthermore, the DEA/DET reaction of $ClONO_2$ is also of particular interest. As expressed in reactions (4.14a)-(4.14d), the DET reaction of this molecule on PSC ice surfaces can directly result in the formation of Cl, ClO, NO$_2$ and NO$_3$ species, which can then lead to ozone destruction. It is surprising that these reactions were not noticed previously.

The Cl, ClO, NO$_2$ and NO$_3$ species resulting from all the above reactions can destroy ozone via the reaction cycles described Sec. 3.3 in Chapter 3. It is worthwhile to note that these reactions can effectively take place without the presence of sunlight and therefore in the winter polar stratosphere. This is actually in good agreement with the observations reviewed in Sec. 3.5.

In addition, it is also well known that the Cl$^-$ ions adsorbed (solvated) at the surface of PSCs can also react with other species to release photochemically reactive Cl$_2$ or ClNO$_2$

$$Cl^- + ClONO_2 \rightarrow Cl_2 + NO_3^- \qquad (4.20a)$$

$$Cl^- + N_2O_5 \rightarrow ClNO_2 + NO_3^- \qquad (4.20b)$$

$$Cl^- + HNO_3 \rightarrow HCl + NO_3^- \qquad (4.20c)$$

Upon photolysis, Cl$_2$ or ClNO$_2$ can release Cl atoms to enhance the destruction of O$_3$ in the springtime polar stratosphere.

Here, we do not exclude the photolysis of ClOOCl (the ClO dimer) that can occur and enhance ozone destruction in the springtime polar stratosphere

$$ClOOCl + h\nu \rightarrow ClOO + Cl \qquad (4.21)$$

Based on the above reactions, the cosmic-ray-driven electron-induced reaction (CRE) mechanism, as schematically shown in Fig. 4.24, has been proposed as an important mechanism for the formation of the O$_3$ hole [Lu and Madey, 1999b; Lu and Sanche, 2001b; Lu, 2009, 2010a, 2013]. It should be noted that given the abundance of photons in the springtime polar stratosphere, the rate-limiting factors for the total ozone-depleting reaction should be not the sunlight intensity but the available electron density and the concentration of *charged-induced adsorbed* (rather than *gaseous*) halogenated molecules on PSC ice surfaces. Both have a linear dependence on the CR intensity.

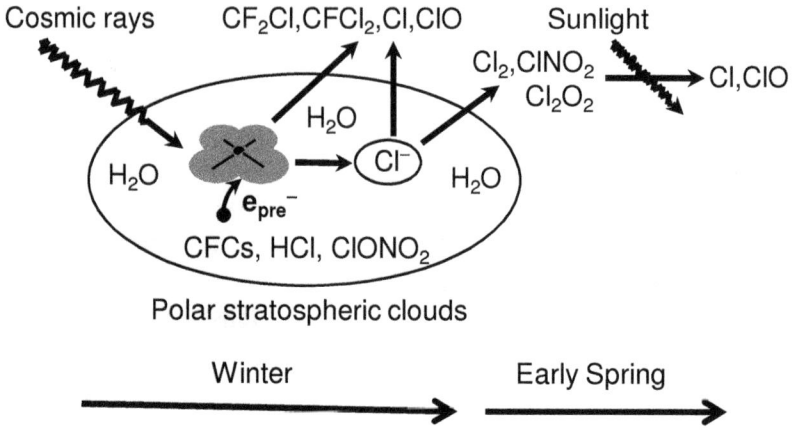

Fig. 4.24. The CR-driven electron-induced reaction (CRE) mechanism of the ozone hole: Cosmic-ray driven electron-induced reactions of halogen-containing molecules in PSCs result in the formation of CF_2Cl, $CFCl_2$, Cl, ClO, and Cl^- ions. The Cl^- ions can either be rapidly converted to reactive Cl atoms or react with other species to release photoactive Cl_2 and $ClNO_2$ in the winter polar stratosphere. The CF_2Cl, $CFCl_2$, Cl and ClO can destroy the O_3 layer in the polar stratosphere during winter. The photolysis of Cl_2, $ClNO_2$, $ClOOCl$ can also enhance ozone destruction upon the return of sunlight in the spring polar stratosphere. Modified from Lu and Madey [1999b], Lu and Sanche [2001a], Lu [2009], Lu [2010a] and Lu [2013].

4.7 An analytic (quantitative) expression of polar ozone loss

A simple quantitative expression of ozone loss due to the CRE mechanism can be derived, in which total ozone loss ($\Delta[O_3]_i$) in the polar stratosphere has been given in the literature [Lu, 2010a; 2013]

$$\frac{\Delta[O_3]_i}{[O_3]_0} = \frac{[O_3]_i - [O_3]_0}{[O_3]_0} \approx -k \times [C_i] \times I_i^2, \qquad (4.22)$$

where $[C_i]$ the equivalent effective chlorine (EECl) in the polar stratosphere, $\Delta[O_3]_i/[O_3]_0$ is the relative total O_3 change, $[O_3]_0$ the total O_3 in the polar stratosphere when $[C_i]=0$, I_i the CR intensity in the year, and k a constant. As mentioned in the above, the CRE reaction depends on the electron density and the *charged-induced adsorbed* (rather than gaseous) concentration of a CFC at the surface of PSC ice, both linearly depending on the CR intensity I. As a result, the ozone loss due to the CRE reaction has a linear dependence on EECl and a quadratic dependence on the CR intensity I_i.

As schematically shown in Fig. 4.24, the CRE mechanism does not exclude any possible contribution of sunlight-related photochemical processes to O_3 depletion. In deriving the CRE equation (Eq. 4.22), however, a simplification was indeed made that the photolysis of gaseous halogens in the spring polar stratosphere is not a limiting factor. This is based on the understanding that the photoactive species (Cl_2, $ClNO_2$, $ClOOCl$) once formed in the winter polar stratosphere can be converted to Cl/ClO to destroy ozone with a conversion efficiency of nearly 100% in the springtime polar stratosphere with plenty of sunlight. Thus, it is expected that only the CRE/DET reaction of halogenated molecules adsorbed on PSC ice is the critical step and the limit factor for calculating total ozone loss in the polar stratosphere.

As will be shown in next Chapter, Eq. 4.22 (the CRE equation) can well reproduce the observed 11-year cyclic variations of not only total ozone but also stratospheric cooling over Antarctica in the past five decades. More quantitative and statistical analyses of observed data from space- and ground-based measurements in terms of Eq. 4.22 will also be presented in Chapter 5.

4.8 Summary

Extremely large DET dissociation cross sections of CFCs adsorbed on ice surfaces at 20~30 K in the orders of magnitude of 10^{-14} to 10^{-12} cm^2 have been measured. The cross sections are enhanced by at least 30 times at the stratospheric temperature of about 190 K.

A surface charging mechanism of ice particles under the polar stratospheric conditions was found. The charging can effectively occur via negative ion resonances of abundant atmospheric molecules (mainly N_2).

Charge-induced adsorption of halogenated molecules on ice surfaces was also demonstrated. This can well be explained by a universal electrostatic coupling mechanism that has been well established in surface physics community.

Electron-induced dissociation rates of halogenated molecules adsorbed on PSC ice particles are extremely large under the winter polar stratospheric conditions at altitudes of 10-20 km. The lifetimes of adsorbed halogenated molecules are estimated to be minutes to hours.

Electron-induced reactions of halogenated molecules can lead to ozone depletion in both the winter and the springtime polar stratosphere. Various radical reactions relevant to ozone depleting reactions have been described in this Chapter. A cosmic-ray-driven electron-induced reaction (CRE) mechanism for the formation of the polar ozone hole has been clearly justified.

Finally, a quantitative description of the CRE mechanism, *i.e.*, the CRE equation, has been given. Ozone loss in the winter and springtime stratosphere is governed by the CR intensity and the equivalent effective chlorine level in the polar stratosphere only, both being the limiting factors to the ozone-depleting reactions.

Chapter 5

The Cosmic-Ray-Driven Theory of the Ozone Hole: Atmospheric Observations

5.1 Introduction

In Chapter 4, we discuss numerous data from laboratory measurements of electron-induced (DEA and DET) reactions of halogenated molecules and the associated cosmic-ray-driven electron-induced reaction (CRE) mechanism for the formation of the polar ozone hole. We will now present and discuss the data from atmospheric measurements in terms of the CRE mechanism in this Chapter.

5.2 Spatial correlation between cosmic rays and ozone depletion

The CRE mechanism has strong *latitude* and *altitude* effects corresponding to the distribution of electrons produced by cosmic rays (CRs) in the atmosphere. Since CRs are composed of charged particles, the earth's magnetic field focuses them onto the South and North Poles. Thus, the CR intensity has a strong latitude variation with maxima at the Earth's south and north poles, as discussed in Sec. 1.3.2 of Chapter 1. Correspondingly, most severe ozone depletion also occurs at the polar stratosphere with high latitudes (Fig. 5.1a). The depletion over Antarctica is much larger than over Arctic in winter due to the colder Antarctic stratosphere and the more stable Antarctic vortex, which produces more

PSC ice particles. In contrast, the solar light intensity has a maximum at the geomagnetic equator and minima at the poles.

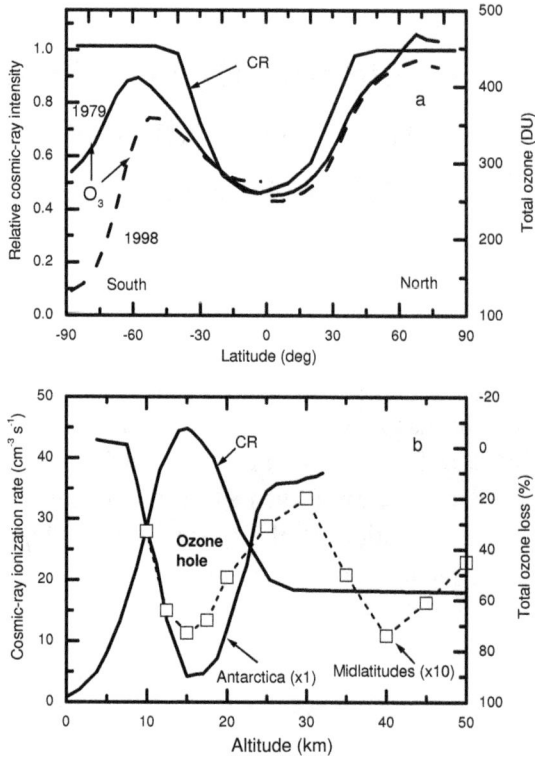

Fig. 5.1. Spatial correlations between cosmic-ray (CR) intensity and ozone depletion. a: Latitude dependences of CR intensity and monthly mean total ozone in pre-O$_3$ hole (Oct. 1979 for Antarctica and March 1979 for Arctic) and O$_3$ hole period (dashed line for Oct. 1998 for Antarctica and March 1998 for Arctic). b: Altitude dependences of the springtime O$_3$ hole over Syowa, Antarctica, and O$_3$ loss per decade from 1979 to 1998 over northern midlatitudes (40°-53°N) (the squares, amplified by 10 times). Adapted from Lu and Sanche [2001a].

Moreover, the altitude dependence of cosmic ray ionization rate shows *a maximum intensity at an altitude of 15–18 km*, due to the production of secondary particles causing multi-ionization. Correspondingly, the ozone hole is observed in the polar stratosphere at

15–18 km in the early spring (Fig. 5.1b). Also shown in Fig. 5.1b is the available data for ozone loss per decade over northern mid-latitudes (40°-53°N) from 1979 to 1998, exhibiting two loss maxima in the upper (~40 km) and lower (~15 km) stratosphere respectively [Randel *et al.*, 1999]. The ozone loss maximum at ~40 km has been attributed to the OH maximum at this altitude or the photodissociation process [Randel *et al.*, 1999].

5.3 Temporal correlation between cosmic rays and ozone depletion

5.3.1 *Eleven-year cyclic variation of polar ozone loss*

The CRE model has predicted an *~11-year cyclic* variation of O_3 loss in the polar hole corresponding to the solar cyclic variation of the CR intensity that has an average periodicity of 11 years (varying in 9-14 years) [Lu and Sanche, 2001a; Lu, 2009, 2010a, 2013]. A time correlation between the annual mean total O_3 in the southern hemisphere (at latitudes 0–65° S) and the CR intensity in the single CR cycle of 1979-1992 was first reported [Lu and Sanche, 2001a]. Unfortunately, Patra and Santhanam [2002] criticized that no such 11-year cyclic variations could exist beyond one solar (CR) cycles. Some atmospheric chemists have kept citing such a criticism to argue against the CRE mechanism. However, the 11-year cyclic variations of total ozone at either the global (at latitudes 60° N–60° S) or the southern hemisphere (0–65° S) have now been well confirmed by the data presented in recent WMO Reports [2006; 2010, 2014] and by the updated NASA satellite data [Lu, 2010a, 2013], as shown in Figs. 3.13a and 3.14a in Chapter 3.

To unravel the CR-ozone correlation, reliable and sufficiently good quality data as well as their correct and scientific presentation are required [Wang *et al.*, 2008c; Lu, 2009, 2010a, 2013, 2014c].

Detailed quantitative and statistical analyses of the observed data for polar stratospheric ozone loss and associated stratospheric cooling in terms of the CRE mechanism have been given recently [Lu, 2010a, 2013]. Atmospheric dynamics is known to cause large fluctuations in total O_3 in the polar hole from year to year. To minimize the unpredictable short-term effect, a three-point (year) adjacent averaging was applied to observed data: $[O_3]_i = \{[O_3]_{i-1} + [O_3]_i + [O_3]_{i+1}\}/3$. This minimal processing can effectively reduce the fluctuation level of measured data. The British Antarctic Survey (BAS) has provided the longest records of ozone and lower stratospheric temperatures over the Antarctica since 1956, which also led to the first discovery of the ozone hole in 1985 [Farman, Gardiner and Shanklin, 1985]. NASA's TOMS and OMI satellite ozone datasets have so far provided the most widely used total O_3 data for the global and Antarctica since 1979, while NOAA's ongoing surface-based observations have provided a measure of ozone-depleting chlorine- and bromine-containing gases in the lower atmosphere. The US NSF-funded cosmic ray measurements at McMurdo ($77.9°$ S, $166.6°$ E) are the only record providing *continuous* time-series CR data over Antarctica since 1960s. Data from these sources have been used in evaluations of the CRE mechanism [Lu, 2009, 2010a, 2013, 2014a, 2014b, 2014c].

Following the Lu-Sanche paper [2001], Lu [2009] showed that there indeed existed 11-year cyclic time correlations of the CR intensity not only with the annual mean total O_3 in the southern hemisphere ($0-60°$ S) but with total O_3 in the spring Antarctic ozone hole ($60-90°$ S) over two CR cycles (1979-2008). The ozone data obtained from the NASA satellites have further confirmed a pronounced ~11-year cyclic correlation between CR intensity and 3-month average total O_3 data in the O_3 hole period over Antarctica ($60-90°$ S) [Lu, 2010a].

Further quantitative and statistical analyses of observed data in terms of the CRE mechanism were given in a recent paper [Lu, 2013]. Here, time-series October monthly mean and 3-month (October-December) mean total O_3 data observed by NASA satellites and fitted by the CRE equation (Eq. 4.22) are shown in Figs. 5.2a and b, respectively. The data show that the Antarctic total O_3 decreased drastically from the end of the 1970s to 1995, following the significant rise of the halogen loading in the stratosphere. From 1995 to the present, total O_3 over Antarctica has exhibited pronounced 11-year cyclic oscillations.

To establish further the reliability of the above conclusion, statistical correlation analyses of the CRE equation and observed ozone data are shown in Fig. 5.2c and d, which respectively plot the October monthly mean and 3-month (October-December) mean total O_3 data versus the product of the equivalent effective stratospheric chlorine $[C_i]$ and the square of CR intensity I_i, *i.e.*, $[C_i] \times I_i^2$. First, as we expect, the 3-point adjacent averaging reduces the fluctuation level of O_3 data significantly. Second, Figs. 5.2a and b show that the CRE equation can well reproduce 11-year cyclic ozone losses in the Antarctic ozone hole. Third, Figs. 5.2c and d show that for the October and the 3-month average total O_3 data, statistical correlation coefficients -R of 0.83 and 0.91 are obtained for the linear fits to the observed data. All the statistical fits were made at a fixed 95% confidence. These results show an excellent linear correlation

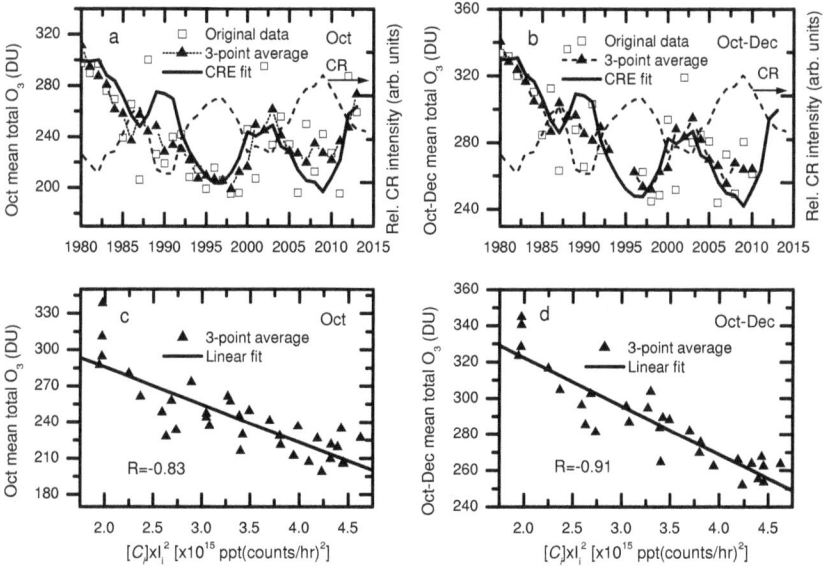

Fig. 5.2. Statistical analyses of the CRE mechanism and the Antarctic (60-90° S) ozone hole during 1980-2013. a and b: Time series data of cosmic ray (CR) intensity (dash line), October mean and 3-month (October-December) mean total O_3 and their 3-point adjacent averaged data, as well as the best fits by the CRE equation (Eq. 4.22). c and d: 3-point average total O_3 data are plotted as a function of the product of $[C_i] \times I_i^2$; linear fits to the data give high linear correlation coefficients -R up to 0.91 and P<0.0001 (for R=0). October mean total ozone data were obtained from NASA TOMS N7/M3/EP/OMI satellites [WMO, 2014]. Adapted from Lu [2013].

between total O_3 and the value of $[C_i] \times I_i^2$ in the Antarctic O_3 hole and thus provide strong evidence of the CRE mechanism (Eq. 4.22).

Furthermore, ozone data at Halley station (75°35' S, 26°36' W), Antarctica obtained from the BAS have also robustly shown pronounced 11-year cyclic variations of polar ozone loss [Lu, 2013], which is shown Fig. 3.15 and Fig. 5.3a. Most remarkably, minima in Antarctic ozone loss (hole) were indeed observed in 2002 and 2013, exactly corresponding to the minima in CR intensity. Also, a statistical correlation analysis of total O_3 at Halley in terms of the CRE equation is shown in Fig. 5.3b. Again, an excellent linear correlation of total O_3 with $[C_i] \times I_i^2$ variations is observed, and a high linear correlation coefficients −R of 0.86 and P(for R=0)<0.0001 is obtained.

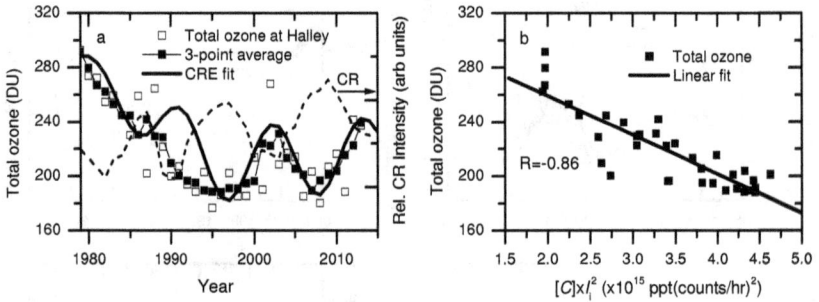

Fig. 5.3. Observed time-series data and statistical analyses of 4 month (September-December) mean total ozone at Halley, Antarctica during 1979-2013. a: Observed time-series CR intensity, total O_3 and 3-point averaged total O_3 data, as well as the best fit by the CRE equation (Eq. 4.22). b: 3-point averaged observed O_3 are plotted as a function of the product of $[C_i] \times I_i^2$; a linear fit to the data is also shown with linear correlation coefficients (R) indicated. Adapted from Lu [2013, 2014c].

5.3.2 *11-year cyclic variation of polar stratospheric cooling*

Moreover, it is also well-known that ozone loss can cause a stratospheric cooling: less O_3 in the stratosphere implies less absorption of solar and infra-red radiation there and hence a cooler stratosphere. Thus, temperature data in the lower polar stratosphere is a direct indicator of polar O_3 loss. Ramaswamy *et al.* [1996] investigated the fingerprint of ozone depletion in the spatial and temporal pattern of lower-stratospheric cooling for the period of one solar cycle (1979-1990). Ten year later, Ramaswamy *et al.* [2006] further studied the stratospheric cooling effect over two solar cycles (1979-2003). Unfortunately, they attributed the non-monotonic decrease in lower stratospheric temperature to the effect of volcanic eruptions. If this were true, then no long-term time correlation between CR intensity and polar stratospheric temperature during any 11-year cycle would be expected.

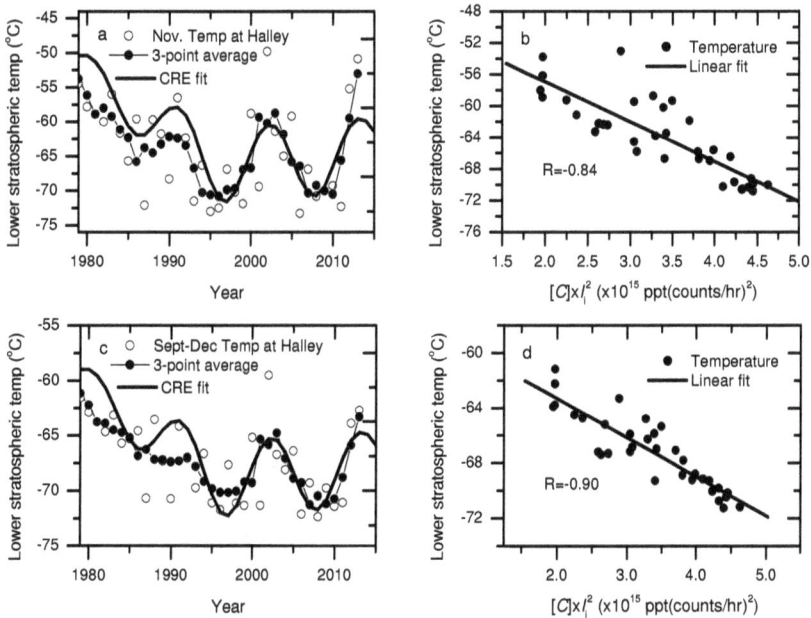

Fig. 5.4. Observed time-series data and statistical analyses of November monthly mean and 4 month (September-December) mean lower stratospheric temperatures at Halley, Antarctica during 1979-2013. a and c: Observed time-series temperatures and 3-point averaged data, as well as the best fit by the CRE equation (Eq. 4.22). b and d: 3-point averaged observed temperature data are plotted as a function of the product of $[C_i] \times I_i^2$; linear fits to the data are also shown with linear correlation coefficients (R) indicated. Adapted from Lu [2013].

The observed fact is, however, that a clear 11-year cyclic correlation between CR intensity and polar stratospheric cooling has been well observed [Lu, 2010a, 2013, 2014c]. Here, the temperatures in the lower stratosphere (100 hPa) at the BAS's Halley station in the later springs (November, immediately after the Antarctic ozone hole peak in October) and in the whole ozone-hole periods (September to December) from 1979 to 2013 are also shown in Fig. 5.4. First, the observed data indeed visibly show in Figs. 5.4a and c that both November mean and 4 month (September-December) mean lower stratospheric temperatures have pronounced *11-year cyclic variations,* which can be well fitted with the CRE equation (Eq. 4.22) derived from the CRE mechanism. It is particularly interesting to note that the observed temperature data have an excellent fit to the CRE equation. Second, statistical correlation analyses of the lower polar stratospheric temperature data at Halley in terms of the CRE equation are shown in Figs. 5.4b and d. Again, excellent linear correlations of stratospheric temperatures with $[C_i] \times I_i^2$ variations are observed, and high linear correlation coefficients $-R$ up to 0.90 and P(for R=0)<0.0001 are obtained.

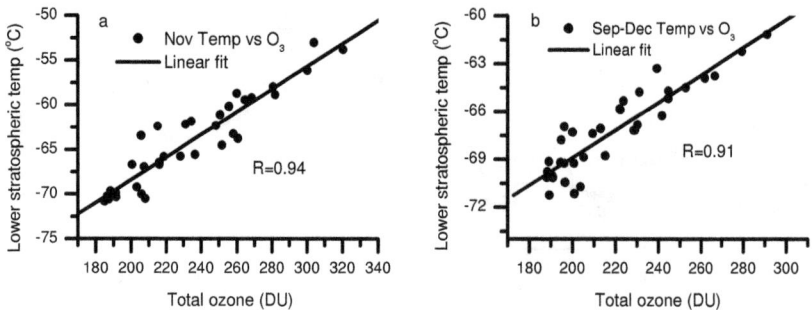

Fig. 5.5. November mean and 4 month (September-December) mean lower stratospheric temperatures versus total ozone at Halley, Antarctica, with linear correlation coefficients R indicated. Adapted from Lu [2010a].

As shown in Fig. 5.5, furthermore, the lower stratospheric temperatures also exhibit excellent linear correlations with measured total ozone at Halley with linear correlation coefficients R as high as 0.94 and P(for R=0)<0.0001 obtained.

The results shown in Figs. 5.2-5.5 have well confirmed that the CRE mechanism can excellently reproduce not only total O_3 but also O_3-loss-induced stratospheric cooling data in the Antarctic hole. The observed pronounced 11-year cyclic variations in total O_3 and lower stratospheric temperature are obviously due to the consequence of the CRE mechanism. These results lead to the important conclusion that both polar O_3 loss and lower stratospheric temperature over the past decades are well described by the CRE equation (Eq. 4.22) with the equivalent effective stratospheric chlorine [C] and the CR intensity *I* as the only two variables.

5.4 Direct effect of solar cycles or cosmic ray cycles

Atmospheric researchers have attempted to include the so-called solar effect in modeling of the *polar* O_3 loss within the photochemical model, while conceding that it is complicate and difficult to understand the O_3 variation in recent years and predict the future trend [WMO, 2006]. Here, it should be pointed out that this might be a pseudo-problem for at least two reasons. First, the direct solar effect, in inverse phase with the CR effect, argues that the maximum solar UV irradiance (corresponding to the lowest CR intensity) would result in the maximum O_3 production via the photolysis of O_2 in the upper stratosphere. This effect predicted small annual O_3 oscillations ($\pm 1.5\%$) in the tropics and mid-latitudes but not in the polar region (especially in the lower polar stratosphere) [WMO, 2006]. Second, the photochemical model would also predict that the maximum solar intensity would produce the largest amount of active Cl to destroy O_3.

Researchers have also proposed the direct cosmic ray effect, which attributes ozone loss to the odd hydrogen (HO_x) and odd nitrogen (NO_y) species generated by CRs in the polar stratosphere [Ruderman *et al.*, 1976], similar to the production of NO_y species from solar particle events proposed by Crutzen *et al.* [1975]. The direct CR effect, however, would predict an 11-year cyclic variation of total O_3 in the polar stratosphere *in any season*. Thus, a direct way to examine the direct solar-cycle and CR-cycle effects is to show the results of time-series total ozone in the summer polar stratosphere: both effects would be most significant to be seen if they exist, while the CRE mechanism is expected not to be effective due to the absence of PSCs. As shown in Fig. 5.6, no such time correlation is observed between total O_3 in the summer polar stratosphere and the CR intensity, in contrast to the winter- and spring-time total ozone over Antarctica [Lu, 2010a]. *Instead, the observed data show a clear and steady increase (recovery) in the summer total ozone over Antarctica since 1995-1996.* These observed data clearly disprove the possibility of linking the direct solar cycle or CR effect to the observed 11-year cyclic variation of total ozone in the polar stratosphere.

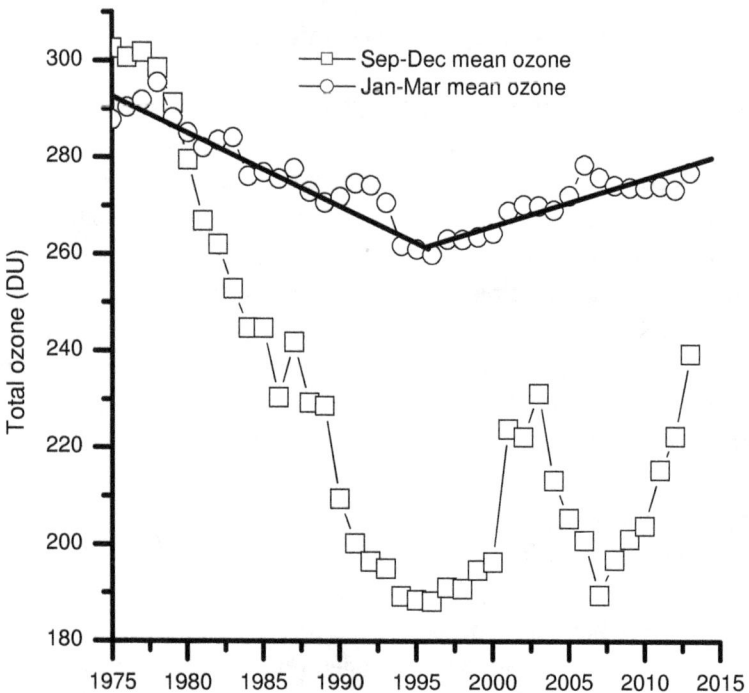

Fig. 5.6. Observed time-series springtime (September-December) and summertime (January-March) total column ozone at Halley, Antarctica during 1975-2013. A 3-point averaged smoothing was applied to the observed data. The summer total ozone over Halley has exhibited no 11-year cyclic variations but a clear and steady increase since 1995-1996. Adapted and updated from Lu [2010a].

There may exist other effects of charged particle precipitation on O_3 loss in the upper stratosphere [Thorne, 1977]. However, solar particle events and energetic electron precipitation are spontaneous frequent events without an 11-year cycle. Thus, they have been ruled out as the possible mechanisms for the observed 11-year cyclic oscillations in total O_3 and stratospheric cooling in the polar regions [Lu, 2010a].

Moreover, it is important to distinguish the observed 11-year cyclic lower polar stratospheric temperature variations directly induced by CRs and indirectly caused by the CR-driven O_3 loss in the polar stratosphere. For this purpose, the temperatures at the lower Antarctic stratosphere (100 hPa) at Halley during the winters (May-August, prior to the large Antarctic O_3 hole season), the springtime (September-December, the large O_3 hole season) over the past 57 years (1956-2013) are shown in Fig. 5.7. Interestingly, the winter polar stratospheric temperature does not show a pronounced oscillation over 11-year CR cycles. In contrast, the springtime polar stratospheric temperature shows *strong 11-year*

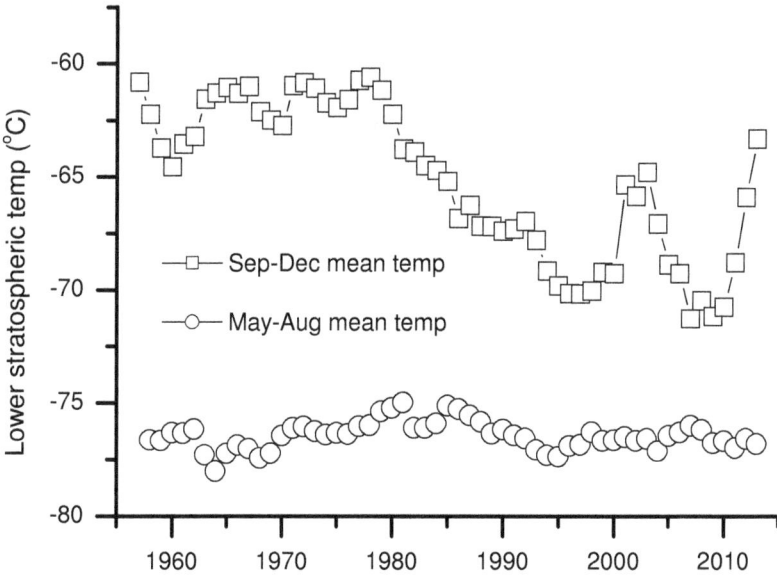

Fig. 5.7. Observed time-series winter (May-August) and springtime (September-December) lower stratospheric temperatures at Halley, Antarctica during 1956-2013. A 3-point averaged smoothing was applied to the observed data. Adapted and updated from Lu [2010a].

cyclic variations. Indeed, the most marked oscillation in lower polar stratospheric temperature occurs in the month (November) right after the largest Antarctic ozone loss usually observed in October [Lu, 2010a, 2013]. Thus, it is observed that the temperature in the polar stratosphere prior to the large O_3-hole season shows no significant time correlation with the CR/solar intensity variation. This observation demonstrates that neither the solar cycle effect nor the pure CR effect (without involving ozone loss) is responsible for the observations of 11-year cyclic ozone loss and stratospheric cooling in the Antarctic O_3 hole over the past about six decades, a conclusion reached in earlier studies [Lu, 2010a, 2013].

As a matter of fact, the observation of pronounced 11-year cyclic oscillations of the total O_3 in the polar ozone hole has forced one to conclude that the CR-driven mechanism must play a dominant role in the polar zone loss. This is because the oscillation amplitude of the CR intensity in 11-year CR cycles is well-known to be small, only about 10% of its mean value, the oscillation amplitudes of polar stratospheric ozone and temperature is observable only if the CRE mechanism plays a major role [Lu, 2009, 2010a]. The observed data have provided strong evidence of the dominance of the CRE mechanism for the polar ozone hole [Lu, 2010a, 2013].

5.5   Seasonal variations of CFCs, N_2O and CH_4

Both N_2O and CH_4 are often used as "tracer gases" in atmospheric chemistry. These molecules are "tracers" of the dynamical motions of stratospheric air masses since they are believed to lack any significant sinks in the troposphere, have well-characterized photochemical rates in the stratosphere and have atmospheric lifetimes much longer than stratospheric transport timescales. Thus, these "trace gases", particularly N_2O, have been used to identify and evaluate the effects of dynamics and chemistry on observed changes in stratospheric air components such as CFCs and ozone. For example, the phenomenon that the concentration of N_2O decreases as that of CF_2Cl_2 in the Antarctic ozone-hole area in the springtime was often taken as evidence to support the air descending model and to disprove the air ascending model in the development of ozone loss mechanisms [Toon and Turco, 1991]. Atmospheric photochemistry models predicted that both CFCs and the tracer gases have similarly long lifetimes and would therefore show a compact linear relationship in their mixing-ratio variations in the stratosphere.

Continuous, time-series data of CFC-12, N_2O and CH_4 over the Antarctic vortex during winter months (June, July and August), which are available in NASA and European Space Agency (ESP) satellite databases, provide important information on the physical process that leads to ozone loss in the polar stratosphere. The concentration distribution of CFCs is generally anti-correlated with the CR intensity distribution, and the decomposition of CFCs is drastically enhanced in the lower polar stratosphere during winter. Lu [2010a] reported one set of such time-series data: the observed monthly average CH_4 data combining the data from both the Cryogenic Limb Array Etalon Spectrometer (CLAES, running over the complete year 1992 only) and Halogen Occultation Experiment (HALOE, running over six years in 1992–1998) aboard the NASA Upper Atmosphere Research Satellite (UARS) were compared with the CF_2Cl_2 data over the winter of 1992 from the CLAES. The data clearly indicate that DET reactions of CFCs but not CH_4 occurred in the lower polar stratosphere during the whole period of winter. Some atmospheric chemists [Grooß and Müller, 2011] criticized that the data for CH_4 from the two instruments (HALOE and CLAES) aboard the UARS would not be comparable with the data of CFC-12 from the CLAES only. However, time-series CH_4 and CFC-12 data from the same instrument CLEAS actually showed a similar result [Lu, 2012a]. This is shown in Fig. 5.8. These data clearly show that for CH_4, no decrease in the *lower* Antarctic stratosphere *below 20 km* from March to August (even an increase in July) was observed, whereas the CF_2Cl_2 level significantly and continuously decreased from ~320 pptv in March to 200 pptv in August. These data provide strong evidence of DET reactions of CFC-12 but not CH_4 in the *winter* lower polar stratosphere.

Moreover, the original ESA's Oxford MIPAS Near Real Time satellite data of CFC-12, N_2O and CH_4 in the lower Antarctic stratosphere (65-90° S) during the winter season (June 23-September 30)

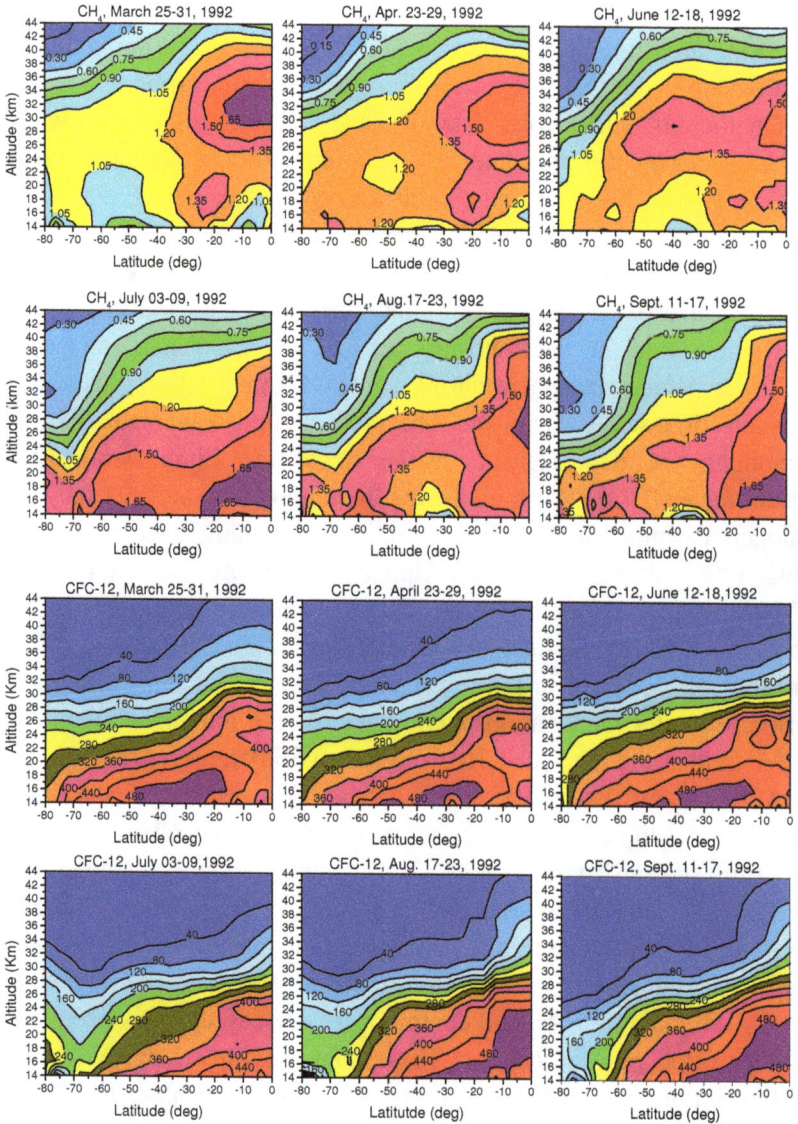

Fig. 5.8. Weekly averaged zonal-mean altitude-latitude maps of CH_4 and CFC-12 mixing ratio in the southern hemisphere from March to September 1992. Data (V9) were obtained from the CLAES aboard the NASA Upper Atmosphere Research Satellite (UARS). Adapted from Lu [2010a, 2012a].

over one decade (2002-2011) have shown a similar observation to the data of NASA's UARS CLAES [Lu, 2013]. As re-plotted as Fig. 5.9, the decadal mean time-series MIPAS data show clearly that the variation of CFC-12 exhibits some similarity to that of N_2O, while the CH_4 data show a very different curve. It is clearly confirmed that the CFC-12 and N_2O levels exhibit a similar continuous decrease since the beginning of winter, while the CH_4 level does not decrease until the end of August. After that, all gases show decreasing trends in September-October and then rising trends in November. Note that in September and October (the early spring), the levels of all gases (CH_4, N_2O and CFC-12) drop in the polar lower stratosphere. This can be well explained either by significant stratospheric cooling and air descending as a result of severe O_3 loss in the springtime lower polar stratosphere or/and by gas-phase reactions of CH_4 with reactive radicals resulting from dissociation of halogenated/N_2O molecules. These data have actually provided solid evidence of CRE (DET) reactions of CFCs and N_2O but not CH_4 in the *winter* polar stratosphere.

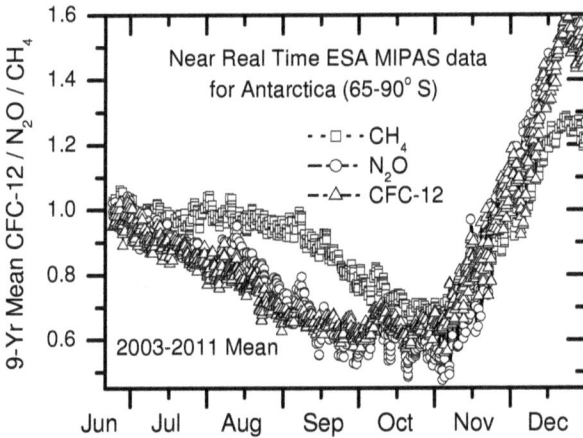

Fig. 5.9. Nine-year mean time-series data of CFC-12, N_2O, CH_4 are averaged from the ESA's MIPAS Near Real Time (daily) satellite data in the lower Antarctic stratosphere (65-90° S) during winter months (June 23-September 30) in 2003-2011, where the data for each gas were normalized to its initial value in the beginning of winter. Adapted from Lu [2013].

In current content of atmospheric chemistry, "air descending" is attributed to cause the concentration decreases of CFCs and some photo-inactive "tracer" gases such as N_2O and CH_4, and it is assumed that no physical or chemical processes contribute to CFC depletion observed in the *winter* lower polar stratosphere [e.g., Müller and Grooß, 2009; Grooß and Müller, 2011]. However, there existed serious problems with their presented "ACE-FTS satellite" CFC/N_2O data [Grooß and Müller, 2013], which will be addressed later in Sec. 5.11. Even ignoring the problems with theie data, there are still at least two major problems with the assumption. First, one must make sure whether DEA/DET processes could occur for these so-called "trace" (photo-inactive) molecules and whether the radicals resulted from the DEA/DET reactions of organic and inorganic halogenated molecules could react with these "trace" gases. For example, the assignment N_2O to be a standard "trace" gas because of its photo-inactivity would be valid only if there were no physical or chemical processes leading to the loss of N_2O in the polar stratosphere, especially in the winter season.

Unfortunately, just similar to CFCs, N_2O is actually a well-known molecule that has DEA resonances at low electron energies at 0.6 and 2.3 eV with an energy threshold of 0.21 eV in the gas phase, as discussed in Sec. 2.4.3 of Chapter 2. The latter is expected to be -1.1 eV on the surface of H_2O ice due to the polarization potential of ~1.3 eV [Lu *et al.*, 2002], and hence DET is effective for N_2O in liquid water or on ice. Indeed, it has been well observed that DET of N_2O can effectively occur with weakly bound electrons, e.g., photoexcited subvacuum hot electrons [Kiss *et al.*, 1991], weakly-bound electrons from Cs-covered surfaces with work-functions of 1.7-2.2 eV [Böttcher and Giessel, 1998] or photoassisted Li atoms in the Li-N_2O complex [Parnis *et al.*, 1995]. Even the latter study showed the reaction of CH_4 with O^- or $O(^1D)$ resulted from the DET of N_2O, and similar reactions might also occur for CH_4 with the DET reaction products (*e.g.*, Cl atoms) of chlorine-containing molecules in the springtime polar stratosphere. Nevertheless, it should also be noted that DET of weakly-bound electrons in PSC ice with CH_4 itself is not expected to occur since CH_4 does not have a DEA resonance at low electron energies below 5 eV in the gas phase (see, Fig. 4.3). Thus, CH_4 might be a better "trace" gas in the *winter* polar stratosphere than N_2O, though its absolute inertness is not guaranteed because of the above reason.

Perhaps, it should be reminded that N_2O is well-known to be reactive with the hydrated electron (e_{aq}^-) and is a source of OH radical in radiation chemistry and biology [Hart and Anbar, 1970; Lehnert, 2008]. It is widely used to convert e_{aq}^- into OH radicals ($OH^\bullet$) in radiolysis of water via the following reactions

$$e_{aq}^- + N_2O + H_2O \rightarrow N_2O^- + H_2O \rightarrow N_2 + OH^\bullet + OH^- \qquad (5.1)$$

This is very well known to radiation chemists and biologists [Hart and Anbar, 1970; Lehnert, 2008]. Thus, N_2O is no longer inactive when there are prehydrated or hydrated electrons trapped at PSC ice surfaces, though its DEA/DET cross section is lower than those of CFCs, HCl and $ClONO_2$.

Second, the "compact" $CFC-N_2O/CH_4$ correlation from the satellite or balloon data taken in the Antarctic springtime (September-November) has been used to support the photochemical models of O_3 depletion. As mention above, however, one should note that in the springtime, the largest O_3 hole has formed and significant stratospheric cooling and air descending occur; all gas concentrations drop. This is very different from the situation in the winter season, as revealed by the observed data presented in Figs. 5.8 and 5.9. Thus, the results cannot be taken as valid evidence to argue against the CR-driven electron reactions of halogenated molecules in the polar stratosphere *in the winter season*.

Third, ironically many observations in the history of ozone research that were in first cases unexplainable by photochemical models were immediately attributed to air motions, from the artificial data processing to erase the observed Antarctic ozone loss in 1979-1985 (before the publication of the ozone hole discovery paper by the BAS team in 1985) and the attribution of the observed Antarctic ozone hole to "air ascending" [see a review by Toon and Turco, 1991], to the attribution of CFC depletion in the winter lower polar stratosphere to "air descending". The former two attributions have certainly been thrown away for decades, while the latter has been disproven by the solid observations shown in Figs. 3.2, 5.8, 5.9 and further by Fig. 5.10 below. As a matter of fact, even the single observation by Dobson in the 1950s (see Fig. 3.2), if being interpreted properly, can counter against all these attributions originating from the expectations of photochemical models. This will be further discussed in next section.

5.6 Ozone loss in the dark polar stratosphere in winter

As shown in Fig. 4.24, the CRE mechanism will lead to the formation of reactive halogen species to destroy ozone in both the *winter* polar stratosphere in darkness and the *springtime* polar stratosphere with sunlight [Lu and Madey, 1999b; Lu, 2010a, 2013]. The British Antarctic Survey (BAS)'s daily total O_3 variations at Rothera, Antarctica during the months from April to November of 1996-2011 are shown in Fig. 3.10 in Chapter 3. This 15-year mean time series total O_3 data have clearly shown that total O_3 starts to drop from a high value of about 300 DU at the beginning of July to about 220 DU (defined as the threshold of the O_3 hole in current atmospheric chemistry texts) at the middle of August and to a minimum value as low as 150 DU in September. This observation shows clear evidence of significant ozone loss in the polar stratosphere during the winter season, consistent with the prediction of the CRE theory. To strengthen this conclusion, more data measured at another Antarctic station, Halley, are further presented as Fig. 5.10 here. It is clearly shown that in the period without significant CFCs in the stratosphere till the middle 1970s, there was no considerable seasonal decrease in total ozone over Antarctica from the end of the fall season to the end of the winter (Fig. 5.10a). This was in fact first observed by Dobson in his famous 1968 paper [Dobson, 1968], mentioned in Sec. 3.2 of Chapter 3 (see, Fig. 3.2). In contrast, there have been significant stratospheric amounts of CFCs since the late 1970s, and correspondingly there has been clear and pronounced polar ozone loss during the winter season. The evidence from the observations shown in Fig. 5.10b is as solid as in Fig. 3.10.

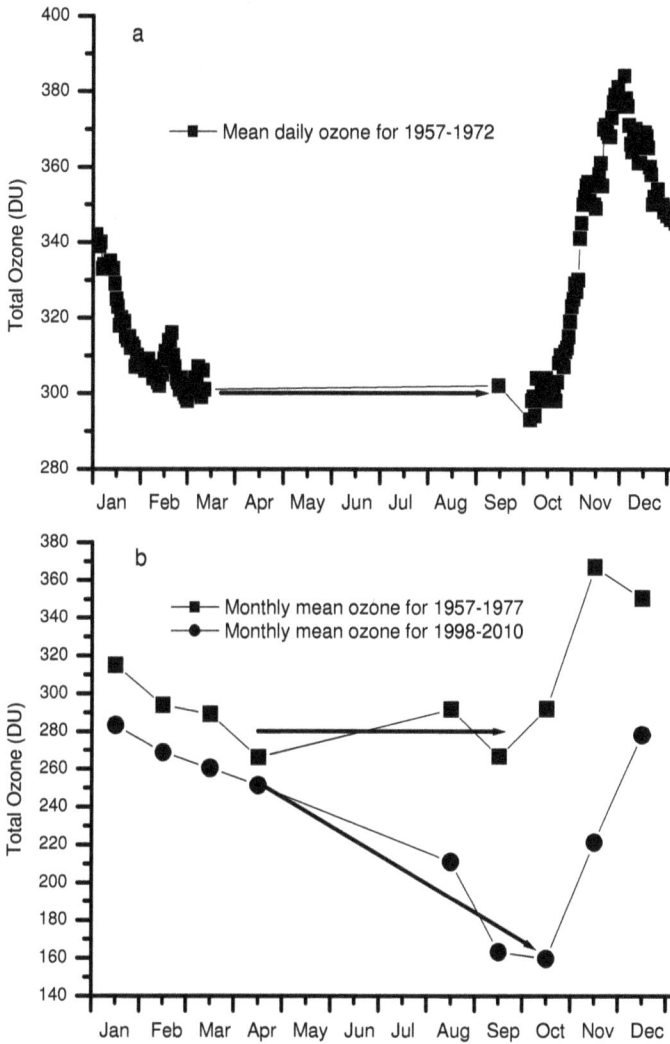

Fig. 5.10. Observed total ozone changes at Halley, Antarctica over the seasons for the periods pre-1970s and post-1970s, corresponding to the periods without and with significant CFCs in the stratosphere. a: The mean daily total ozone for the period of 1957-1972. b: The mean monthly total ozone for the periods of 1957-1977 and 1998-2010. The arrows are added to guide the eyes. Based on data from the British Antarctic Survey (BAS).

Note that this polar ozone loss in the dark winter in the past decades cannot be attributed to any air motions because the observed data for the period before the mid-1970s show no such change in polar total ozone (Figs. 3.2 and 5.10a). Indeed, the real-time (daily) variation of total O_3 over Antarctica (Fig. 3.10) generally follows well that of CFCs (Fig. 5.9) during the winter season, indicating that the CRE mechanism plays an important role in causing severe O_3 loss over Antarctica. Due to the lack of sunlight in the lower polar stratosphere during early and mid-winter, the significant decreases of CFCs and total O_3 in June-August cannot be explained by photochemical models. By contrast, these observations (Figs. 3.2, 3.10, 5.9 and 5.10) agree with the CRE theory quite well, giving strong evidence of the latter.

5.7   Evaluation of the Montreal Protocol

As reviewed in Sec. 3.5 of Chapter 3, a total column ozone increase (recovery) either in the global mean or in the springtime Antarctica and Arctic has not yet been observed [WMO, 2014], despite the decreases of major ozone-depleting substances (CFCs) controlled by the Montreal Protocol. This somewhat disagrees with the expectations from the photochemical models.

In contrast to the photochemical models, the observations mentioned above have shown that the summertime Antarctic total ozone has exhibited a clear and steady recovery since ~1995 (see Figs. 3.16 and 5. 6). Moreover, both total O_3 loss and stratospheric cooling in the winter and springtime Antarctica can be well reproduced by the CRE equation (Eq. 4.22) that leads to the dependence of O_3 loss on the stratospheric EECl level [C_i] and CR intensity I_i only, as shown in Figs. 5.2-5.4. The CR intensities have been well recorded since 1960s, showing a rising trend in the past four solar cycles [Lu, 2010a, 2013]. This means that no sign in recovery of recorded Antarctic O_3 losses is most likely due to rising CR intensities, which compensates the declining EECl levels in the polar stratosphere. Thus, the *real* change of the stratospheric EECl levels can be revealed by correcting measured total O_3 data or temperature data in the lower polar stratosphere with the CR-factor of $1/I_i^2$. A recent study used the CR-factor of $1/I_iI_{i-1}$ [Lu, 2013], but there is no considerable difference between the two factors. Therefore the real effectiveness of the Montreal Protocol can be evaluated, provided that reliable data of stratospheric O_3 / temperatures and CRs are used.

Here, the CR data at McMurdo and NOAA's Equivalent Effective Chlorine (EECl) data measured in the lower atmosphere at Antarctica and mid-latitudes are further plotted in Figs. 5.11A and B, respectively, which show that the tropospheric EECls measured on the Earth's surface have declined since its peak observed around 1994. Indeed, the observed tropospheric EECl at Antarctica, normalized to the 1980 value (as 100%), has declined to 143% in 2013 by about 14% from the peak value of 157% in 1994, while the observed tropospheric EECl at mid-latitudes has declined to 129% in 2013 by about 23% from the peak value of 152% in 1994. The NASA's October monthly mean total O_3 data over the South Pole (60-90° S) and annual mean total O_3 in low- and mid-latitudes (65° S-65° N) in 1979-2013 (2010), as well as their 3-year averaged data, are shown in Figs. 5.11C and D, respectively. The observed data in Fig. 5.11C show that from 1995 to the present, total O_3 over Antarctica has exhibited pronounced 11-year cyclic oscillations. Total ozone at mid-latitudes has shown a much smaller magnitude of *continuous* decrease and clear 11-year cyclic modulations since 1979 up to the present (Fig. 5.11D).

The observed Antarctic and non-polar O_3 data after correction by the CR-factor of $1/I^2$ given in the CRE equation (Eq. 4.22) are shown in Fig. 5.11E and F, respectively, in which polynomial fits to the data give $R^2=0.80$ and 0.70 (coefficient of determination) with the probability P<0.0001 for $R^2 = 0$ (no trend). Most strikingly, Fig. 5.11E shows that *O_3 losses in the Antarctic hole have had a clear and steady recovery since around 1995.* This is consistent with the directly observed recovery in summer Antarctic ozone without the correction by the CR-factor ($1/I^2$) (shown in Fig. 5.6). *Quantitatively, the October mean total O_3 loss over Antarctica has recovered from the peak loss of ~37% at ~1995 to a loss of ~25% in 2013.* Comparing the O_3 data in Fig. 5.11E with the EECl data in Fig. 5.11A, one can clearly see that the corrected O_3 loss over Antarctica follows the NOAA surface-measured EECl closely, with a short delay of only 1~2 years in the polar stratosphere. Indeed, the projected EECl with a 2-year delay has shown an approximately 15% decline from its peak at 1995-1996. This result solidly indicates that CFCs are indeed one of the main causes of the Antarctic O_3 hole, which

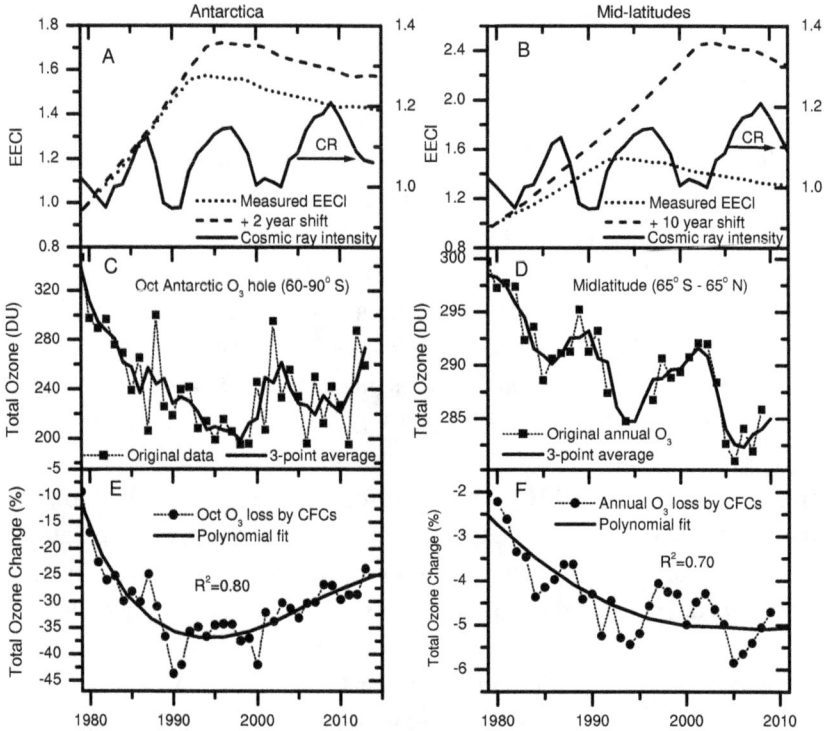

Fig. 5.11. Observed and corrected time-series total ozone in Antarctica (60-90° S) and mid-latitudes (65° S-65° N) during 1979-2013. A and B: NOAA's equivalent effective chlorine (EECl) data measured in the troposphere (solid lines) and projected EECl in the stratosphere (dash lines) at Antarctica and mid-latitudes, as well as CR intensities I measured at McMurdo, Antarctica. Both stratospheric EECl and CR data are normalized to their values in 1980. C and D: October monthly mean total O_3 over Antarctica and annual mean total O_3 at mid-latitudes, obtained from NASA TOMS N7/M3/EP/OMI satellites; also shown are the 3-point adjacent averaged smoothing. E and F: 3-point mean total O_3 data are corrected by the CR-factor of $1/I^2$; polynomial fits to the corrected data give coefficients of determination R^2 indicated and P<0.0001 for R^2=0 (no trend). It is clearly shown that after the CR effect is removed, the O_3 hole over Antarctica has a clear recovery, while no recovery in O_3 depletion at mid-latitudes has been seen. Adapted from Lu [2013] with the October Antarctic ozone data updated from WMO [2014].

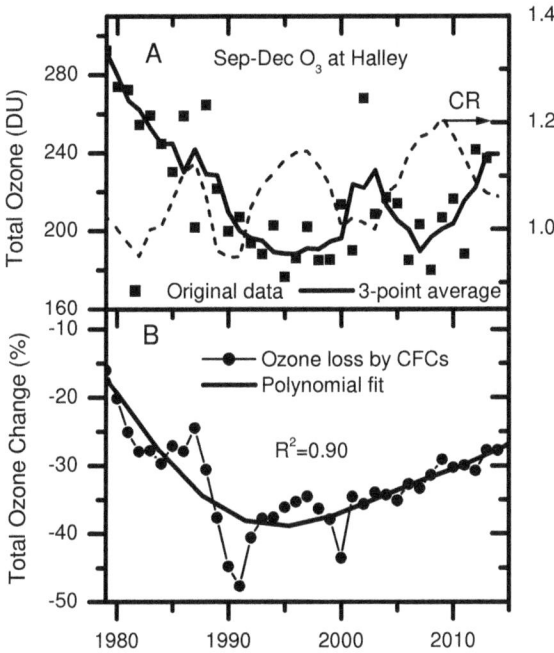

Fig. 5.12. Observed and corrected time-series total ozone at Halley, Antarctica during 1979-2013. A: 4 month (September-December) averages of observed total ozone and the 3-point adjacent averaged smoothing, as well as CR intensities I measured at McMurdo, Antarctica. B: 3-point mean total O_3 data are corrected by the CR-factor of $1/(I^2)$; a polynomial fit to the corrected data give a high coefficient of determination $R^2=0.90$ and $P<0.0001$ for $R^2=0$ (no trend). It is clearly revealed that after the CR effect is removed, the total ozone over Halley during the Antarctic O_3-hole period has exhibited a clear recovery. Adapted from Lu [2013, 2014c].

has shown a sensitive response to the decrease in the tropospheric CFCs regulated by the Montreal Protocol. In contrast, Fig. 5.11F shows no sign in recovery for O_3 loss at mid-latitudes. This means that a time delay of ≥ 10 years between the surface-measured and stratospheric EECl in mid-latitudes is required, as shown in Fig. 5.11B. Thus, it is obvious that the stratospheric EECl decline and associated O_3 recovery in mid-latitudes are significantly delayed, compared with those in the polar stratosphere.

Moreover, the BAS observed data of total O_3 at Halley before and after correction by the CR-factor of $1/I_i^2$ are shown in Figs. 5.12A and B, in which a polynomial fit to the corrected data give $R^2=0.90$ (coefficient of determination) with the probability $P<0.0001$ for $R^2 = 0$ (no trend). Consistent with the NASA satellite O_3 data for the Antarctica (60-90° S) in Fig. 5.11, total O_3 data at Halley again show a pronounced recovery since ~1995. Quantitatively, the total O_3 at Halley has exhibited a pronounced recovery from ~38.8% in loss in 1994-1995 to ~27.5% in 2013, after the CR effect is removed. Therefore, a rapid response to the decrease of ozone-depleting substances controlled by the Montreal Protocol is clearly revealed.

In a short summary, excellent statistical correlations of total O_3 and polar stratospheric cooling with $[C_i] \times I_i^2$ shown in Figs. 5.2-5.4 have strongly shown the validity of the CRE equation (Eq. 4.22) to describe the Antarctic O_3 hole. Eq. 4.22 can unravel the direct effect of human-made CFCs on O_3 loss after the removal of the CR effect. By correcting the observed data with the CR-factor of $1/I_i^2$, O_3 loss in the polar O_3 hole have shown a clear recovery since around 1995, closely following the decrease of the surface-measured tropospheric EECl that peaked in 1993-1994. *A pronounced ozone recovery from the loss of ~38±2% in 1994-1995 to ~26±2% in 2013 in the Antarctic O_3 hole has now been clearly revealed.* This result not only shows the validity of the CRE theory, but places the Montreal Protocol inhibiting the use of CFCs on a global scale on a firm scientific ground.

5.8 Comparison of the CRE model with the photochemical model

The CRE mechanism significantly differs from the photochemical model for stratospheric ozone depletion. The latter assumes that the sunlight photolysis of CFCs in the upper tropical stratosphere, air transport and the subsequent heterogeneous chemical reactions of transported inorganic halogens on ice surfaces in PSCs are the three major processes for the activation of halogenated compounds into photoactive halogens. In contrast, the CRE model believes that the *in-situ* CR-driven electron-induced reaction of halogenated molecules including organic and inorganic molecules (CFCs, HCl, ClONO$_2$, *etc*) adsorbed or trapped at

PSC ice in the winter polar stratosphere is the key step to form active halogen species that then lead to ozone loss in both winter and springtime polar stratosphere.

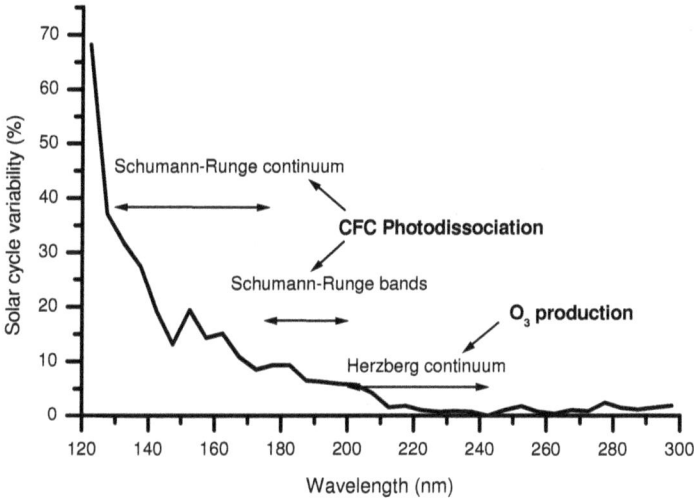

Fig. 5.13. Solar cycle variability (percent) defined as 100×[flux(solar max)-flux(solar min)]/flux(solar min) as a function of wavelength. Based on data from Brasseur *et al.* [1999].

First, the observed data shown in Figs. 5.6 and 5.7 have in fact ruled out any possibility that the solar effect would be responsible for the 11-year periodic variations of polar ozone and stratospheric cooling. Second, the photochemical models proposed that the 11-year cyclic variation of total O_3 in the tropics and mid-latitudes (but no polar regions) was attributed to the pure solar cycle effect: maxima in UV solar irradiance cause maxima in photochemical O_3 production [WMO, 2006]. This explanation, however, ignores another aspect of the photochemical models: maxima in UV solar irradiance would lead to maxima in activation of halogens for O_3 destruction. It should be noted that in the past 3 solar cycles, UV solar irradiance at the Herzberg continuum (200–242 nm) relevant to O_3 production varied by ~3%, which is far less than ~8% at the Schumann–Runge bands (175–200 nm) and 18% at the Schumann–Runge continuum (130–175 nm) relevant to CFC photodissociation leading to O_3 destruction [Krivova *et al.*, 2009b]. This is shown in Fig. 5.13. Furthermore, photochemical models give the maximum 11-year O_3 variation in the *upper* stratosphere at ~40 km, while the observed total O_3 cyclic variation originates mainly in the *lower* stratosphere at altitudes below 25 km for mid-latitudes and the origin remains uncertain in terms of photochemical models [WMO, 2006, 2010].

Another important observation to reveal the true underlying mechanism for O_3 loss lies in the data shown in Figs. 5.6, 5.11 and 5.12. In current context of atmospheric chemistry, the photodissociation mechanism proposed that CFCs would mainly decompose in the upper tropical stratosphere; air carrying the photoproducts (inorganic species) is then transported to the lower Antarctic stratosphere. Thus, a long delay (~6 or 5.5 years) from the troposphere peak was projected for the EECl to destroy O_3 in the Antarctic stratosphere [WMO, 2010, 2014]. The situation for the mid-latitudes of both hemispheres is different from the Antarctica primarily because it was thought that air in the mid-latitude stratosphere would have a younger mean 'stratospheric age' (~3 years) compared to air above Antarctica. As a result, CFCs in the mid-latitude stratosphere would need less time to become degraded by UV sunlight, and hence the mid-latitude stratospheric EECl was shifted by ~3 years only from the values measured at the troposphere [WMO, 2010, 2014]. This understanding of CFCs in the atmosphere is just opposite to the observed data shown in Figs. 5.6, 5.11 and 5.12. In contrast, the CRE mechanism gives that CFCs are mainly *in-situ* destroyed in the *polar* stratosphere and therefore the EECl to destroy O_3 in the polar region should be more sensitive to the change of CFCs in the troposphere. That is, a short delay (1~2 years) is expected between the polar stratospheric EECl change and the CFC change observed in the troposphere. Differently due to the low electron density in the mid-latitude stratosphere (Fig. 5.1), the electron-induced dissociation of CFCs to destroy O_3 at mid-latitudes is much slower. In other words, CFCs have a much longer residence time in the mid-latitude stratosphere than in the polar stratosphere. As a result, a much longer lag time from the tropospheric CFC change is expected for the EECl and resultant O_3 recovery in the mid-latitude stratosphere. This is exactly observed in the data of Figs. 5.6, 5.11 and 5.12, showing strong evidence of the CRE mechanism.

5.9 Future trends of the ozone hole

The excellent agreements between the CRE mechanism and the observations presented in Sec. 5.2-5.6 give high confidence in applying the CRE equation (Eq. 4.22) to predict the future recovery of the ozone hole with the projected variations of human-made EECl and natural CRs. The Montreal Protocol has been effective in regulating ozone-depleting halogen-containing molecules, so that the EECl in the polar stratosphere is expected to continue the decreasing trend observed in the past decade [WMO, 2010, 2014].

The CR-intensity variation with an average periodicity of 11 years and its modulation of ~10% are well known, which can generally be expressed as [Lu, 2010a]

$$I_i = I_{i0}\left\{1 + 10\% \sin[\frac{2\pi}{11}(i - i_0)]\right\}, \tag{5.2}$$

where I_{i0} is the median CR intensity in an 11-year cycle. Note that I_{i0} at Antarctica in the past three solar (CR) cycles has an increasing rate of ~2% per 11-year cycle. The best fit to all the observed CR data at the Antarctica (McMurdo) from 1960s-2009 yielded I_{i0}=8800[1+2%(i-1979)/11] (10^2 count/hr) [Lu, 2010a]. Eq. 5.2 can be used to calculate future CR intensities.

The observed and analyzed data shown in Figs. 5.6, 5.10 and 5.11 have soildly established that the stratospheric EECl data have *a lag time of only 2 years* from the tropospheric EECl data over Antarctica. With the above projections of future CR change and projected EECl data obtained with a lag time of only 2 years from the tropospheric EECl data, the observed and calculated 4-month (September-December) mean total O_3 and lower stratospheric temperatures at 100 hPa over Halley for 1979-2080 by Eq. 4.22 are shown in Fig. 5.14. Also indicated in Fig. 5.14 are the observed data confirming the predictions of 11-year cyclic Antarctic ozone loss [Lu and Sanche, 2001a] and associated stratospheric cooling

[Lu, 2010a]. The Antarctic O_3 hole is predicted to recover to the 1980 level around 2058($\pm$5), depending on the variations of not only the halogen loading but also CRs in the stratosphere [Lu, 2013]. The photochemical models reviewed in the earlier WMO Report [2010] gave a slower recovery by about 30~40 years, which predicted a recovery to the pre-1980 value by the end of this century. But the latter has now been revised to around 2075 in the newest Report [WMO, 2014], closer to 2058 given by the CRE mechanism [Lu, 2013].

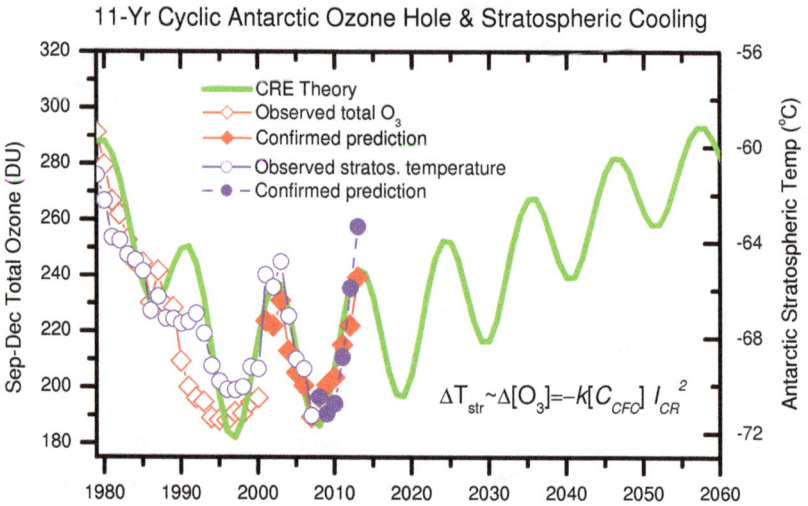

Fig. 5.14. Observed and theoretical total ozone and lower stratospheric temperature (100 hPa) at Halley (75°35' S, 26°36' W), Antarctica during the ozone hole months (September-December) in 1979-2013, and predicted future change. The observed data (open and solid diamond symbols for total ozone and open and solid circle symbols for temperature) are the averages of the data for 4 months (September-December), measured by the British Antarctic Survey (BAS); only a minimum 3-point smoothing was applied to the observed data. The data presented as solid symbols (solid diamonds and solid circles) were observed *after* the predictions of 11-year cyclic variations in polar ozone loss and stratospheric cooling were made by Lu and Sanche [2001a] and Lu [2010a], respectively. The theoretical data were calculated from the CRE equation (Eq. 4.22) with the equivalent effective chlorine (EECl) and the cosmic-ray (CR) intensity in the polar stratosphere as variables. Updated and adapted from Lu [2013].

5.10 Any effect of non-halogen greenhouse gases on polar ozone loss and climate change?

The observed data of ozone and stratospheric temperatures over the Antarctica in the past 50 years have also great implications for global climate change. The data plotted in Figs. 5.2, 5.3, 5.4 and 5.14 show that the 11-year cyclic variations of total ozone and lower stratospheric temperature over Antarctica are nearly completely controlled by the stratospheric EECl and the intensity of cosmic rays as the only two variables. This fact indicates no effects of non-halogen greenhouse gases (CO_2, N_2O and CH_4) on stratospheric climate of Antarctica over the past five decades. This is in striking contrast to the predictions of previous climate models that the large magnitude of polar stratospheric cooling due to the increase of non-CFC greenhouse gases would be observed, even as large as that induced by O_3 loss, which would in turn enhance polar ozone loss [Ramanathan *et al.*, 1985, 1987; Austin *et al.*, 1992; Shindell *et al.*, 1998].

For instance, those models predicted that the Antarctic and Arctic temperatures would several K colder due to stratospheric cooling induced by non-CFC greenhouse gases. It was also predicted that the ozone loss over Arctic would become very severe, nearly doubling the ozone depletion area and the ozone loss by 2000-2009, relative to the observed values in 1990-1999 [Shindell *et al.*, 1998]. Lower-stratospheric cooling induced by increasing greenhouse gases was also predicted to expand the ozone-hole area over Antarctica in 2000-2009, though the changes would be less drastic than the Arctic ozone. Remarkably, it was also noted that without climate forcing from increasing non-halogen greenhouse gases, the average Arctic ozone loss would only change due to changes in chlorine amounts. Since the equivalent effective stratospheric chlorine (EESC) level in the 2000-2010 was close to those observed in the late 1990s (1998-1999) [WMO, 2010], the predicted much greater mean Arctic ozone loss would then display the distinct signature of stratospheric cooling induced by increasing non-halogen greenhouse gases [Shindell *et al.*, 1998]. Unfortunately, none of these predictions agree with the observed data of ozone over Arctic and Antarctica in the past 16 years. In fact, the current Arctic ozone loss is no greater than that one decade ago, while the Antarctic ozone loss has been well reproduced by the CRE equation (Eq. 4.22). These facts, in turn, indicate no climate forcing from increasing non-halogen greenhouse gases. Moreover, the effect of stratospheric cooling induced by non-halogen greenhouse gases, if had existed, would have been observed clearly in the stratospheric ozone and temperature data over Antarctica, with significant increases of these gas concentrations over the past five to six decades. The observed data in Figs. 5.2-5.4 and 5.14 have clearly shown no such effect of rising non-halogen greenhouse gases.

5.11   Responses to the criticisms by some atmospheric chemists

Despite the substantial and convincing observations reviewed in the previous Sections, Rolf Müller and Jens-Uwe Grooß have long argued that no correlation would exist between CRs and O_3 loss in the polar region [Müller, 2003, 2008; Müller and Grooß, 2009; Grooß and Müller, 2011; Müller and Grooß, 2014a, 2014b]. Müller even argued that no further studies of the CR-related mechanism for O_3 depletion should have been motivated [Müller, 2008]. As revealed previously [Wang *et al.*, 2008c; Lu, 2009, 2012a, 2014a, 2014c], however, the scientific facts have been that either the fluctuation levels (up to 20%) of the O_3 data presented in those criticizing papers [Müller, 2003, 2008] were too large to allow examination of the effect of the CR intensity modulation (typically about 10%) on O_3 depletion over 11-year cycles, or the data used by those authors [Müller and Grooß, 2009; Grooß and Müller, 2011] are impossible to be found in the given source (namely the "ACE-FTS satellite databases").

For example, no one could obtain *time-series monthly-averaged "ACE-FTS data" of* N_2O and CFC-12 in *the Antarctic polar region* for *EIGHT months (points) per year*, which were unfortunately reported in Fig. 2 of Müller and Grooß [2009]. Similarly, one cannot find the "ACE-FTS data" shown in Fig. 1 of Grooß and Müller [2011]. This is because the Canadian satellite has never covered the Antarctic region in February and October and has had little coverage to the Antarctica in November (see the satellite latitude cover plot in Fig. 1 of Lu [2014a]). In fact, after the publication of my paper [Lu, 2012a] revealing the serious problems with their "ACE-FTS data", they published a Corrigendum admitting that "The months for which the (ACE-FTS) data were shown were not correctly indicated. ... the data do not cover this complete latitude range especially *they do not extend to the South Pole*" [Grooß and Müller, 2013]. In their recent Note [Müller and Grooß, 2014b], they attempted to argue again that their presented data were "original data, obtained directly from the ACE-FTS data base", albeit having "incorrect labels regarding the coverage of the ACE-FTS satellite measurements". Such an argument, however, cannot change the fact that it is impossible to obtain the "ACE-FTS data" used in their papers [Müller and Grooß, 2009; Grooß and Müller, 2011] from the real ACE-FTS database and that their argument against the CRE mechanism is invalid, as clearly and evidently shown by the satellite cover plot. As a physicist, I have been quite surprised that the above-listed papers by Müller and Grooß have been published in scientific journals. For more details about the facts and my responses, the readers are referred to my recent publications [Lu, 2012a, 2014a, 2014c].

5.12 Summary

The proposed CRE mechanism of the ozone hole has been well confirmed by substantial and convincing observations from both laboratory and field measurements. According to this theory, ozone loss in the polar O_3 hole can be well calculated with the simple CRE equation: $-\Delta[O_3]=k[C]I^2$, where $[C]$ is the equivalent effective chlorine level in the polar stratosphere and I the CR intensity in the year.

In this Chapter, we have reviewed and analyzed comprehensive data sets of halogenated gases, CRs, total ozone and O_3-loss-induced stratospheric cooling over Antarctica. These analyses have given excellent quantitative and statistical results consistent with the CRE theory. Indeed, the CRE mechanism has been well proven by substantial observations of spatial and temporal variations of CRs, CFCs, and total ozone.

In particular, the CRE equation has well reproduced 11-year cyclic variations of the Antarctic O_3 hole and associated stratospheric cooling, and significantly improved our predictive capabilities for future polar ozone loss. The results have shown that the CRE/DET reaction, rather than the photoactivation of halogen species in the gas phase, is indeed the critical step and the limit factor leading to ozone loss in the polar stratosphere.

After the removal of the CR effect, a pronounced recovery in the Antarctic O_3 hole since ~1995 has been clearly discovered, while no sign in recovery of O_3 loss in mid-latitudes has been observed. The polar O_3 hole has shown a sensitive response to the decline in total halogen burden in the low troposphere since 1994 due to the regulation by the Montreal Protocol. This result has not only shown the validity of the CRE theory but placed the Montreal Protocol on a firm scientific ground.

Unfortunately, nearly all the experimental and observational findings reviewed in this Chapter and Chapter 4 have been ignored in all the WMO Reports [2002, 2006, 2010, 2014]. Almost none of the main findings have been objectively discussed in these Reports. However, the predictions from the CRE theory have indeed been well confirmed by the substantial observed data, as summarized in this Chapter. Another example is the precise prediction by the CRE theory of an ozone loss maximum (a deepest ozone hole) in 2008 and an ozone loss minimum in 2013 over the Antarctica [Lu, 2009], though it is always challenging to predict polar ozone loss in a particular year. Both have been well confirmed. As shown in Fig. 5.15, a historical total ozone minimum of 75 DU was indeed observed at the Belgrano station (77° S, 35° W), Antarctica on 5th October 2008. This exactly confirms the prediction made from the CRE theory prior to the observation [Lu, 2009; also, a press release was indeed made by the University of Waterloo on the 16th September 2008].

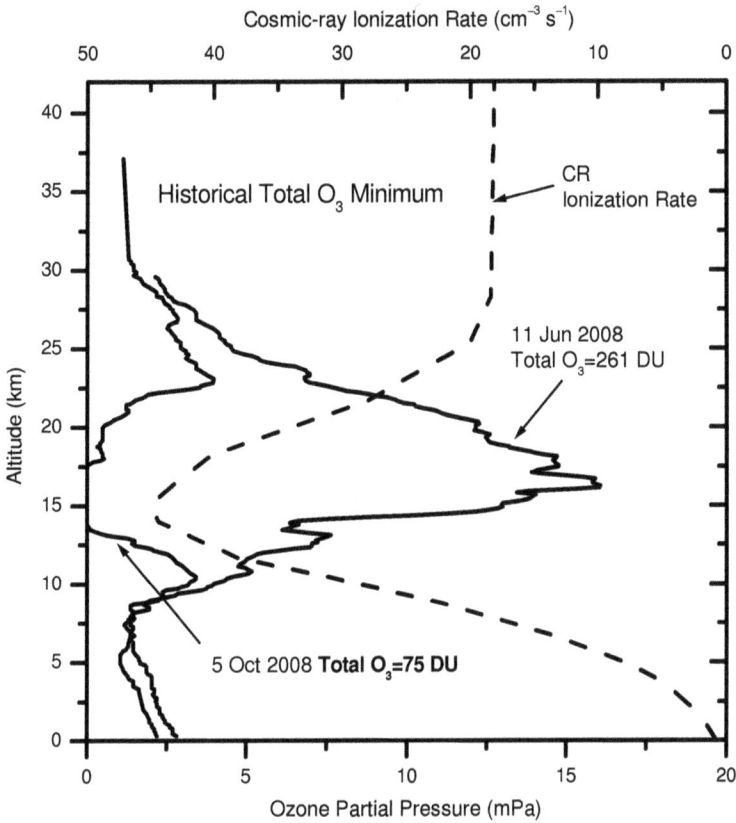

Fig. 5.15. Observation of a historical total ozone minimum of 75 DU at the Belgrano station (77° S, 35° W), Antarctica on the 5th October 2008. This is the lowest total ozone value in record. For comparison, the ozone profile on the 11th June 2008 and the altitude profile of cosmic-ray (CR) ionization rate are also shown.

The newest WMO Report [2014] states that the current best estimates (from the state-of-the-art photochemical models) for when effective equivalent stratosphere chlorine will return to its 1980 values are around 2050 for the midlatitudes and around 2075 for the Antarctic. It is also predicted that the evolution of the O_3 layer in the late 21st century would largely depend on the atmospheric abundances of CO_2, N_2O, and CH_4 [WMO, 2014]. In view of the fact that the substantial observations reviewed in this Chapter have been completely ignored in the 2014 Report, it is highly unlikely that these statements/predictions would eventually be true. This conclusion can also be inferred from those predictions made in the previous Reports [WMO, 1994, 1998, 2002, 2006, 2010], which turned out to agree rarely with the observations.

As a matter of fact, observed data have shown no effect of increasing non-halogen greenhouse gases on polar stratospheric ozone loss and associated stratospheric cooling over the past five to six decades. The high linear correlation coefficients (up to 0.94) observed in Figs. 5.4 and 5.5 indicate that the long-term temperature variation in the lower polar stratosphere is nearly completely controlled by the CRE equation with the total halogen level and CR intensity as only two variables. Thus, the observed data have shown no sign in greenhouse effect of increasing non-halogen gases (CO_2, CH_4, N_2O) on the stratospheric climate of Antarctica over the past five to six decades. This fact motivated us to look in the real effect of these gases on global surface temperature, which will be particularly addressed in Chapter 8.

Chapter 6

Conventional Understanding of Climate Change

6.1 Introduction to climate change

In the 1820s, the French scientist Joseph Fourier proposed that the atmosphere traps the outgoing heat radiation from the Earth's surface as a box with a glass cover—the box's interior warms up with entering sunlight while the heat cannot escape from the box. This was certainly an over-simplified explanation. It is now known that comparing the atmosphere to a greenhouse is not precisely correct. Nevertheless, Fourier's analogy indeed led to the calling of heat trapping by the atmosphere as "*the greenhouse (GH) effect*", implying that changes in atmospheric compositions could affect the Earth's climate.

In the 1860s, the Irish physicist John Tyndall discovered in his laboratory measurements that certain gases, including water vapor and carbon dioxide (CO_2), are opaque to infrared (terrestrial) radiation. He suggested that such gases high in the atmosphere keep our planet warm by trapping outgoing heat (terrestrial) radiation emitted from the surface.

The elementary ideas suggested by Fourier and Tyndall were developed further by the Swedish chemist Svante Arrhenius in 1896. He investigated how changes in the amount of CO_2 could affect the climate. He calculated that a reduction in the natural level of CO_2 in the air by 50% would lead to an ice age and a doubling of the CO_2 level would increase the global mean surface temperature by 5-6 °C. This result is actually close enough to the results obtained by state-of-the-art computer climate models given in recent IPCC Reports—the current estimate is that a doubling of CO_2 would lead to about 3 °C of global warming

[IPCC, 2001, 2007]. In the simple model calculations by Arrhenius, he calculated the warming effect of CO_2 in a straightforward way, starting with measured data (of solar radiation reaching the earth's surface) and all the obvious physics assumptions similar to those used in current climate models. Consequently, it is not surprising that both preliminary and state-of-the-art climate models reached roughly the same result.

Arrhenius also postulated that the significant rise of the CO_2 level due to the rapid industrial revolution would likely lead to the amount of warming he calculated. Then the British electrical engineer Guy Stewart found in 1938 that the CO_2 level had increased by about 10% since the 1890s, and argued that it could explain the rise in temperature recorded over 1890-1938. Since 1958, direct atmospheric measurements of CO_2 have been made at Mauna Loa Observatory in Hawaii by the US NOAA's Global Monitoring Division, and the potential seriousness of global warming due to the observed continuous rise in CO_2 has received much attention across the globe. Fig. 6.1 shows the measured atmospheric CO_2 concentrations and global mean surface temperatures (GMST) during 1958-2012 [IPCC, 2013].

The co-incidence of the observed rapid global warming and the increase in the CO_2 from 1970 to around 2000 (see Fig. 6.1) strongly boosted the argument that anthropogenic CO_2 would be the main driver of modern climate change. Since the early 2000s, however, the subject of global warming has led to intense debate and large controversies. It seems now generally agreed that there exist large discrepancies between predictions from climate models and observations [e.g., Fyfe *et al.*, 2013; IPCC, 2013].

A true calculation of warming caused by the GH effect requires measurements more accurate and complete than the simple measurement on solar radiation reaching the earth's surface such as used in Arrhenius' model calculations. The details of precise radiation physics of the Earth, exactly what bands of radiation are absorbed in the atmosphere by *greenhouse gases* (CO_2, CFCs, water molecules, *etc.*), what degrees of the absorption are saturated, and what climate sensitivity and feedback factors are precisely determined to be, are all needed to produce a more accurate measure of warming.

This chapter gives a brief description of basic radiation physics of the Earth-atmosphere system and presents a simple model of the greenhouse effect. The discussion will be focused on the CO_2 theory of modern global warming. Finally, the response of the climate system to radiative forcing caused by an increase in the atmospheric abundance of a greenhouse gas (GHG) will be addressed.

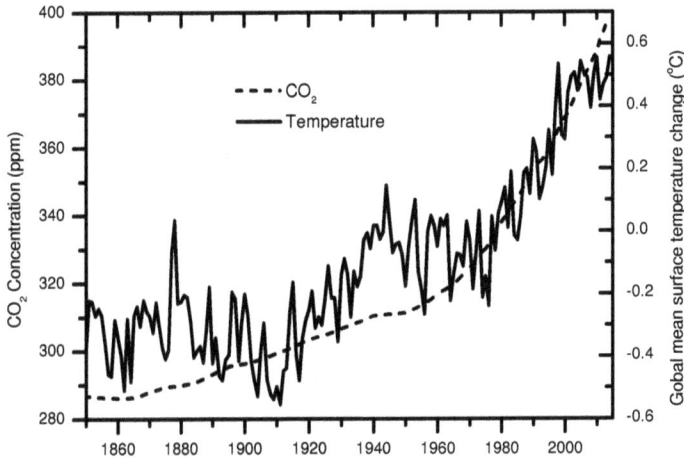

Fig. 6.1. Time-series atmospheric CO_2 concentrations and global mean surface temperatures from 1850 to 2012. CO_2 data were obtained from the IPCC Report AR5 [2013]. Annual global (near) surface temperatures were from the UK Met Office's HadCRUT4 dataset with temperature anomalies (°C) relative to 1961-1990 [Morice *et al.*, 2012].

6.2 Solar radiation

6.2.1 *Solar radiation spectrum*

The Sun can approximately be described as a blackbody with a temperature of about 6000 K. The spectrum of solar radiation can be theoretically given by Planck's formula, as shown in Fig. 6.2. The spectrum of the solar electromagnetic radiation striking the Earth's atmosphere covers a wavelength range of 100 nm to about 1 mm.

At the top of the atmosphere (TOA) of the Earth, the sunlight has a power of 1361-1362 W/m^2 [IPCC, 2013] which is called the solar constant F_S. At ground level, the Sun's power decreases significantly, with the apparent sharp dips at certain wavelengths, which are due to the absorption by atmospheric gases, mainly molecular oxygen (O_2), ozone (O_3), and water (H_2O) vapor (see Fig. 6.2).

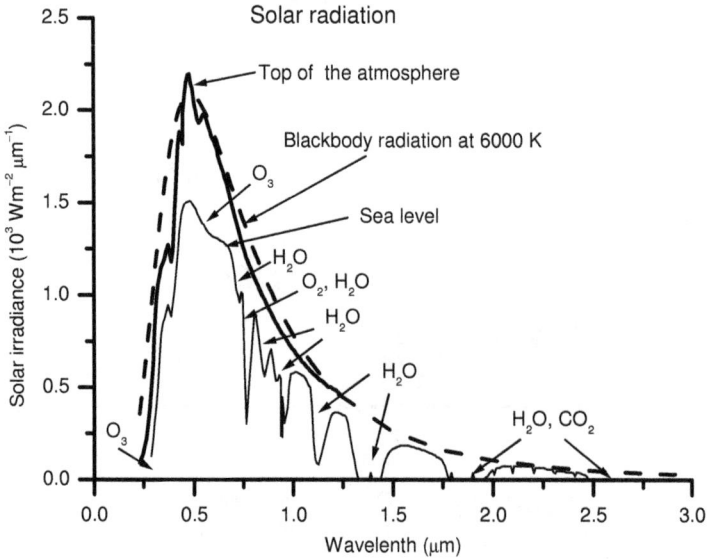

Fig. 6.2. Theoretical blackbody radiation spectrum of the Sun at 6000 K (dash line), and measured solar irradiance spectra at the top of atmosphere (thick solid line) and at sea level (thin solid line), respectively. Based on the Planck formula and data from Gast [1961] and Seliger [1977]. Adapted from Brasseur *et al.* [1999] and Jacob [1999].

6.2.2 *Solar variation*

The longest record of solar variations is the changes in the number of sunspots recorded since ~1610, while direct measurements of total solar irradiance (TSI) have only been available since the mid-1970s and are based on a composite of many different observing satellites. The correlation between TSI and other proxies of solar activity make it possible to estimate past solar activity. Most important among these proxies is the record of sunspot observations, as sunspots are closely correlated with changes in solar output. Fig. 6.3 shows time series variation of annual mean sunspot number (SSN) from 1700 to 2014, where the SSN modulation with an 11-year periodicity is marked. Direct measurements of radio emissions from the Sun at 10.7 cm can also provide a proxy of solar activity that can be measured from the ground as the Earth's atmosphere is transparent at this wavelength. Lastly, solar flares are another index of solar activity, which usually occur in the presence of sunspots, and hence the two are correlated. As mentioned in Chapter 1, however, solar flares make only little perturbations to the solar luminosity.

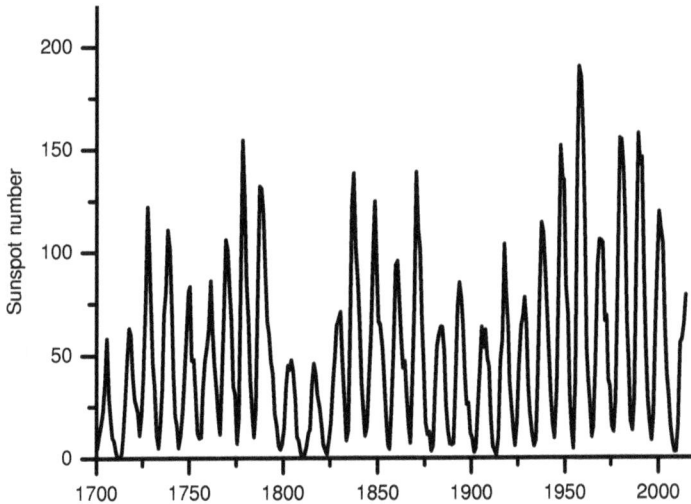

Fig. 6.3. Time series variation of annual mean sunspot number (SSN) from 1700 to 2014. Based on data from the Royal Observatory of Belgium's Solar Influences Data Analysis Center (http://sidc.oma.be/).

There were no reliable measurements of solar constant changes before the mid-1970s. Satellite-based observations of total solar irradiance started in 1978. These measurements show that the average solar constant has had only small variations since the 1970s.

6.3   Earth's blackbody radiation (Terrestrial radiation)

The heat radiation of the Earth surface after absorbing solar radiation can approximately be treated as a blackbody radiation. The latter, in history, was referred to cavity radiation, which is the radiation emitted from a small hole at the wall of an opaque enclosure kept at temperature T. The radiation energy density (the energy per unit volume) in the wavelength interval $d\lambda$ or the frequency interval dv is given by the Planck formula

$$u_\lambda(T)d\lambda = \frac{8\pi hc}{\lambda^5(e^{hc/\lambda kT} - 1)}d\lambda, \qquad (6.1)$$

$$u_v(T)dv = \frac{8\pi h v^3}{c^3(e^{hv/kT} - 1)}dv, \qquad (6.2)$$

where c is the speed of light, $h=6.63\times10^{-34}$ J s is the Planck constant, and $k = 1.38\times10^{-23}$ J K^{-1} is the Boltzmann constant.

In the unit volume, photons carrying the radiation energy move isotropically in all directions. An observer actually measures the energy flux $\Delta\Phi$ (energy per unit time, per unit area) at the speed c of the photons within a small solid angle $\Delta\Omega$. Moreover, since an irradiating object emits electromagnetic (EM) radiation with a continuous spectrum of wavelengths (frequencies), the observer using a spectrometer measures the radiation energy flux in a wavelength interval from λ to $\lambda+\Delta\lambda$, where $\Delta\lambda$ is the so-called resolution of the spectrometer. Thus, the measured *radiance spectrum* over the entire spectrum of wavelengths corresponds to the radiation energy flux per unit solid angle, per unit wavelength (frequency) interval for a blackbody at temperature T, given by

$$B_\lambda(T)d\lambda = (\frac{c}{4\pi})u_\lambda(T)d\lambda = \frac{2hc^2}{\lambda^5(e^{hc/\lambda kT} - 1)}d\lambda, \qquad (6.3)$$

$$B_v(T)dv = (\frac{c}{4\pi})u_v(T)dv = \frac{2hv^3}{c^2(e^{hv/kT} - 1)}dv, \qquad (6.4)$$

The radiance spectra of the Sun at ~5800 K and the Earth at 288 K are shown in Fig. 6.4.

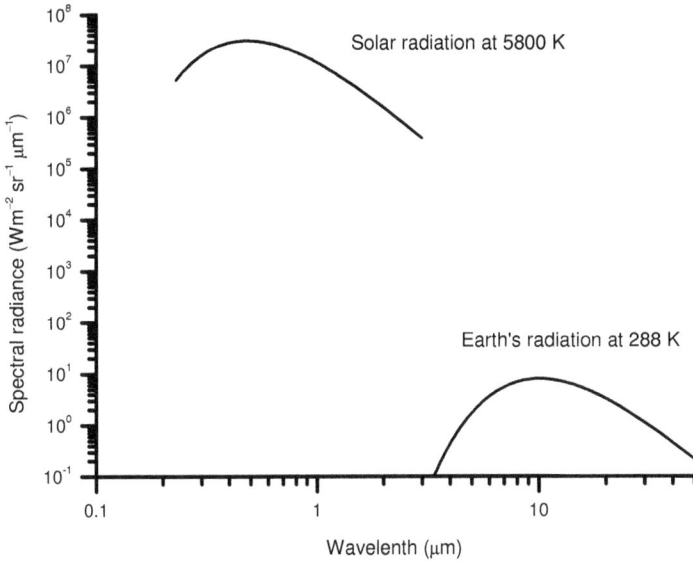

Fig. 6.4. The radiance spectra of the Sun and the Earth.

The total radiation energy flux emitted by a blackbody, obtained by integrating $u_\lambda(T)$ over all wavelengths or $u_\nu(T)$ over all frequencies, is

$$F = \frac{c}{4}\int_0^\infty u_\lambda(T)d\lambda = \sigma T^4 \qquad (6.5)$$

where σ is the Stefan-Boltzmann constant.

The wavelength λ_{max} at which $B_\lambda(T)$ is at the maximum, obtained by solving $\partial B_\lambda(T)/\partial\lambda = 0$, is given by Wien's displacement law

$$T\lambda_{max} = hc/4.965k = 2898\mu m \cdot K \qquad (6.6)$$

Note that by solving $\partial B_\nu(T)/\partial\nu = 0$, one obtains the frequency ν_{max} at which $B_\nu(T)$ is at the maximum: $\nu_{max}/T = 2.821k/h$, corresponding to $\lambda_{max} = hc/2.821kT$. It follows that the wavelength corresponding to the $B_\nu(T)$ peak is 1.76 times the wavelength λ_{max} at which $B_\lambda(T)$ peaks for any blackbody temperature, as shown in Fig. 6.5.

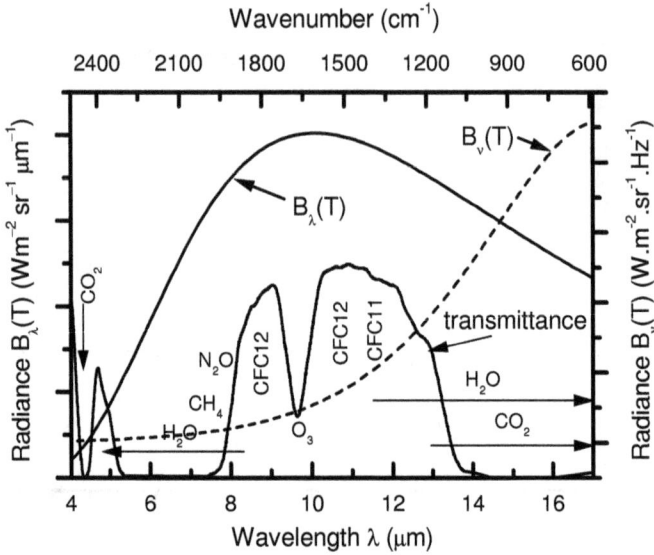

Fig. 6.5. Earth's blackbody radiance spectrum and atmospheric transmittance spectrum. The radiation energy flux per unit solid angle, per unit wavelength — $B_\lambda(T)$ as a function of wavelength λ and the radiation energy flux per unit solid angle, per unit frequency v —$B_v(T)$ as a function of wavenumber. Also shown is a theoretical atmospheric transmittance spectrum obtained with Modtran4 calculations [Ratkowski *et al.*, 1998].

Given that the Earth surface temperature is about 288 K, $B_\lambda(T)$ will have a peak at $\lambda_{max}=10.06$ μm in the infrared (IR) wavelength range. Since IR absorption spectra of molecules are usually measured by an IR spectrometer that uses a Fourier transform method, the radiance spectrum of the Earth is often presented as a function of frequency or wavenumber $(1/\lambda)$. A terrestrial radiation spectrum measured at the top of the atmosphere by the Nimbus-3 IRIS instrument in a satellite in 1971 [Hanel *et al.*, 1972] is shown in Fig. 6.6. Note that when $B_v(T)$ is applied, one obtains that $B_v(T)$ peaks at the wavenumber around 565 cm^{-1}, corresponding to the wavelength of 17.71 μm. Since CO_2 happens to have a strong absorption band at 600-770 cm^{-1} (13-17 μm) centering at ~667 cm^{-1} (~15 μm), it was often showed that the blackbody radiation spectrum of the Earth centers at the absorption band of CO_2. The solution to this paradox will be described in Chapter 8.

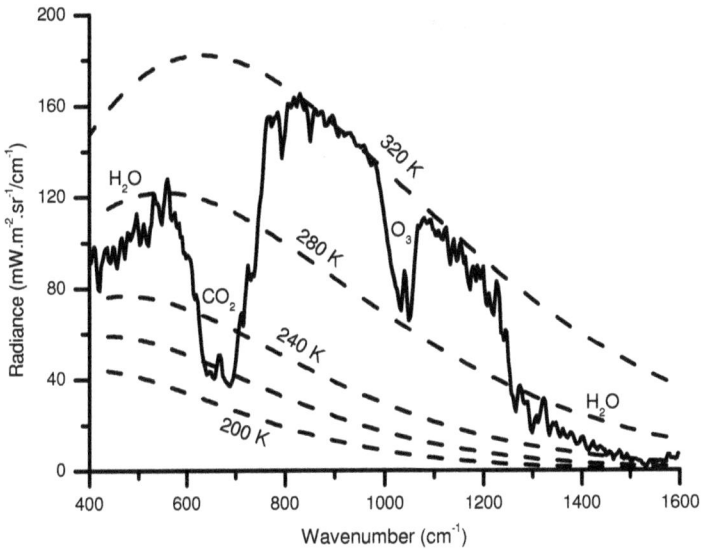

Fig. 6.6. Radiance spectrum measured by a satellite over North Africa at noon. For comparison, blackbody radiance spectra at selected temperatures are included. The absorption bands of H_2O, CO_2, and O_3 are indicated. Based on data from Hanel *et al.* [1972].

A theoretical atmospheric transmittance spectrum obtained with Modtran4 calculations is also shown in Fig. 6.5 [Ratkowski *et al.*, 1998]. It can be seen from Figs. 6.5 and 6.6 that besides the 15 μm band of CO_2, there are absorbing bands of O_3 at 9.6 μm and other molecules such as H_2O, CH_4 and N_2O. These radiative gases are called greenhouse gases. It is worth noting that CO_2 contributes to the strong absorption band at 13-17 μm, but H_2O is the effective absorber in the entire infrared spectral range and has two major bands at 5-8.3 μm and 11-17 μm. CH_4 and N_2O also have strong IR absorption band strengths at 7.6 μm and 7.8 μm, respectively. These gases are therefore main greenhouse gases.

Owing to its abundance, water vapor is the most important greenhouse gas and absorbs a large fraction of the terrestrial radiation in the lower atmosphere [Clark, 1999].

6.4 Absorption of radiation by the atmosphere

As discussed in Chapter 1, the absorption of radiation energy by a molecule may lead to an excitation of the molecule, that is, a transition between electronic or vibrational or rotational energy levels of the molecule. An electronic transition to a higher electronic state generally requires the highest excitation energy (UV radiation at <400 nm). A rotational transition requires the lowest excitation energy (far-IR radiation at >20 μm). A vibrational transition requires an intermediate energy (IR radiation at 700 nm–20 μm), corresponding to the wavelength range of maximum longwave radiation from the earth's surface. In contrast, little absorption by atmospheric molecules occurs in the range of visible radiation (400-700 nm) which lies in the energy gap between electronic and vibrational transitions, corresponding to the wavelength range of main solar radiation. Therefore, the atmosphere is essentially transparent to the visible sunlight but is an effective 'blanket" (heat trap) to longwave (terrestrial) radiation emitted from the Earth's surface.

Some molecules absorb radiation in the wavelength range (5-50 μm) of terrestrial radiation leading to vibrational and vibrational-rotational excitations and are hence called *greenhouse gases (GHGs)*, while others do not absorb the longwave radiation and hence are not GHGs. For example, gases such as H_2O, CO_2, O_3 and CFCs have strong absorption lines in the IR and are therefore effective GHGs, while gases such as N_2 and O_2 do not absorb in the IR and are not GHGs. The impact of GHGs on the earth's climate depends on their abundances and absorption strengths in the wavelength range of terrestrial radiation. Amongst atmospheric gases, H_2O, CO_2, CH_4, N_2O, O_3, and CFCs are important GHGs because they absorb a significant fraction of longwave radiation emitted from the surface. Water vapor is the most important GHG due to its atmospheric abundance and extensive IR absorption bands. Similarly, CO_2 is also a main GHG. However, CFCs may have far-reaching importance for defining the climate of the Earth. This is because there is an *atmospheric window* at 8-13 mm, where the unpolluted atmosphere is essentially transparent except for the strong absorption band of O_3 at 9.6 μm, as shown in Fig. 6.5. Therefore, terrestrial radiation mainly propagates to space through this spectral window. As a consequence, any polluting molecule with a strong absorption in the 8 to 13 μm region is highly effective GHGs. The importance of CFCs and other halogen-containing GHGs will particularly be addressed in Chapter 8.

6.5 The radiative equilibrium of the Earth

In climate research, there are essentially two types of models: *conceptual models* and *'quasi-realistic' models* [*e.g.*, von Storch, 2010]. The former describe the basics of the climate system and the effect of physical processes in a simplified language, often taking the form of simple equations that allow for analytical solutions. In contrast, the *'quasi-realistic' models* seek to maximize complexity for optimizing different aspects, such as spatial resolution and the number of parameters. The latter usually take the form of complex and lengthy programming codes to be run on an advanced computer. General circulation models (GCMs), particularly coupled atmosphere-ocean GCMs (AOGCMs, *e.g.*, CMIP5) represent the largest complexity in climate models and internalize as many processes as possible. Indeed, the procedures of calculating surface temperature changes from GCMs are very complicated and there are many uncertainties in performing the calculations including many parameters. The greenhouse effect is best illustrated from the global average conceptual models, as shown in many texts [*e.g.*, Jacob, 1999; Brasseur *et al.*, 1999; Andrews, 2000; Salby, 2012].

In this section, we introduce a simple radiative balance model to explain why the global mean surface temperature is about 288 K and to understand the greenhouse effect, given the input of solar radiation and some basic radiation physics.

As shown in Fig. 6.7, we have to notice that the total solar power intercepted by the Earth's surface is $F_S \pi R_E^2$, where R_E is the Earth's radius and F_S the solar constant. Also, the Earth-atmosphere system has an albedo A approximately equal to 0.3, that is, ~30% of the incoming solar radiation is reflected back to space without penetrating through the Earth's atmosphere. Therefore, only $(1-A)F_S \pi R_E^2$ of the incoming solar power can be absorbed by the earth.

If there were no atmosphere or if the atmosphere would not have any absorption of any incoming solar radiation and outgoing thermal radiation from the Earth, then the fraction of the solar power, $(1-A)F_S\pi R_E^2$, would hit on the earth surface. It is a good approximation that the earth is a black body, which absorbs all this fraction of solar power and then emits as a black body in all directions at a uniform temperature T_s. By the Stefan-Boltzmann law, the blackbody radiation energy flux (power per unit area) is σT_s^4. When the earth is in thermal equilibrium, we have

$$(1-A)F_S\pi R_E^2 = 4\pi R_E^2 \sigma T_s^4 \tag{6.5}$$

That is, the mean unreflected incoming solar flux at the top of the atmosphere (TOA)

$$F_0 = (1-A)F_S/4, \tag{6.6}$$

which is about 240 Wm^{-2}, must balance the thermal emission flux from the earth surface

$$F_g = \sigma T_s^4 \tag{6.7}$$

So that

$$F_0 = (1-A)F_S/4 = \sigma T_s^4 \tag{6.8}$$

With $A=0.3$, $F_S=1365$ Wm^{-2}, and the Stefan-Boltzmann constant $\sigma=5.67 \times 10^{-8}$ $Wm^{-2}K^{-4}$, we calculate $T_s=255$ K. This value is significantly lower than the observed global mean surface temperature of about 288 K. This is clearly due to the unrealistic assumption that there would be no absorption in the Earth's atmosphere.

Now, we consider the effect of adding a non-transparent/absorbing atmosphere to the above model. A *gray* atmosphere is first assumed: the atmosphere is transparent to short-wave radiation from the sun but absorbs long-wave thermal radiation from the ground with a constant absorption coefficient that is independent of wavelength λ. Under equilibrium at the TOA, the incoming energy flux F_0 from the Sun is now balanced by the outgoing energy flux from the Earth *for an observer looking down the atmosphere from space*

$$F^\uparrow = \varepsilon \sigma T_s^{\,4} \qquad\qquad (6.9)$$

So that

$$F_0 = (1-A)F_S/4 = \varepsilon \sigma T_s^{\,4}, \qquad\qquad (6.10)$$

where ε is called the effective *emissivity* of the atmosphere. A unit effective emissivity (ε=1) would give T_s=255 K corresponding to the absence of an atmosphere, while ε=0.615 gives the observed surface temperature T_s=288 K.

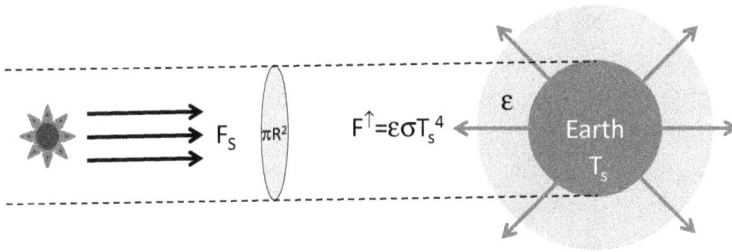

Fig. 6.7. A simple radiative model of the greenhouse effect. The atmosphere as a vertically continuous absorbing medium is transparent to incoming (unflected) solar radiation F_0=(1-A)F_S/4 but has an absorption efficiency ρ for long-wave (thermal) radiation from the ground. The ground at temperature T_s is assumed to be a blackbody with a radius R emitting upward flux F_g=$\sigma T_s^{\,4}$. The atmosphere is not a blackbody but emits fluxes both upwards and downwards.

It is worthwhile to note that the equilibrium condition, *i.e.*, Eq. 6.10, gives a higher surface temperature for a lower effective emissivity ε of the atmosphere. This appears counter-instinctive for an observer on the ground. *Kirchhoff's law* states that if an object absorbs radiation of wavelength λ with an efficiency ρ_λ, then it emits radiation of that wavelength at a fraction ρ_λ of the corresponding blackbody emission at the identical temperature. Thus, the effective emissivity ε of the atmosphere is equal to the effective *absorptivity* of long-wave radiation by the atmosphere. Correspondingly, $\varGamma$=1-ε is the effective transmittance of long-wave radiation in the atmosphere. An observer on the ground would expect that the more atmospheric absorption of terrestrial radiation from the surface, the more significant the greenhouse effect of the atmosphere and hence the higher the resultant surface temperature T_s. For an observer from space, on the contrary, the less efficient the absorption of terrestrial radiation by the atmosphere, the brighter the surface (the hotter the surface). This is exactly observed in the outgoing longwave radiation (OLR) spectrum measured by satellites at TOA. For example, the OLR flux measured by a satellite in the atmospheric window at 8-13 μm, in which the atmosphere is only weakly absorbing except for the absorption band of O_3 at 9.6 μm, corresponds to a blackbody at the temperature of the earth's surface, about 320 K (see Fig. 6.6). By contrast, the measured OLR flux in the strong absorption bands of CO_2, CH_4, N_2O and H_2O correspond to much lower blackbody temperatures. Although the surface temperature T_s is generally measured by an observer on the ground, the effective emissivity (ε) or effective transmittance ($1-\varepsilon$) of the atmosphere in long-wave regions is actually only measurable by an observer (satellite) from space. It should be cautious to link the two variables measured from two observing systems.

One could improve the above simple radiative balance model by considering the atmosphere as a vertically continuous absorbing medium and the temperature distribution in the atmosphere. It can be refined in the vertical to *radiative-convective models*, which consider upwelling and downwelling radiative transfer through atmospheric layers that both absorb and emit longwave radiation, and upward transport of heat by convection (especially in the lower troposphere). The most sophisticated are the *general circulation models* (GCMs). Three-dimensional GCMs resolve the horizontal heterogeneity of the surface and its atmosphere by solving globally the 3-dimensional equations for fluid motion and energy transfer and integrate these over time. Atmospheric GCMs (AGCMs) model the atmosphere and impose sea surface temperatures as boundary conditions, while coupled atmosphere-ocean GCMs (AOGCMs) combine the two models.

6.6 Radiative forcing

In the previous section, we discuss how climate models can be used to estimate the surface temperature change in response to a change in the atmosphere. In these models, a radiative perturbation arising from an increase in a greenhouse gas, called *radiative forcing,* induces an initial rise in surface temperature. This is followed by complex responses including enhanced evaporation of water vapor into the atmosphere (a positive feedback), changes in cloud cover, and changes in the atmospheric or oceanic circulation. It has been shown in various climate models from the global energy balance models to GCMs that the equilibrium surface temperature has a linear relationship with the initial radiative forcing [IPCC, 2001, 2007, 2013]. Thus, the radiative forcing provides a useful metric to access and compare the impacts of various anthropogenic and natural variations on the earth's climate. In this section, we present the definition of radiative forcing, and discuss the utility and limitations of radiative forcings for various GHGs.

6.6.1 *The concept of radiative forcing*

Radiative forcing (ΔF) is the net change in the energy balance of the Earth system arising from a certain imposed perturbation. It is usually expressed in Watts per square meter (W/m^2) averaged over a particular period of time and quantifies the energy imbalance caused by the occurrence of the imposed change. Radiative forcing is difficult to observe, but it is widely calculated in climate models as it provides a simple quantitative basis for assessing and comparing the potential responses of the climate to different imposed drivers, particularly global mean surface temperature changes. A positive forcing (more incoming energy) causes a warming of the climate system, while a negative forcing (more outgoing energy) causes a cooling of it. As will be discussed below and in Chapters 7 and 8, radiative forcing can be caused by anthropogenic and natural drivers such as changes in solar radiation and concentrations of greenhouse gases.

6.6.2 *Defining radiative forcing*

The radiative forcing caused by a change in concentration of a specific greenhouse gas in the atmosphere is defined as the resulting energy flux imbalance in the radiative budget for the Earth system. Consider the simple radiative balance model of the Earth, shown in Fig. 6.7. At the TOA, the global mean net radiation flux F_z (positive for downward) can be expressed as

$$F_z = F_0 - F^\uparrow \tag{6.11}$$

For the system to be at equilibrium, the net radiation flux $F_z = 0$, that is,

$$F_0 = F^\uparrow \tag{6.12}$$

Consider now that the climate system is imposed by a small energy perturbation, for example, by an increase in the atmospheric concentration of a GHG or in the incoming solar radiation. Before the surface temperature changes, this energy imbalance translates into an initial radiative flux perturbation ΔF at the TOA: $\Delta F = F_0 - F^\uparrow$. ΔF is called the *radiative forcing* caused by an increase in the atmospheric concentration of the GHG or in the incoming solar radiation.

In climate models, radiative forcing is most commonly computed in terms of the radiative perturbation at the tropopause rather than at the TOA. That is, F_0 and $F^\uparrow$ after the energy perturbation are retrieved from the model at the tropopause after stratospheric temperatures have been allowed to readjust to radiative equilibrium while surface and tropospheric temperatures and state variables such as water vapor and cloud cover are fixed at the initial unperturbed values. The rational for this treatment is that in view of the weak stratospheric-tropospheric dynamical coupling, a radiative perturbation in the stratosphere could have relatively little effect on the earth's surface temperature. This definition of radiative forcing (ΔF, also called stratospherically-adjusted radiative forcing, as distinct from instantaneous radiative forcing) has been adopted in the IPCC Third Assessment Report (TAR) [IPCC, 2001], Fourth Assessment Report (AR4) [IPCC, 2007] and Fifth Assessment Report (AR5) [IPCC, 2013]. Radiative forcing (ΔF) is generally more indicative of the surface and tropospheric temperature responses than instantaneous radiative forcing, especially for climate drivers such as ozone change that substantially alter stratospheric temperatures.

In the IPCC AR5 Report, alternative definitions of radiative forcing have been developed: Another measure of radiative forcing is the so-called *effective radiative forcing* (ERF). ERF is defined as the change in net downward radiative flux at TOA after allowing for atmospheric temperatures, water vapour, and clouds to adjust, while keeping surface temperature or a portion of surface conditions unchanged. The primary methods to calculate ERF are (1) fixing sea surface temperatures and sea ice cover at climatological values while allowing all other parts of the system to respond until reaching steady state or (2) analyzing the transient global mean surface temperature response to an instantaneous perturbation and using the regression of the response extrapolated back to the start of the simulation to derive the initial ERF [IPCC, 2013].

Both the radiative forcing (ΔF) and ERF concepts have strengths and weaknesses. Sophisticate climate model simulations required to calculate the ERF can be more computationally demanding than those for ΔF or instantaneous ΔF since many years are required to reduce the influence of climate variability. The presence of meteorological variability can also make it difficult to isolate the ERF of small forcings that are easily isolated in the pair of radiative transfer calculations performed for ΔF. For ΔF, on the other hand, a definition of the tropopause is required, which may be ambiguous.

However, analyses of current-generation GCMs (e.g., Coupled Model Intercomparison Project Phase 5 — CMIP5) have shown that ΔF and ERF are nearly equal in many cases. In particular, ΔF and ERF are nearly equal for well-mixed greenhouse gases (WMGHGs) such as CO_2, N_2O, CH_4, O_3, and CFCs [IPCC, 2013]. This will be shown in Fig. 6.9 below.

6.6.3 *Radiative forcing due to greenhouse gases*

It is relatively simple to calculate the radiative forcing. The climate models used for ΔF calculations may be as simple as a 1-dimensional (vertical) formulation of radiative equilibrium, or as complicated as

AOGCMs. Since the results from various climate models are quite similar [Dufresne and Bony, 2008; IPCC, 2013], it is of less importance to select a specific model.

For a greenhouse gas, radiative forcing ΔF values as a function of changing concentration can be calculated using radiative transfer codes that examine each absorption spectral line of the molecule under atmospheric conditions. Greenhouse gases absorb and emit longwave (IR) radiation in bands composed of discrete lines with extended wings. For gases such as CFCs and HCFCs that are present in very small atmospheric concentrations in parts per trillion by volume (ppt), it is secure to use a linear relationship of the absorbance (and hence the radiative forcing) with gas concentration. In contrast, for gases such as CH_4 and N_2O that have orders of magnitude larger atmospheric concentrations in parts per billion by volume (ppb), the absorption at the center of the IR bands has already been saturated so that it is thought in climate models that the absorbance would continue to increase with rising gas concentrations, mainly occurring in the edge wings of the bands. Hence, the absorbance is assumed to be approximately proportional to the square root of gas concentration for CH_4 and N_2O. For the most abundant greenhouse gases such as CO_2 in in parts per million by volume (ppm), the atmosphere is completely opaque in the center of the absorption bands; increasing absorption with rising concentrations could only occur in the edge wings of the IR bands. In this case, a logarithmic relationship between the absorbance and CO_2 concentration is assumed in climate models [IPCC, 2001, 2007, 2013].

In fact, the calculations of radiative forcing are often simplified into an algebraic formulation that is specific to a greenhouse gas. Since the late 1990s, the IPCC Reports have used simplified analytical expressions that are derived from atmospheric radiative transfer models to calculate the ΔFs for the well-mixed greenhouse gases. Such ΔF calculations have either shown excellent agreement (within 5%) with high spectral resolution radiative transfer calculations used in GCMs [Myhre *et al.*, 1998; IPCC TAR, 2001] or resulted in the exactly same value (*e.g.*, 3.71 W/m^2 for a doubling of CO_2 concentration) as that given by more advanced AOGCMs (see an excellent review by Dufresne and Bony, 2008). This is not surprising as the above-mentioned relationships between the absorption and atmospheric concentration for the main greenhouse gases are implicitly assumed in the climate models. The simple functional forms of the relationships, and their agreement with explicit radiative transfer calculations used in GCMs, are strong bases for their continued usage, ranging from the IPCC TAR [2001] and AR4 [2007] to AR5 [2013] Reports, as well as the WMO [2006], [2010] and [2014] Reports. Such simple formulas of ΔF used to calculate the radiative forcings (ΔFs) in Wm^{-2} from carbon dioxide (CO_2), methane (CH_4), nitrous oxide (N_2O), and halogen-containing greenhouse gases are the following:

CO_2: $$\Delta F = 5.35 \times \ln(C/C_0) \tag{6.13}$$

CH_4: $$\Delta F = 0.036(\sqrt{M} - \sqrt{M_0}) - (f(M,N_0) - f(M_0,N_0)) \tag{6.14}$$

N_2O: $$\Delta F = 0.12(\sqrt{N} - \sqrt{N_0}) - (f(M_0,N) - f(M_0,N_0)) \tag{6.15}$$

CFCs and HCFCs: $$\Delta F = \chi(X - X_0) \tag{6.16}$$

Here, $f(M,N) = 0.47\ln[1 + 2.01 \times 10^{-5}(MN)^{0.75} + 5.31 \times 10^{-15}M(MN)^{1.52}]$; C is CO_2 in ppm; M is CH_4 in ppb; N is N_2O in ppb; X is CFCs or HCFCs in ppb. The subscript 0 denotes the unperturbed (1750) concentration: $C_0 = 278$ ppm, $M_0 = 722$ ppb, $N_0 = 270$ ppb, and $X_0 = 0$. The same expression is used for all halogenated greenhouse gases, but with a different value of χ (*i.e.*, the radiative efficiency) for each gas. The constant in the simplified expression for CO_2 is based on radiative transfer calculations with three-dimensional climatological meteorological input data, while for CH_4 and N_2O, the constants are derived with radiative transfer calculations using one-dimensional global average meteorological input data [IPCC, 2001].

According to the IPCC AR5, the above simple functional forms of ΔF generally have a total uncertainty of approximately 10% arising mainly from the uncertainty in the radiative transfer modeling for well-mixed GHGs [see Sec. 8.SM.3 in IPCC, 2013]. The IPCC AR5 also notes that uncertainty in ΔF calculations in many GCMs is actually substantially higher both due to radiative transfer codes and meteorological data such as clouds adopted in the simulations. Therefore, there is strong basis to use the simple functional forms of ΔF, instead of complex GCMs, to access and compare the potential climate effects of various well-mixed GHGs (WMGHGs), as performed in the IPCC AR5.

With Eqs. 6.13-6.16, the IPCC AR4 (2007) assessed the radiative forcing from 1750 to 2005 of the WMGHGs to be 2.63 Wm^{-2}. The four most important gases were CO_2, CH_4, CFC-12 and N_2O in that order. Halogen-containing gases, including chlorofluorocarbons (CFCs), hydrochlorofluorocarbons (HCFCs), hydrofluorocarbons (HFCs), perfluorocarbons (PFCs) and SF6, contributed 0.337 Wm^{-2} to the total ΔF. Since AR4, N_2O has overtaken CFC-12 as the third largest contributor to ΔF. The IPCC AR5 [2013] has given a calculated total ΔF of 2.83 Wm^{-2} arising from all WMGHGs in 2011, with the uncertainties (90% confidence ranges) assessed to be approximately 10%.

The tropospheric mixing ratio of CO_2 has increased globally from 278 ppm in 1750 to 390.5 ppm in 2011. Using Eq. 6.13, the calculated ΔF of CO_2 from 1750 to 2011 is 1.82 Wm^{-2}. The IPCC AR5 [2013] concludes that since AR4 (2007), the ΔF of CO_2 has increased by 0.16 Wm^{-2} and continues the rising rate of almost 0.3 Wm^{-2} per decade, and that the rising rate in the ΔF from the WMGHGs over the last 15 years has been dominated by CO_2. CO_2 has accounted for over 80% of the total ΔF increase from all WMGHGs.

Fig. 6.8 shows the radiative forcing values since the preindustrial period at 1750, calculated by Eqs. 6.13-6.16 for the five major WMGHGs (CO_2, CH_4, N_2O, CFC-11 and CFC-12) and a group of 26 minor halogenated WMGHGs, including the sixteen Montreal Protocol regulated halogen GHGs (CFCs, HCFCs, halons, CCl_4, CH_3CCl_3, CH_3Br and CH_3Cl) [WMO, 2010; IPCC, 2013] and the twelve fluorinated GHGs [HFCs, PFCs and SF_6, averaged from the 4 Representative Concentration Pathways (RCPs)] [IPCC, 2013]. Except for the HFCs, PFCs and SF_6 containing no chlorine or bromine, these gases are also ozone-depleting gases and have been regulated by the Montreal Protocol and its amendments. The five major GHGs account for about 96% of the radiative forcing by WMGHG increases since 1750. The remaining 4% is contributed by the 26 minor halogenated gases.

Between 1970 and 1990, halogenated GHGs made a significant contribution to the rate of change of ΔF. Since the Montreal Protocol and its amendments, the rate of change of ΔF from halogenated GHGs has been much less. The total ΔF of halogenated GHGs is dominated by four gases, namely CFC-12, CFC-11, HCFC-22 and CFC-113 in the order. As expected from Eqs. 6.13-6.16, CO_2 dominates the total radiative forcing, while CFCs become relatively smaller contributors to the total ΔF over time, especially since around 2000.

It is interesting to note that the results in Fig. 6.9 show that there are essentially no differences between radiative forcings (ΔFs) calculated by Eqs. 6.13-6.16 and effective radiative forcings (ERFs) obtained from current-generation GCMs and given in the newest IPCC Report [2013].

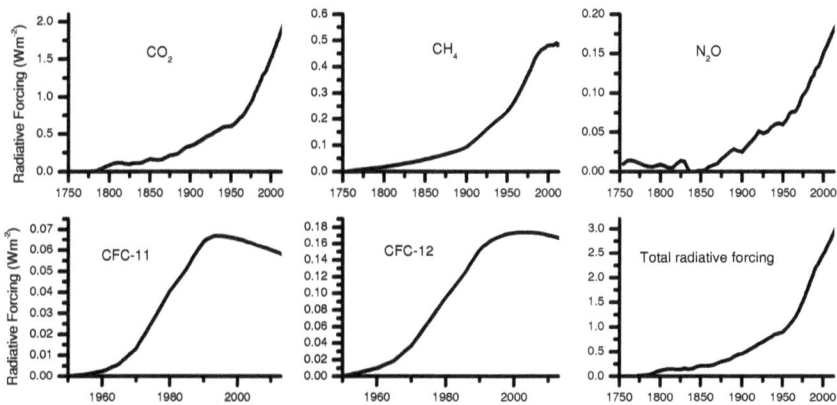

Fig. 6.8. Calculated radiative forcings (ΔFs) of five main well-mixed greenhouse gases (WMGHGs) between 1750 and 2013, relative to the pre-industrial period at 1750. Calculated by Eqs. (6.13)-(6.16) with the gas abundances obtained from Table AII.1.1 in IPCC [2013]. The total radiative forcing includes the contributions from CO_2, CH_4, N_2O, CFC-11 and CFC-12, as well as 26 other minor, long-lived halogenated GHGs, including the 16 halogen GHGs controlled under the Montreal Protocol (CFCs, HCFCs, Halons, CCl_4, CH_3Br, CH_3Cl, etc.) and the 12 fluorinated GHGs controlled under the Kyoto Protocol (HFCs, PFCs, and SF_6, obtained by averaging 4 RCPs).

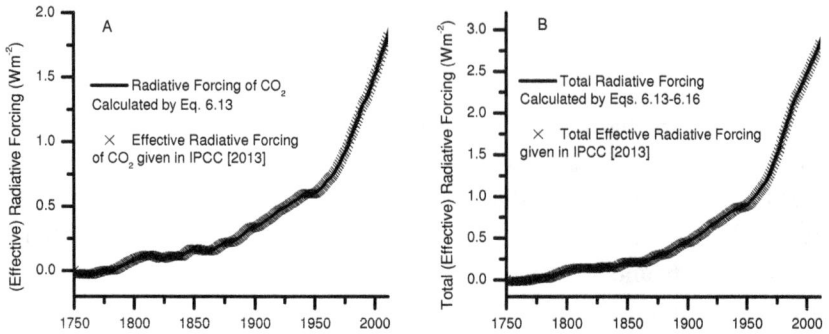

Fig. 6.9. Comparison of calculated radiative forcings (*ΔFs*) and effective radiative forcings (ERFs) of well-mixed greenhouse gases (WMGHGs) between 1750 and 2011, relative to the pre-industrial period at 1750. A. For CO_2; B. for the total (effective) radiative forcing for all GHGs. *ΔFs* were calculated by Eqs. (6.13)-(6.16) with the gas abundances obtained from Table AII.1.1 in IPCC [2013], whereas the data of ERFs were obtained directly from Table AII.1.2. in IPCC [2013].

6.7   Climate sensitivity factor

Climate change occurs when the climate system responds in order to counteract the radiation flux changes, and all such responses are explicitly excluded from the definition of radiative forcing. *Climate sensitivity factor* is a measure of how responsive the temperature of the earth's climate system is to a change in radiative forcing ΔF. The simplest model to derive the relationship between a sustained forcing ΔF and the equilibrium global mean surface temperature response (ΔT_s) is based on the energy-balance model [*e.g.*, Brasseur *et al.*, 1999].

In the definition described in Section 6.6.2, radiative forcing is equal to an initial radiative flux perturbation dF at the TOA caused by a change in a greenhouse gas concentration. In response to this forcing dF, the system will adjust to reach a new equilibrium, leading to a change dT_s in surface temperature. From Eq. 6.9, the OLR flux for the initial atmosphere in radiative equilibrium at the TOA is generally expressed as $F^{\uparrow}=\varepsilon F_g$. A small increment dC of a GHG concentration induces an initial *increase* of the absorption of terrestrial radiation in the atmosphere. This leads to a corresponding *decrease* in OLR flux at the TOA: $-F_g d\varepsilon$. By definition of radiative forcing dF, we have

$$dF = -F_g d\varepsilon \tag{6.17}$$

Note that for an observer (satellite) from space, $d\varepsilon$ is negative, as discussed in Sec. 6.5. Given that the perturbation is maintained for some time, a new equilibrium state is eventually reached where the earth surface temperature T_s is increased by dT_s from its initial state. From Eq. 6.10, the new radiative equilibrium is subject to

$$d[\varepsilon F_g] = dF_0 = 0,$$

or

$$d\varepsilon F_g + \varepsilon dF_g = 0.$$

So that Eq. 6.17 becomes

$$dF = \varepsilon dF_g = 4\varepsilon\sigma T_s^3 dT_s. \tag{6.18}$$

Thus, we obtain the climate sensitivity factor (α_c):

$$\alpha_c \equiv \frac{dT_s}{dF} = \frac{1}{4\varepsilon\sigma T_s^3}, \tag{6.19}$$

where α_c is defined as the change in surface temperature T_s in response to a radiative forcing dF. This equation can widely be found in the literature [*e.g.*, Jacob, 1999; Brasseur *et al.*, 1999; Salby, 2012].

With $T_s = 288$ K and an estimated effective emissivity $\varepsilon = \sim 0.612$, Eq. 6.19 gives an $\alpha_c = 0.30$ K/(Wm^{-2}), which is often called the *reference climate sensitivity* or the Planck feedback factor in climate models. This gives a surface temperature increase ΔT_s by 0.30 K in response to a radiative forcing ΔF of 1.0 Wm^{-2} or a rise of about 1.1 K for the *calculated* ΔF of 3.71 Wm^{-2} that would arise from a doubling of CO_2 in climate models (Eq. 6.13). It also gives a significantly lower temperature rise than the observed global mean surface temperature rise of about 0.6 K during 1970-2000. This discrepancy is due to the neglect of the climate feedback factors of water vapor, lapse rate, albedo, clouds, *etc.* that amplify the response to a change in forcing ΔF. In climate models such as AOGCMs, feedback parameters include those of water vapor (w), lapse-rate (L), cloud (c), and surface albedo (A) [Dufresne and Bony, 2008; IPCC, 2013]. Complex climate models (CMIP5) reviewed in the IPCC AR5 [2013] give that these feedbacks increase the climate sensitivity to a value of 1.0 ± 0.5 (0.5-1.5) K/(Wm^{-2}) with wide ranges, varying from model to model. This is equivalent to a total amplification factor β of 1.67-5.0. So that the global mean equilibrium surface temperature can finally be given by

$$\Delta T_s = \beta \alpha_c \Delta F = \lambda_c \Delta F \qquad (6.20)$$

where $\lambda_c = \beta \alpha_c$ is the *equilibrium climate sensitivity factor*. Comparing the observed global surface temperature rise $\Delta T_s = \sim 0.6$ K during the period of 1970 to 2000 with the total radiative forcing calculated by Eqs. 6.13-6.16 gives rise to a λ_c value of approximately 0.8 K/(W/m^2) [IPCC, 2001], which is nearly the same as $\lambda_c = 0.79$ K/(W/m^2) obtained in AOGCMs [Dufresne and Bony, 2008].

In IPCC Reports and climate models, however, climate sensitivity is often expressed as the temperature change associated with a doubling of the atmospheric concentration of CO_2. The *equilibrium climate sensitivity* (*ECS*) refers to the equilibrium change in global mean near-surface air temperature that would result from a sustained doubling of the atmospheric (equivalent) CO_2 concentration (ΔT_{x2}). The IPCC AR4 estimated the *ECS* value as "likely to be in the range 2 to 4.5 °C with a best estimate of about 3 °C". The newest AR5 (2013) assures that "The CMIP5 model spread in equilibrium climate sensitivity ranges from 2.1 °C to 4.7 °C and is very similar to the assessment in the AR4."

However, the value of ΔT_{x2} (*ECS*) is well-known to have large discrepancies: it varies largely among models, is sensitive to model parameters, and is not well constrained by observations. Estimates for modern climate conditions give the ΔT_{x2} values in the range from 0.1 to about 10 °C [e.g., Idso, 1980, 1998; Andronova and Schlesinger, 2001; IPCC, 2001; Forest, *et al.*, 2002; Gregory, *et al.*, 2002; Ring *et al.*, 2012]. For example, Idso [1980] made interesting calculations based on eight natural experiments, giving a climate sensitivity of 0.1 °C/(Wm^{-2}) that results in a ΔT_{x2}=0.4 °C for a doubling of CO_2. He also noted that several of the cooling forces estimated have equivalent magnitudes, but of opposite sign, to the predicted warming effect of CO_2, suggesting that little net temperature change would ultimately result from the ongoing buildup of CO_2 in Earth's atmosphere. Andronova and Schlesinger [2001] concluded that ΔT_{x2} could lie between 1 and 10 °C, with a 54 percent likelihood that it lies outside the IPCC range. Forest, *et al.* [2002] using patterns of change and the MIT EMIC method estimated a 95% confidence interval of 1.4–7.7 °C for ΔT_{x2}, and a 30% probability that sensitivity was outside the 1.5 to 4.5 °C range. Gregory, *et al.* [2002] estimated a lower bound of 1.6 °C by estimating the change in Earth's radiation budget and comparing it to the global warming observed over the 20th century. Recently, Ring *et al.* [2012] also concluded that the ΔT_{x2} is on the low side of the range given in the IPCC Report (AR4). In summary, there are obviously large uncertainties in estimated ΔT_{x2}.

6.8 Global surface temperature changes given by climate models

There exist large controversies about the ability of GCMs to simulate true climate changes and there are indeed large discrepancies between modeled results and observations [*e.g.*, Fyfe *et al.*, 2013; IPCC, 2013]. Despite these problems, all GCMs seem to give a linear relationship between the initial radiative forcing and the ultimate perturbation to the surface temperature [IPCC, 2013]. This is expressed in Eq. 6.20 obtained from the global energy balance models, *i.e.*, $\Delta T_s = \lambda_c \Delta F$.

The radiative forcings ΔF calculated by the functional expressions, *i.e.*, Eqs. 6.13-6.16, are nearly identical to those by GCMs through calculations of the spectrally, line-by-line absorptions of greenhouse gases in the atmosphere. This is shown in Fig. 6.9. Moreover, Dufresne and Bony [2008] gave an excellent review on various (totally 12) AOGCMs. For the 12 AOGCMs reviewed therein, the multimodel mean values are: the net radiative forcing ΔF is 3.71±0.2 W/m^2 for a doubling of CO_2, λ_c (=1/η, where η is the sum of the Planck, water vapor, lapse-rate, cloud, and surface albedo feedback parameters) is $1.27^{-1}=0.79\pm0.19$ $K/(W/m^2)$, and ΔT_{x2} is 3.1 ± 0.7 K. These results are nearly the same as those calculated from $\Delta F=5.35\times\ln2=3.71$ W/m^2 by Eq. 6.13, and $\Delta T_s=\lambda_c\Delta F=3.0$ K by Eq. 6.20 with an $\lambda_c=0.8$ $K/(W/m^2)$ mentioned above. Complex climate models (CMIP5) reviewed in the IPCC AR5 [2013] give similar results: $\Delta F=3.7\pm0.8$ W/m^2 and $\Delta T_{x2}=3.2\pm1.3$ K, and $\lambda_c=1.0\pm0.5$ $K/(Wm^{-2})$.

From the above results, it can be concluded that although GCMs (*e.g.*, CMIP5) are far more complex and have more parameters and equations, their results have no essential differences from those given by their first generation (energy-balance) climate models. This is understandable as there is no major difference in conceptual physics between the two generations of conventional climate models.

It might be interesting to calculate global surface temperature changes ΔT_s induced by main GHGs including CO_2, CH_4, N_2O, halogenated gases (CFCs, HCFCs, HFCs, etc.) using the ΔF formulas (Eqs. 6.13-6.16) that have been used in IPCC Reports [2001, 2007, 2013]. To make the calculations, the observed atmospheric abundances of non-halogen GHGs (CO_2, CH_4, N_2O), sixteen Montreal Protocol regulated halogen GHGs (CFCs, HCFCs, halons, CCl_4, CH_3CCl_3, CH_3Br and CH_3Cl) and twelve fluorinated GHGs (HFCs, PFCs and SF_6) are obtained from the WMO Report [2010] and the IPCC AR5 [2013]. An equilibrium climate sensitivity factor $\lambda_c=0.8$ $K/(W/m^2)$ was used. The calculated results are shown in Fig. 6.10.

First, a drastic continuous warming trend is shown. This is consistent with the results of various climate models and is because of the dominance of CO_2 in the calculated radiative forcing ΔF, which accounts for over 80% of the ΔF increase since 1750 [IPCC, 2013]. Second, the surface temperature would keep rising as long as the assumed ΔF relationships of non-halogen GHGs (Eqs. 6.13-6.15) are used. This is due to the fact that the concentrations of non-halogen GHGs are orders of magnitude higher than those of halogenated GHGs. This conclusion is consistent with that of AOGCMs reviewed in the IPCC AR5 [2013].

However, it is also worthwhile to note that the calculated ΔT_s arising from the total forcing ΔF has been significantly larger than the observed ΔT_s since 1970. To fit to the observed surface rise during 1970-2002,

AOGCMs introduced the so-called transient temperature change ΔT_s^t [also called the transient climate response (TCR), instead of equilibrium temperature change ΔT_s], in which an additional parameter, the ocean heat uptake efficiency κ ($=0.69$ $K^{-1}Wm^{-2}$) is added to the total feedback parameter η [see *e.g.*, Dufresne and Bony, 2008]

$$\Delta T_s^t = \Delta F/(\eta + \kappa) = \lambda_c^t \times \Delta F \qquad (6.21)$$

where $\eta = 1/\lambda_c$ ($=1.27$ $K^{-1}Wm^{-2}$) and $\lambda_c^t = 0.51$ $K/(Wm^{-2})$. The calculated results are also shown in Fig. 6.10 (dash line), which show an improved agreement with the observed data between 1970 and 2000. Nevertheless, the calculated ΔT_s^t data do not agree with the observed data in the period prior to 1970 and still gives a continued warming trend over the past 16 years, which is also inconsistent with the observed surface temperature trend. These results lead to a conclusion that if the assumed radiative forcing relationships [Eqs. 6.13-6.16] were valid, a continued warming trend would inevitably result. This is indeed the general conclusion in all the IPCC Reports and current climate models [2001, 2007, 2013]. This important issue will be further addressed in Chapter 8.

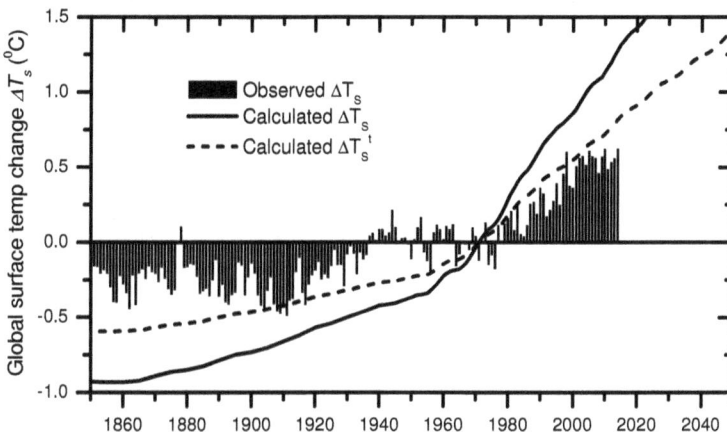

Fig. 6.10. Observed and calculated global mean surface temperature changes ΔT_s for 1850-2050, with respect to the temperature in 1970. ΔT_s are calculated by $\Delta T_s = \lambda_c \times \Delta F$ with $\lambda_c = 0.8$ $K/(Wm^{-2})$ and transient temperature change $\Delta T_s^t = \lambda_c^t \times \Delta F$ with $\lambda_c^t = 0.51$ $K/(Wm^{-2})$ (dash line) for all GHGs (see text). Annual global (near) surface temperatures were from the UK Met Office's HadCRUT4 dataset [Morice *et al.*, 2012].

6.9 Concluding remarks

It has been shown in various climate models from the global energy balance models to GCMs that the equilibrium surface temperature has a linear relationship with the radiative forcing. Thus, the radiative forcing provides a useful metric to access and compare the impacts of various anthropogenic and natural variations on the earth's climate.

The radiative forcings ΔF calculated by the functional expressions (Eqs. 6.13-6.16) are nearly identical to those by advanced GCMs (*e.g.*, CMIP5) through calculations of the spectrally, line-by-line absorptions of greenhouse gases in the atmosphere.

There exist large discrepancies between modeled results and observations, indicating that the assumptions in current climate models are likely incorrect.

Researchers, policy-makers and the public should be aware of the above fact. An open opinion about different models may be required in accessing the impacts of human activities and climate change and making decisions about future policies.

A simple physics model could be far more comprehensible than the state-of-the-art three-dimensional general circulation models. The former could be far more useful in estimating the impacts of various natural and anthropogenic drivers on global climate change, though the latter have the most advanced computations, different parameters and great complexity. A simple physics model could also provide far more reliable and convincing results once it is constrained with observations. Such models should play a more important role in delivering outcomes of climate research to the world's leaders and policy-makers.

Chapter 7

Natural Drivers of Climate Change

7.1 Introduction

As shown in last Chapter, there is a robust observation that the Earth's global surface temperature increased by about 0.6 K over the last three decades of the 20th century. This change has mainly been attributed to anthropogenic origin [IPCC, 2013]. However, a proper understanding of the relative roles of various climate drivers, both natural and human-made, is a prerequisite for a quantitative assessment of the human-made contribution to the climate change.

The most obvious natural influence is due to direct or indirect effects of the Sun, which is the source of the external energy input into the Earth's climate system. The Sun's effects can include three possible mechanisms: (1) variations in the Sun's radiative output (direct effect); (2) variations in the astronomical alignment of the Sun and the Earth (Milankovitch cycles) (indirect effect); and (3) the influence of the Sun's activity on galactic cosmic rays (GCRs) proposed to affect cloud cover.

The second mechanism is believed to be the prime cause of ice ages and the interglacial warming periods that have dominated the longer term evolution of the climate over the past few million years. Various parameters of the Earth's orbital and rotational motion change at periods of 23-100 kyr. This effect is substantial only at millennial and longer timescales, and therefore it is unlikely to have contributed to the drastic global warming observed in the second half of the 20th century. It will not be discussed further here.

Thus, only the first and third mechanisms of the Sun's influence will be considered further below. The first mechanism is generally considered to be the main cause of the solar contribution to global climate change and will be described in some detail below. These include changes in total solar irradiance (TSI), i.e., in the energy input to the Earth system, and variations in solar spectral irradiance (SSI), particularly in UV irradiance, which can enhance the Sun's effect by influencing the chemistry in the Earth's middle atmosphere. There are a number of excellent reviews on solar irradiance variability and its effect on Earth's climate in the literature [*e.g.,* Solanki and Krivova, 2003; Foukal *et al.*, 2006; Fröhlich, 2006; Lockwood and Fröhlich, 2007; Krivova *et al.*, 2009a; Fröhlich, 2012; Krivova and Solanki, 2013; Solanki *et al.*, 2013].

In addition, volcanic eruptions are another natural contributor to climate change and may have dramatic, rapid impacts on climate. However, their contribution to the sharp increase in global surface (warming) in 1970 to 2000 is very small. It will only be discussed briefly in this Chapter.

7.2 Variations in the Sun's radiative output

The main natural driver of the Earth's climate system is the solar radiative output, which has been a subject of intense and continued interest [*e.g.,* Eddy, 1976; Reid, 1987; Friis-Christensen and Lassen, 1991; Hoyt and Schatten, 1993; Willson, 1997; Solanki and Krivova, 2003; Foukal *et al.*, 2006; Fröhlich, 2006; Lockwood and Fröhlich, 2007; Krivova *et al.*, 2009a, 2009b; Fröhlich, 2012; Krivova and Solanki, 2013; Solanki *et al.*, 2013]. Variations in TSI and SSI are the two primary physical quantities. TSI integrated over all wavelengths is the total solar energy flux entering the TOA, and therefore any changes in the TSI disturb the overall energy balance of the climate system. Variations in SSI (given in Watts per meter squared per nanometer), reflecting the change in the spectral distribution of the solar irradiance, particularly in the UV (and visible and IR spectral ranges as well), could also have an influence on the chemistry and dynamics of the Earth's atmosphere [Foukal *et al.*, 2006; Fröhlich, 2012; Krivova and Solanki, 2013; Solanki *et al.*, 2013].

There were controversies about the reconstructed composites of historic TSI and SSI time-series data. These arose mainly from the fact that TSI and SSI data from direct satellite-based measurements have only been available since 1978 and are based on a composite of many different observing satellites [*e.g.*, Willson, 1997; Willson and Mordvinov, 2003; Dewitte *et al.*, 2004; Fröhlich, 2006; Krivova *et al.*, 2009a, 2009b; Fröhlich, 2012; Krivova and Solanki, 2013].

7.2.1 *Variations in TSI*

As reviewed recently [Fröhlich, 2012; Krivova and Solanki, 2013; IPCC, 2013], there are some differences in constructed TSI composites in not only absolute values but also time-series trends of TSI during the last three solar cycles or the contribution of different spectral ranges to the irradiance variation. Three main constructed composite series are referred to as the Active Cavity Radiometer Irradiance Monitor (ACRIM) [Willson, 1997; Willson and Mordvinov, 2003], the Physikalisch Meteorologisches Observatorium Davos (PMOD) [Fröhlich and Lean, 1998; Fröhlich, 2006], and the Royal Meteorological Institute of Belgium (RMIB) [Dewitte *et al.*, 2004] series.

There are two main differences between ACRIM and PMOD. The first is the rapid drift in calibration between PMOD and ACRIM before 1981. This difference arises as both composites employ the Hickey–Frieden (HF) radiometer data, while PMOD has implemented a re-evaluation of the early HF degradation but ACRIM has not. The second difference (involving the RMIB composite as well) is the bridging of the gap between the end of ACRIM I (mid-1989) and the start of ACRIM II (late 1991) observations. The possibility of a change in HF data during this gap is not taken into account in ACRIM and thus its TSI values are larger by more than 0.5 Wm^{-2} during solar cycle 22. These differences lead to different long-term TSI trends in the composites, as shown in Fig. 7.1. The ACRIM composite shows an upward trend of 0.04-0.05% per decade between consecutive solar activity minima in 1986 and 1996, and a declining trend from 1996 to 2008. In contrast, the PMOD composite shows the opposite, that is, a declining trend in solar minima since 1986, following the solar-cycle-averaged sunspot number [Fröhlich, 2006, 2012].

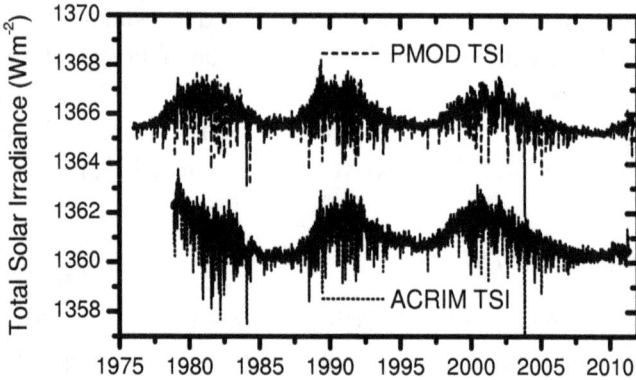

Fig. 7.1. Time series PMOD and ACRIM TSI composites constructed from measurements since 1976/1978.

Krivova *et al.* [2009a] recently re-analyzed the ACRIM composite, employing the more appropriate Spectral And Total Irradiance Reconstructions for the Satellite era (SATIRE-S) model. Their 'mixed' ACRIM–SATIRE-S composite showed no increase in the TSI from 1986 to 1996, in contrast to the ACRIM composite; a slight decrease by approximately 0.011–0.05% was actually found though it could not be estimated accurately. Independent models assuming irradiance variability to be driven by the evolution of the surface magnetic field have confirmed the necessity for correction of HF data and agreed better with the PMOD long-term trend. It has been concluded that the PMOD composite is the most accurate among the existing composites [Ball *et al.*, 2012; Krivova and Solanki, 2013; IPCC, 2013].

The TSI measured by the Total Irradiance Monitor (TIM) on the space borne Solar Radiation and Climate Experiment (SORCE) is 1360.8 ± 0.5 Wm^{-2} in 2008 [Kopp and Lean, 2011], which is ~4.5 Wm^{-2} lower than the PMOD TSI value in 2008 [Fröhlich, 2009]. This difference is probably due to instrumental biases in measurements prior to TIM. The latter has been taken as the most accurate [IPCC, 2013].

All the constructed composite series have now been standardized to the TIM series [IPCC, 2013], as shown in Fig. 7.2. And it has also been noted that the few tenths of a percent bias in the absolute TSI value has minimal consequences for climate change, though the TIM TSI values are lower than other composites [IPCC, 2013].

In all the four TSI composites mentioned above, TSI variations between solar maxima and minima of 11-year solar cycles were observed to be approximately 0.1% [IPCC, 2013]. This change is well-constrained from various observations. Different mechanisms have been proposed to explain the observed change, which have mostly been attributed to the modulation by the solar surface magnetic field. This variation is mainly due to an interplay between relatively dark sunspots, bright faculae and bright network elements [Krivova and Solanki, 2013; Solanki *et al.*, 2013].

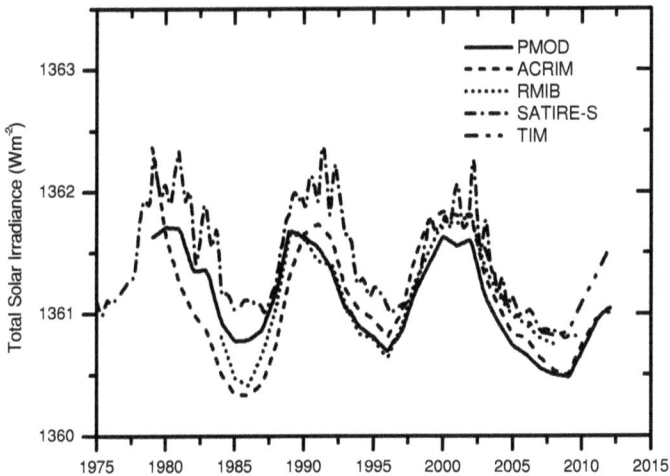

Fig. 7.2. Comparison of the three smoothed total solar irradiance (TSI) composites (PMOD, ACRIM and RMIB) as well as TSI reconstructed by the SATIRE-S, all standardized to the TIM series. Based on data from IPCC [2013] and Solanki *et al.* [2013].

Since direct time series measurements have covered only ~35 years, longer-term reconstructions are important for a reliable evaluation of the connection between solar variability and the Earth's climate change.

Unfortunately, an extrapolating of reconstructed TSI composites back to preindustrial time is a very challenging task, and reconstructed TSI composites have large uncertainties. The magnitude of the secular trend in TSI in long-term reconstructions has remained the most uncertain [Lean *et al.*, 1992; Hoyt and Schatten, 1993; Krivova *et al.*, 2010; Krivova and Solanki, 2013]

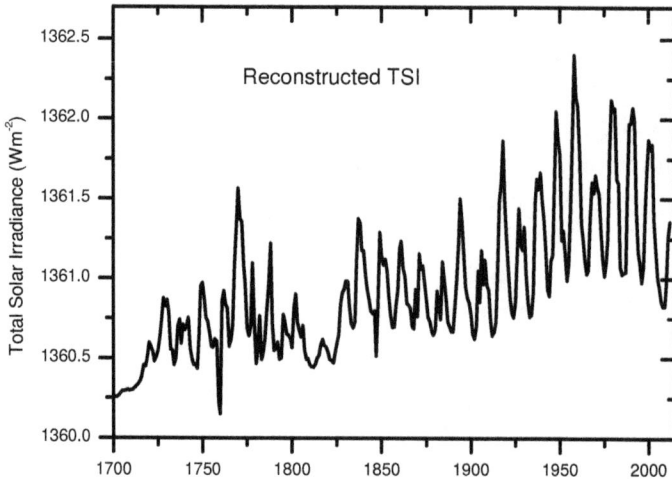

Fig. 7.3. Time series total Solar Irradiance (TSI, Wm⁻²) reconstruction since 1750. The series are standardized to the PMOD measurements of solar cycle 23 (1996–2008). The latter has been standardized to the TIM series. Based on data from Ball *et al.* [2012] and Krivova *et al.* [2010].

Fig. 7.3 shows a recent TSI reconstruction by Krivova *et al.* [2010] between 1745 and 1973 and from 1974 to 2012 by Ball *et al.* [2012], which has been recommended as the best estimate in the newest IPCC Report [2013]. The reconstruction is based on physical modeling of the evolution of solar surface magnetic flux, and its relationship with sunspot group number (before 1974) and sunspot umbra and penumbra and faculae afterwards. This is believed to provide a more detailed reconstruction than other models. The best estimate given in the IPCC Report [2013] from the assessment of the most reliable TSI reconstruction gives a 7-year running mean radiative forcing between the minima of 1745 and 2008 of 0.05 W m⁻², corresponding to a TSI increase of 0.29 W m^{-2}. The range of radiative forcing from TSI changes is 0.0 to 0.10 W m^{-2} in several updated reconstructions using the same 7-year running mean past-to-present minima years, corresponding to a TSI increase of 0-0.57 W m^{-2}.

Krivova *et al.* [2007] have also estimated the possible range of the secular TSI increase as 0.9–1.5 W/m^2 since the Maunder minimum, which is consistent with most other recent estimates derived under various assumptions. Similarly, Krivova and Solanki [2013] show that between the end of the Maunder minimum and the end of the 20th century, the cycle-averaged TSI has increased by 1.25 W/m^2 or about 0.9 %. Nevertheless, Solanki *et al.* [2013] also conclude that current estimates of the TSI change since the end of the Maunder minimum (the end of the 17th century) range from 0.8 W m^{-2} to about 3 W m^{-2}, i.e., over nearly a factor of four. These authors also note that the time dependence is also different in various reconstructions and is rather uncertain. All reconstructions rely on indirect proxies that inherently do not give consistent results. Thus, there exist large discrepancies among the reconstruction models.

7.2.2 *Variations in solar spectral irradiance*

Although regular monitoring of solar spectral irradiance (SSI) in the UV also began in 1978, a consistent time series does not exist [Krivova and Solanki, 2013]. This is because, apart from differences in absolute levels, degradation trends and other problems in data from various instruments depend strongly on the wavelength, making it essentially impossible to have a proper self-consistent cross-calibration of data measured by different instruments [DeLand and Cebula, 2008]. Regular observations in a broader spectral range covering the visible and the near-IR only started in 2002/2003 [Krivova and Solanki, 2013].

Multiple space-based measurements made in the past 35 years indi-cated that UV variations at wavelengths below 400 nm account for about 30-63% of the TSI variations in solar cycles, while about 37-70% were produced within the visible and infrared [Krivova and Solanki, 2013; IPCC, 2013]. Measurements from the Spectral Irradiance Monitor (SIM) on board SORCE over the solar cycle 23 declining phase shows large discrepancies with prior measurements, indicating that additional validation and uncertainty estimates are needed [IPCC 2013].

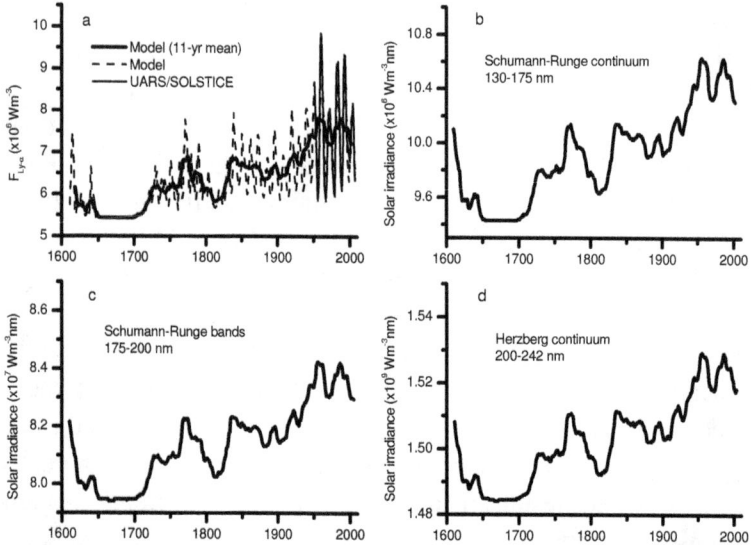

Fig. 7.4. Reconstructed solar spectral irradiance. a. Solar irradiance in Ly-α reconstructed using the SATIRE-T model (dash line), the 11-year smoothing line (thick solid line) by Krivova *et al.* [2000], and the composite (thin solid line) of UARS/SOLSTICE measurements and proxy models by Woods *et al.* [2000]. b, c and d: Reconstructed solar irradiance at the Shumann-Runge oxygen continuum, the Schumann-Runge oxygen bands and the Herzberg oxygen continuum, respectively, based on data from Krivova *et al.* [2010].

Reconstructions of preindustrial UV variations have also been made recently. The reconstruction by Krivova *et al.* [2010] is based on what is known about spectral contrasts of different surface magnetic features and the relationship between TSI and magnetic fields. The authors interpolated backwards to the year 1610 based on sunspot group numbers and magnetic information. As shown in Fig. 7.4, Krivova *et al.* [2010] showed reconstructed solar irradiance in Ly-α and several other spectral ranges of special interest for climate models, the available measurements by the UARS Solar Stellar Irradiance Comparison Experiment (SOLSTICE) and a composite time series compiled by Woods et al. [2000]. The results show that the UV SSI appears to have generally increased over the past four centuries, with larger trends at shorter wavelengths. However, there is a slight declining trend in SSI since around 1950, consistent with the TSI variation shown in Figs. 7.2 and 7.3.

Variations in SSI in the UV may have impacts on the chemistry of the stratosphere. Ozone is the main gas involved in stratospheric radiative heating. Observations showed that stratospheric ozone responds to solar activity. The vertically integrated ozone column varies by 1–3% in phase with the 11-year solar cycle, with the largest signal in the upper stratosphere in the subtropics [Solanki *et al.*, 2013]. This is often attributed to variations in ozone production rate due to changes in solar UV irradiance. As discussed in Sec. 5.8 of Chapter 5, however, ozone destruction rate variations due to UV variations are completely ignored in current atmospheric photochemical models. The latter would expect that an increase in UV solar irradiance would lead to an increase in photodissociation of CFCs and activation of halogens for O_3 destruction [Lu, 2013]. In any case, however, the radiative forcing due to solar-induced ozone changes is only a small fraction of the solar radiative forcing [IPCC, 2013].

7.2.3 *Variations of other indices of solar activity*

A further analysis of other indices of solar activity might also be helpful for a more reliable evaluation of the TSI composites since the 1970s [Lu, 2013]. Observational records of the sunspot number (SSN) began about 300 years ago, and there has been little disagreement about the observed data of SSN. Approximately every 11 years, a maximum of solar activity is reached, with a large number of sunspots present on the solar surface. Solar cycle activity maxima are separated by minima during which only a few or no sunspots are present on the solar surface. As an indicator of solar activity, the number of sunspots is expected to have a consistent behaviour with TSI. Also, the cosmic-ray intensity modulated by the strength of the Sun's open magnetic field is another indicator of solar activity and has been well recorded by independent detection stations in many parts of the world since the 1950s; the data are available from the Network of Cosmic Ray Stations. Thus, with combined measured data

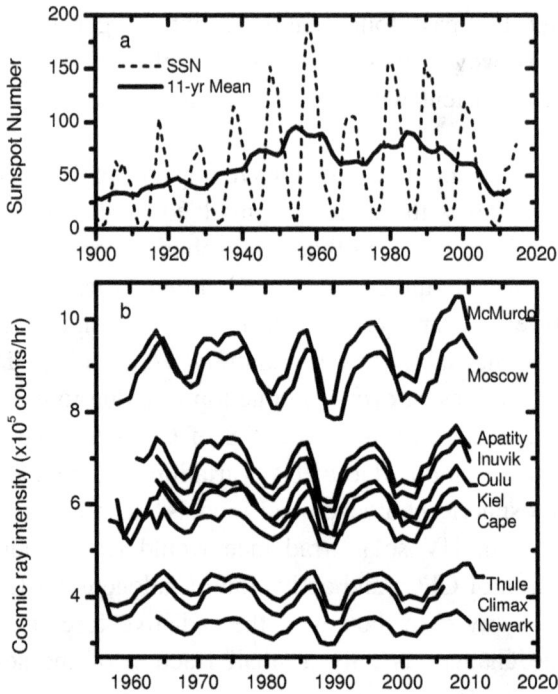

Fig. 7.5. Time series variations of the indicators of solar activity. a. Annual mean sunspot number (SSN) and 11-year average sunspot number (SSN) from 1900 to 2014. b: Annual Mean cosmic ray intensity data measured at ten neutron detector stations (McMurdo, Moscow, Apatity, Inuvik, Oulu, Kiel, Cape, Thule, Climax and Newark) from the 1950s to 2011. Updated from Lu [2013].

of TSI, SSI, SSN and the CR intensity from multiple sources, it is possible to obtain a reliable evaluation of the solar effect on the Earth's climate since 1970. Recorded time-series SSNs since 1900 are plotted in Fig. 7.5a, while Fig. 7.5b shows the CR intensities from measurements at ten stations (McMurdo, Moscow, Apatity, Inuvik, Oulu, Kiel, Cape, Thule, Climax and Newark) at various altitudes (0-3 km) and latitudes from the polar regions to midlatitudes since the 1950s. It can clearly be seen that the mean SSN had a rising trend in the first half of the 20th century and a declining trend after that, particularly in the past 3 solar cycles. The variation of the CR intensity should be anti-correlated with that of the solar activity: the solar activity minimum (maximum) corresponds to the CR maximum (minimum). As shown in Fig. 7.5b, indeed, the observed CR data from multiple stations show that the overall mean CR intensity has had an increasing trend during the period of 1970 up to the present. Clearly, the observed CR maximum at 1997 is slightly larger than that in 1986. This is consistent with the PMOD TSI composite [Fröhlich, 2006, 2012], the ACRIM–SATIRE-S composite [Krivova *et al.*, 2009, 2010] and the IPCC-recommended TSI composite.

7.3 Effects of cosmic rays on clouds

There is also strong interest in studying the effects of galactic cosmic rays (GCRs) on cloud cover at low troposphere ($\leq$3km), following the first report of the correlation between GCRs and cloud cover by Svensmark and Friis-Christensen [1997]. High solar activity means a stronger heliospheric magnetic field and thus a more efficient screening of cosmic rays to reach the earth's atmosphere. It has been hypothesized that atmospheric ionization by cosmic rays create ions which facilitates aerosol nucleation and new particle formation with a further impact on cloud formation [Svensmark, 1998; Marsh and Svensmark, 2000; Kirkby *et al.*, 2011]. Changing amount or properties of clouds will modify the Earth's albedo and hence affect climate. Thus, the increased GCR flux would form more clouds amplifying the cooling effect expected from reduced solar activity.

There are a number of observational studies and model simulations to investigate the physical mechanism for the correlation [Svensmark, 1998; Marsh and Svensmark, 2000; Yu and Turco, 2001]. The GCR-cloud correlation has been reviewed by Carslaw *et al.* [2002]. Since then, other researchers have reported both positive and negative results about the correlation. Similarly, researchers have also studied the possible effects of CRs on the formation of PSCs [Yu, 2004; Kasatkina1 and Shumilov, 2005]. Simulations by Pierce and Adams [2009] showed that changes in cloud condensation nuclei concentrations from changes in CRs during a CR cycle are two orders of magnitude too small to account for the observed changes in cloud properties. They therefore concluded that the CR-cloud effect is too small to play a significant role in current climate change. In contrast, Svensmark *et al.* [2009] showed the correlation between CRs and atmospheric aerosols and clouds. The CLOUD experiment at CERN has so far returned only equivocal results on the effectiveness of cosmic rays in producing clouds [Kirkby *et al.*, 2011]. Overall, the direct GCR-cloud (PSC) correlation remains the subject of significant controversies, and the debate is expected to continue.

7.4   Other natural contributions to climate change

Here we briefly discuss volcanic eruptions as another natural contributor to climate change. Volcanic eruptions that inject large amounts of SO_2 gas into the stratosphere can result in an externally forced change in climate on the annual and multi-decadal time scales. This can explain part of the pre-industrial climate change. Particularly small size sulphate aerosols cause impacts on climate because they are effective scatters of sunlight and have long lifetimes. To be significant, sulphate aerosols must be injected into the stratosphere, where they from tropical and high-latitude eruptions have a lifetime of about one year and several months, respectively. This is much longer than the lifetime of only about one week for aerosols in the troposphere [IPCC, 2013].

However, there have been no major volcanic eruptions since Mt Pinatubo in 1991, though there were several smaller eruptions. Thus, the CMIP5 simulations did not include the recent small volcanic forcing in their calculations. Even the impact of the 1991 Mt Pinatubo eruption on global climate is rather limited [Frolicher *et al.*, 2011]. Although volcanic eruptions can have impacts on the climate, their contribution to the rapid global warming in the late half of last century is generally insignificant [Solanki *et al.*, 2013].

7.5   Influence of solar variability on climate

The rising trend in the ACRIM TSI in 1986 and 1996 was once proposed to account for the global temperature rise in recent decades [Willson,

1997]. However, the PMOD TSI series indicates that the measured TSI variations since 1978 are too small to have contributed appreciably to accelerated global warming over 1970-2000 [Foukal *et al.*, 2006; Fröhlich, 2006; Lockwood and Fröhlich, 2007; Fröhlich, 2012]. These studies and other studies [*e.g.*, Solanki and Krivova, 2003; Krivova *et al.*, 2009] have shown that the solar activity cannot have had a significant influence on global climate since ~1970, irrespective of the specific process dominant in determining Sun-climate interactions: TSI changes or solar UV changes or cloud coverage changes by CR flux variations. Lockwood and Fröhlich [2007] have even concluded that all the trends in the Sun for either a direct TSI effect on climate or an indirect effect via CR-regulated cloud coverage or for a combination of the two have been in the opposite direction to that required to explain the observed rise in global temperature in the late 20th century. Overall, the combined time series observations of TSI (and SSI), SSN and CR intensity have unambiguously shown that the natural influence on climate is indeed in the opposite direction to that required to explain the observed rise in global temperature in the decades of 1970-2002 [Lu, 2013]. That is, the natural effect is negligible and the human effect must be dominant. This conclusion, consistent with most of the studies, seems fairly established by substantial observations [Fröhlich, 2012; Solanki *et al.*, 2013; Lu, 2013; IPCC, 2013].

Furthermore, the above conclusion can be based on a more quantitative evaluation of the impact of solar variability. Here, the observed global surface temperatures and the observed and constructed TSI composites since 1850 are shown in Fig. 7.6. Some studies showed that the global temperature indeed closely followed the TSI variation up to 1970 [Hoyt and Schatten, 1993; Solanki and Krivova, 2003; Lu, 2013]. Even an excellent linear correlation with coefficients of 0.83~0.97 between 11-year mean TSI and temperature was obtained [Solanki and Krivova, 2003]. The observed data seemed to indicate that the solar effect played the dominant role in climate change prior to 1970.

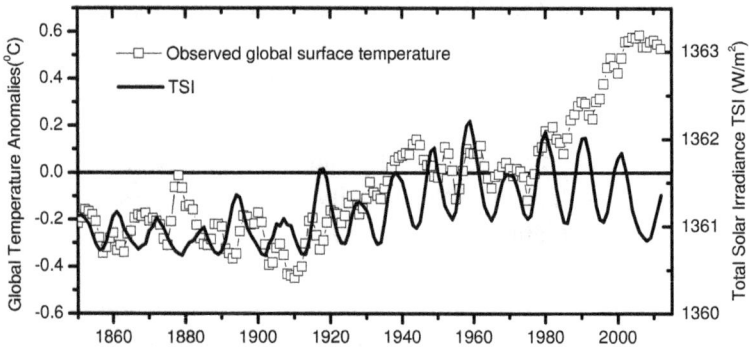

Fig. 7.6. Observed global mean surface temperatures and the observed and constructed TSI composites since 1850. Annual global surface temperatures were from the UK Met Office's HadCRUT4 dataset, relative to with the temperature at 1970; the TSI data have the same source as in Fig. 7.3. A 3-point smoothing was applied to the observed data.

Solanki *et al.* [2013] recently reviewed an approach by Lean and Rind [2008], using a multiple linear regression analysis to separate different contributors to global mean surface temperature over the past century. The results showed that the Sun may have a contribution of approximately 0.07 K (neglecting the 11-year cyclic oscillation) to an overall global warming from 1889 to 1960, and has had little effect since 1960. Thus, the Sun's contribution to the global surface temperature rise of ~1.0 K over the last century was suggested to be ~7% (note that this is somewhat contradictory to the last sentence, since most of the warming was observed after 1960!). However, the authors also noted that the index of solar variability used as the regression index had a small long-term trend and the result depended fundamentally on the assumed temporal variation of the solar forcing, which has some uncertainty.

Solar radiative forcing has been used to evaluate the effect of solar variability on global surface temperature. This is based on the fact that nearly all climate models including AOGCMs have shown that the change in equilibrium global mean surface temperature is linearly proportional to the radiative forcing and is essentially independent of the drivers of the forcing (Chapter 6). The constant of proportionality is called the climate sensitivity parameter (λ_c). Given that plausible estimates for the TSI (F_S, the so-called solar constant) increase from the Maunder minimum to the present lie in the range 0.8–3.0 W m^{-2} [Solanki *et al.*, 2013], this gives a solar radiative forcing of ΔF_0=(1-0.3)ΔF_S/4=0.14-0.525 (Wm^{-2}). Using an estimated λ_c=0.6 K/(Wm^{-2}), Solanki *et al.* [2013] suggested that a solar-driven global surface temperature increase lies in the range of 0.08 to 0.30 K since the 17th century, and that the Sun may have contributed 8–30% of the observed global warming of about 1.0 K till the end of 20th century.

The above-mentioned estimations of the solar contribution to global surface temperature change are instructive but are incorrect in some sense. It is now generally agreed that the PMOD composite is most accurate among the existing TSI composites, showing either a slight *declining* trend or nearly constant of TSI since 1978 up to the present [Foukal *et al.*, 2006; Fröhlich, 2006; Fröhlich, 2012; Solanki *et al.*, 2013; IPCC, 2013]. Thus, all the solar-driven global surface temperature *increase* in the range of 0.08 to 0.30 K since the 17th century must have occurred prior to 1978. Since the observed global mean surface temperature increase was 0.2-0.3 K only from the 17th century to 1978, the Sun should have contributed 27–100% of the observed global warming from the 17th century to 1978, prior to the drastic warming of about 0.6 K from ~1975 to 2000.

It is generally agreed that there are large uncertainties in various reconstructions of solar radiative output prior to the start of direct measurements in 1978, and therefore evaluations of solar contribution to global climate change prior to the 1970s have also large uncertainties. Perhaps, the most valuable and reliable information obtained from measurements of solar variability to understanding global climate change lies in the following relatively reliable observations made after 1978:

(1) In all TSI composites, the TSI variation of 0.1% (equivalent to about 1.36 Wm^{-2}) between solar maxima and minima of 11-year solar cycles is well-constrained from observations [IPCC, 2013].

(2) The observed global surface temperature variation between solar maxima and minima of the past several 11-year solar cycles is also well-constrained to be about 0.11 K from observations [Lean and Rind, 2008; Solanki *et al.*, 2013].

(3) The long-term TSI variation has shown a nearly zero trend or a slightly declining trend since 1978. In any TSI composites either ACRIM or PMOD, the magnitude of long-term (secular) TSI change has been far less than the TSI variation of 0.1% (about 1.36 Wm^{-2}) between solar maxima and minima of 11-year solar cycles since 1978 [Fröhlich, 2012; Krivova and Solanki, 2013; IPCC, 2013].

These well-constrained data from reliable and direct observations indicate at least three important aspects of global climate change, described as follows.

(1) The solar contribution to the drastic global warming by about 0.6 K in 1975-2000 is negligibly small (far less than 0.1 K). This conclusion has generally been agreed [Fröhlich, 2012; Krivova and Solanki, 2013; Lu, 2013; IPCC Report, 2013].

(2) The earth surface temperature has an instantaneous response to a change in radiative forcing generated at the TOA as the annual mean surface temperature variation follows the TSI variation instantaneously in 11-year solar cycles. This implies that the introduction of the so-called impulse response function in climate models, which allows the climate system to take centuries to millennia to reach an equilibrium surface temperature, is most likely unnecessary and artificial.

(3) When reliable observed data are used, solar irradiance variability can provide a most direct and reliable approach to obtain the climate feedback factor and the solar climate sensitivity.

The last point will be discussed in detail in the next section.

7.6 Derivation of solar climate sensitivity factor and total feedback factor from solar irradiance variability

From the radiative equilibrium of the Earth, expressed in Eq. 6.10, $F_0=(1-A)F_S/4=\varepsilon\sigma T_s^4$, that is, the incoming energy flux $F_0=(1-A)F_S/4$ from the Sun is balanced by the outgoing energy flux $F^{\uparrow}=\varepsilon\sigma T_s^4$ from the Earth, we can readily determine the effect of the change in solar output F_S on the surface temperature T_s. This can be deduced as follows. From Eq. 6.10, we obtain $(1-A)dF_S=16\varepsilon\sigma T_s^3 dT_s$, leading to the following expression

$$dF_S / F_S = 4dT_s / T_s. \tag{7.1}$$

Since $\Delta F_S/F_S$ during solar cycles has well been measured to vary by approximately 0.1% from solar maximum to solar minimum in solar cycles, Eq. 7.1 gives the direct surface temperature variation $\Delta T_s \approx 0.1\% \times 288/4$ K=0.072 K. This is smaller than the observed variation of 0.11 K in global mean equilibrium surface temperature in 11-year solar cycles [Lean and Rind, 2008; Solanki *et al.*, 2013]. This is due to lack of the feedback factors of water vapor, lapse rate, albedo, clouds, *etc.* in Eq. 7.1. Indeed, the ratio of the observed temperature variation to the one calculated by Eq. 7.1 gives a direct measure of the total (feedback) amplification factor

$$\beta=0.11/0.072=1.53. \tag{7.2}$$

From $F_0=(1-A)F_S/4$, we can obtain solar radiative forcing arising from a change in TSI (solar constant) F_S

$$\Delta F_0=(1-A)\Delta F_S/4, \tag{7.3}$$

and the direct surface temperature change due to a solar radiative forcing is given by the solar climate sensitivity factor (α_c^s)

$$\alpha_c^s \equiv \frac{dT_s}{dF_0} = 4\frac{dT_s}{(1-A)dF_S}. \tag{7.4}$$

Substituting Eq. 7.1 into Eq. 7.4, we obtain

$$\alpha_c^s \equiv \frac{dT_s}{dF_0} = \frac{T_s}{(1-A)F_S}. \tag{7.5}$$

With the well observed T_s=288 K, F_S=1360-1365 Wm^{-2}, and A=0.3 (albedo), we obtain the solar climate sensitivity factor α_c^s=0.30 K/(Wm^{-2}). The latter is the same as the reference climate sensitivity or the Planck feedback factor in climate models, obtained in Chapter 6 (Sec. 6.7).

With the observed β=1.53 (Eq. 7.2), the *solar equilibrium climate sensitivity factor* λ_c^s is further given by

$$\lambda_c^s = \beta\alpha_c^s = 0.46 \text{ K/(Wm}^{-2}), \tag{7.6}$$

and the equilibrium global mean surface temperature change due to a solar radiative forcing is given by

$$\Delta T_s = \lambda_c^s \times \Delta F_0 . \tag{7.7}$$

Since it is well observed that the radiative forcing ΔF_0=0.24 Wm^{-2} (corresponding to the TSI variation of 1%, *i.e.*, ΔF_S=1.36 Wm^{-2}) in 11-year solar cycles, Eq. 7.7 reproduces the equilibrium surface temperature variation ΔT_s=0.11 K observed in the past several solar cycles, as shown in Fig. 7.7.

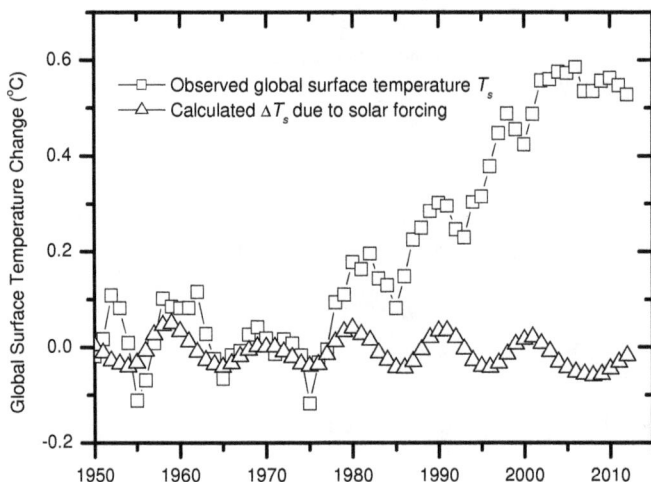

Fig. 7.7. Observed global mean surface temperatures and calculated global surface temperature change due to solar forcing since 1950 (see text). Annual global surface temperatures have the same source as Fig. 7.6.

Using a larger surface temperature variation $\varDelta T_s$=0.15 K in 11-year solar cycles, Douglass and Clader [2002] obtained a larger feedback factor β=2.1 and a larger solar equilibrium climate sensitivity factor λ_c^s =0.63 K/(Wm^{-2}). However, the best estimate of observed equilibrium surface temperature variations in 11-year solar cycles has given $\varDelta T_s$=0.11 K in the most recent literature [Lean and Rind, 2008; Solanki et al., 2013]. Thus, the present values of λ_c^s =0.46 K/(Wm^{-2}) and β=1.53 given here should be more realistic.

7.7 Conclusions

Observations have shown strong evidence that natural climate drivers have played a negligibly small role in long-term climate change since 1950, leading to a long-term global mean surface temperature rise of far less than 0.1 K. Research on the potential climate impacts of solar variability should be focused on reliable measurements available since 1978, which could provide valuable information on global climate change.

With the observed modulations of 0.11 K in global mean equilibrium surface temperature during 11-year solar cycles, we obtain a total climate (feedback) amplification factor β=1.53 and a solar equilibrium climate sensitivity factor λ_c^s=0.46 K/(Wm^{-2}). These results will play an important role in quantifying human-made contribution to global climate change, which will be addressed in the next Chapter.

Chapter 8

New Theory of Global Climate Change

8.1 Introduction

As discussed in Chapter 6, there are long discrepancies about the scientific origins and perspectives of global climate change [e.g., Idso, 1980, 1998; Friis-Christensen and Lassen, 1991; Lindzen, 1997; von Storch and Stehr, 2000; Veizer *et al.*, 2000; Soon *et al.*, 2001; Pielke Jr., 2001; Douglass *et al.*, 2004; Lu, 2010a, 2010b, 2012b, 2013; Fyfe *et al.*, 2013; IPCC, 2013; Wyatt and Curry, 2014].

In late 2008, this author made an (probably the first) observation and prediction of the long-term reversal in global warming and proposed an anthropogenic cause. The manuscript reporting the finding and prediction, entitled "New Observations of the Effects of CFCs on Ozone Depletion and Climate Change", was first submitted on 11th May 2009 to a journal, which had published one of the first few papers on the CRE mechanism of the ozone hole (addressed in Chapters 4 and 5). Unfortunately the paper did not go through the review process; it was subsequently submitted to several other journals but received all rejections. Here is the abstract of the first submitted manuscript:

"Following the recent finding of 11-year cyclic polar ozone depletion [Lu, Phys. Rev. Lett. 102, 118501(2009)], a clear 11-year cyclic variation in polar stratospheric cooling as a result of cosmic-ray-driven ozone depletion from 1956 to 2008 is revealed in this study. It is also strikingly found that the polar stratospheric temperature is solely determined by total ozone in the polar stratosphere, indicating that non-CFC greenhouse gases play a negligible effect. More striking, it is also observed that CFCs co-operated with cosmic rays are responsible for not only stratospheric ozone depletion but the global warming observed in the late 20th century dominantly. This means that global warming may be reversed in the 21st century due to the decreasing emission of CFCs into atmosphere."

Fortunately, the paper was finally accepted and published in *Physics Reports* online on December 3, 2009 [Lu, 2010a], in which this author predicted a long-term global mean surface temperature declining trend that started around 2002, corresponding to the declining of halogen-containing molecules (mainly chlorofluorocarbons—CFCs) in the atmosphere. The *Phys Rep* paper was followed by an invited paper published in *Journal of Cosmology* [Lu, 2010b], which made a theoretical calculation of the global surface temperature increase as a result of anthropogenic emission of CFCs, HCFCs and CCl_4. The calculation with only one parameter (the equilibrium climate sensitivity factor λ_c=1.8) showed that these gases could indeed be responsible for the observed global surface temperature rise of 0.6 °C from 1950 to around 2000, and since then a global cooling trend started. More recently, an in-depth statistical and quantitative analysis of comprehensive observed datasets has further strengthened the conclusion and shown strong evidence of the warming 'hiatus' associated with the declining halogenated gases regulated by the Montreal Protocol [Lu, 2012b, 2013].

Today it is generally accepted that there has been a 'hiatus' in warming since 1998 [Fyfe *et al.*, 2013; IPCC, 2013; Trenberth and Fasullo, 2013; Kosaka and Xie, 2013; Wyatt and Curry, 2014]. As recently noted by researchers [*e.g.*, Fyfe *et al.*, 2013], there is striking discrepancy between observed and model-simulated global mean surface temperature (GMST) trends since 1998. For this period, the observed trend of 0.05 ± 0.08 °C per decade is much smaller than the average simulated trend of 0.21 ± 0.03 °C per decade. The divergence between observed and simulated global warming by global climate models (*e.g.*, CMIP5) began in the early 1990s, as is seen from the comparison of observed and simulated running trends from 1970–2013 [Fyfe *et al.*, 2013]. All historical simulations by the current generation of climate models (*e.g.*, CMIP5) do not reproduce the observed global temperature trends, particularly the recent 'hiatus' in global warming over the past 16 years [Fyfe *et al.*, 2013; IPCC, 2013].

It is worthwhile to note that researchers have proposed various mechanisms (mainly natural climate variability) for the stopping of global warming [IPCC, 2013; Trenberth and Fasullo, 2013; Kosaka and Xie, 2013; Wyatt and Curry, 2014], but natural climate variability in climate models has its limits [Knight *et al.*, 2009]. The model simulations set up a duration of an observed 15-year absence of warming as indication of a discrepancy with the prediction of climate models, and thus predicted in 2009 that "no sort of natural variability could hold off greenhouse warming much longer" and "(rapid) warming will resume in the next few years", as truly documented in interviews with climate researchers by the *Science* reporter Richard Kerr [2009].

In climate research, there are essentially two types of models: conceptual models and 'quasi-realistic' models [*e.g.*, von Storch, 2010; Hegerl and Zwiers, 2011]. The former have often been used to explain the greenhouse effect, as presented in Chapter 6. The 'quasi-realistic' models seek to maximize complexity for optimizing different aspects, such as spatial resolution and a large number of parameters. The latter usually require complex and lengthy programming codes to be run on an advanced computer. It seems that efforts in recent climate research have been focused on 'quasi-realistic' models [*e.g.*, von Storch, 2010; IPCC, 2007, 2013]. However, the assumptions in any climate model must be validated with observations. The failure of models to explain real world observations often indicates that the assumptions in the models are most likely incorrect.

This Chapter will introduce a new theory of modern global warming, which was developed since 2008 [Lu, 2010a, 2010b, 2012, 2013]. This theory has explained the 'hiatus' in global warming as due to an anthropogenic cause, and predicted a long-term cooling trend that started around 2002, corresponding to the declining of halogen-containing molecules (mainly chlorofluorocarbons—CFCs) in the atmosphere. In Sec. 8.2, the scientific basis of the new theory will be given. The observations to support the conclusion strongly that the greenhouse effect of increasing non-halogen greenhouse gases (GHGs) has been completely saturated will be presented in Sec. 8.3. It will be followed by discussion of early calculations of the greenhouse effects of halogenated gases in Sec. 8.4. A refined physics model of global climate change will be presented in Sec. 8.5. In Sec. 8.6, *no-parameter* conceptual model calculations will be presented to reproduce the observed data of global mean surface temperature since 1950, including the observed ending of global warming over the past 16 years, and to show a long-term cooling trend for the coming five to seven decades.

8.2 Revisiting the earth's blackbody radiation spectrum

Historically finding an explanation for blackbody radiation in physics was one of the two major problems around the turn of 20th century. The explanation via the revolutionary concept of *energy quanta* put forward by Max Planck in 1900 is generally regarded as the dawn of 20th century quantum theory [Klein, 1962; Kragh, 1999]. Today we know that radiation from both the Earth and Sun can approximately be treated as blackbody radiation.

The radiation energy flux in the wavelength interval $d\lambda$, $B_\lambda(T)$, or the frequency interval dv, $B_v(T)$, of a blackbody is given by Planck's formula, Eq. 6.3 or Eq. 6.4 in Chapter 6. Although CO_2 has a strong absorption band centered at 667 cm^{-1} (15 μm), it is actually a misconception to compare either $B_\lambda(T)$ or $B_v(T)$ with the atmospheric absorption (transmittance) spectrum, since they are different in physical nature and units. This is analogous to the argument that the Sun has a radiation peak at 880 nm (in the near IR region) derived from $B_v(T)$ rather than at the well-known visible wavelength of ~500 nm obtained from $B_\lambda(T)$ using Wien's displacement law. Like the previous solution to the solar Wien peak [Soffer and Lynch, 1999; Overduin, 2003], this paradox can be solved with the integral of either $B_\lambda(T)$ over a wavelength interval of $\Delta\lambda$ or $B_v(T)$ over a frequency interval of Δv ($=-c\Delta\lambda/\lambda^2$), which must give the same radiation intensity I [Lu, 2010b]. This comes directly from their definitions: $B_\lambda(T)d\lambda=B_v(T)dv$. The intensity spectrum

$I(\lambda)$ can be obtaining by integrating $B_\lambda(T)$ over a certain wavelength interval $\Delta\lambda$ (*e.g.*, 3 μm)

$$I(\lambda) = \int_{\lambda-\Delta\lambda/2}^{\lambda+\Delta\lambda/2} B_\lambda(T)d\lambda. \tag{8.1}$$

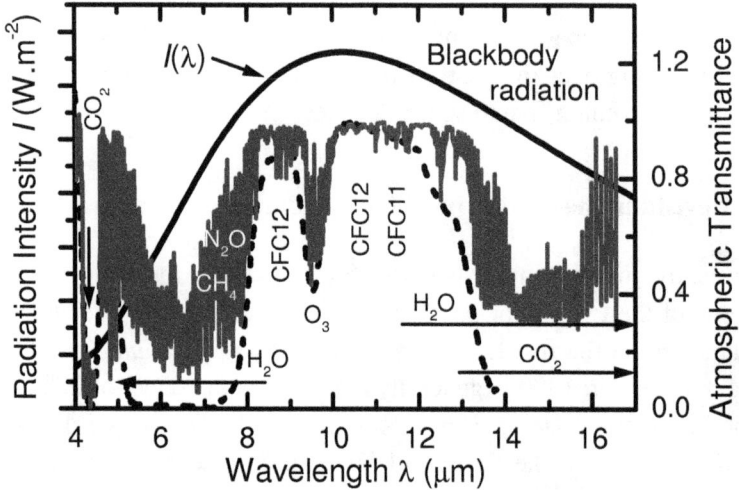

Fig. 8.1. Earth's blackbody radiation intensity spectrum $I(\lambda)$ and atmospheric transmittance spectrum $\Gamma(\lambda)$. $I(\lambda)$ (thick solid line) obtained by integration of radiation power density $B_\lambda(T)$ at T=288 K over a wavelength interval (3 μm); outgoing long-wave radiation transmittance spectrum was measured at the TOA by a satellite in 1997 (thin solid line) [Clerbaux *et al.*, 2003]; and transmitted radiance spectrum measured at the ground with the Sun as a blackbody source (short dash line) [Ratkowski *et al.*, 1998], normalized to the maximum transmittance of measured OLR in the 8-13 μm atmospheric window. The absorption band positions of CFCs, O_3, H_2O, CO_2, CH_4 and N_2O are indicated. Modified from Lu [2010b; 2013].

The intensity spectrum $I(\lambda)$ over the wavelength range of 4-17 μm is shown in Fig. 8.1, which gives a peak around 10.1 μm [Lu, 2010b, 2013], along with the satellite-measured atmospheric transmittance spectrum [Clerbaux *et al.*, 2003]. Also shown in Fig. 8.1 is the atmospheric transmittance spectrum measured at the ground with the Sun as a blackbody source, which closely resembles the theoretical atmospheric transmittance spectrum obtained with Modtran4 calculations shown in Fig. 6.5 [Ratkowski *et al.*, 1998]. The spectral region of 8-13 μm is generally called the atmospheric "window", because the unpolluted atmosphere in this spectral region is quite transparent, except for absorption by the ozone layer in the 9.6 μm band. One can see the "ozone dip" in Fig. 8.1. Another, more critical, reason for calling it a "window" is that the majority of Earth's radiation energy is emitted into space at wavelengths between 8 and 13 μm, where the maximum intensity of Earth's blackbody radiation is emitted.

The conclusion is that over 80% of the total radiation energy from Earth's surface and clouds is emitted into space through the atmospheric window. As a consequence, any polluting molecule that strongly absorbs radiation in the atmospheric window is highly effective in generating a greenhouse effect. Unfortunately, many halogenated molecules such as CFCs and HCFCs are not only ozone-depleting molecules but are also highly effective greenhouse gases, since they have strong absorption bands in the atmospheric window at 8-13 μm.

As also shown in Fig. 8.1, CO_2 contributes to the strong absorption bands at 4-5 μm and 13-17 μm, while CH_4 and N_2O also have strong IR absorption band strengths at 7.6 μm and 7.8 μm, respectively. However, H_2O (water) is the most effective absorber in the entire infrared spectral range, having two major bands at 5-8.3 μm and 11-17 μm. These gases are therefore strong GHGs.

It is important to note that CO_2, CH_4 and N_2O have high atmospheric concentrations in ~400 ppm, 1.9 ppm and 327 ppb, respectively, which are 10^6, 10^4 and 10^3 times those of CFCs and HCFCs in 100-500 ppt [IPCC, 2007, 2013; WMO, 2011]. Although the CH_4 and N_2O levels are lower than CO_2, IR absorption band strengths of CH_4 at 7.6 μm and N_2O at 7.8 μm are 1~2 orders of magnitudes higher than that of CO_2 at 15 μm. As mentioned in Chapter 6, there is no controversy that the absorption of terrestrial radiation emitted from the earth surface by these non-halogen GHGs at the centers of their IR bands has been entirely saturated. Indeed, the current debate lies on whether the absorption at the edge wings of their IR bands would continue to increase with rising gas concentrations and contribute considerably to the observed global surface temperature change. This must be examined closely by observations, as will be shown in next Section.

8.3 Evidence of the null warming effect of rising non-halogen greenhouse gases

8.3.1 *No effects of increasing non-halogen gases on polar ozone loss and stratospheric cooling*

First, as noted in Sec. 5.10 of Chapter 5, the absence of the greenhouse effect of increasing non-halogen GHGs has been revealed in the well-observed 11-year cyclic variations of polar ozone loss and

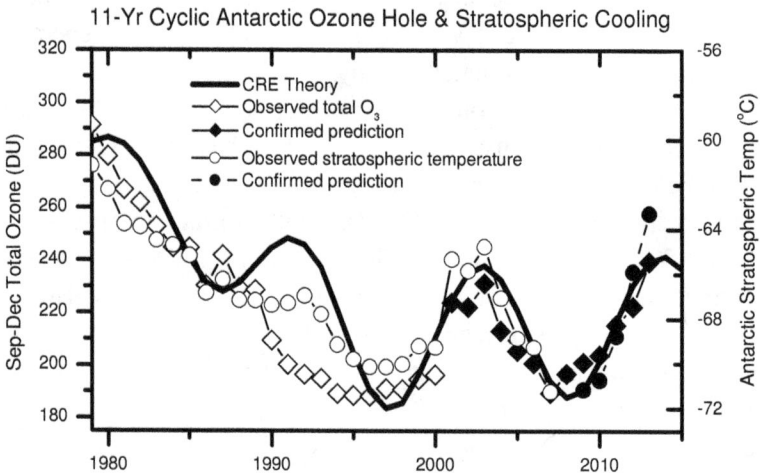

Fig. 8.2. Observed and theoretical total ozone and lower stratospheric temperature (100 hPa) at Halley (75°35' S, 26°36' W), Antarctica during the ozone hole months (September-December) in 1979-2013. The observed data (open and solid diamond symbols for total ozone; open and solid circle symbols for temperature) are the averages of the monthly mean data for 4 months (September-December), measured by the British Antarctic Survey (BAS); only a minimum processing by 3-point smoothing was applied to the observed 4-month averaged data. The data presented as solid symbols (solid diamonds and solid circles) were observed *after* the predictions of 11-year cyclic variations in polar ozone loss and stratospheric cooling were made by Lu and Sanche [2001a] and Lu [2010a], respectively. The theoretical data were calculated from the CRE equation (Eq. 4.22) with the equivalent effective chlorine (EECl) and the cosmic-ray (CR) intensity in the polar stratosphere as variables. Updated and adapted from Lu [2012b, 2013].

stratospheric cooling over Antarctica in the past five decades [Lu, 2010a, 2010b, 2013, 2014c]. The observed data have shown the solid fact that 11-year cyclic changes of both total ozone loss and temperature in the winter and springtime lower stratosphere over Antarctica are governed nearly completely by the simple CRE equation with the polar stratospheric level of effective equivalent chlorine and the intensity of cosmic rays as the only two variables. The detailed and updated results are shown in Fig. 8.2 [Lu, 2014c]. This is in striking contrast to the predictions of climate models simulating the effect of increasing non-halogen GHGs (CO_2, CH_4 and N_2O) on polar stratospheric ozone loss and associated stratospheric cooling [Ramanathan *et al.*, 1985, 1987; Austin *et al.*, 1992; Shindell *et al.*, 1998]. Based on the model predictions, the observed data, in turn, indicate no climate forcings from increasing non-halogen GHGs on stratospheric climate of Antarctica over the past decades.

8.3.2 *Global surface temperature vs CO_2 or CFCs*

Any possible correlation between global surface temperature and atmospheric CO_2 concentration from 1850 to 1930 was analyzed by Lu [2010b]. This analysis is particularly meaningful, as in this duration, the anthropogenic CO_2 level in the earth's atmosphere had increasingly enlarged due to the well-known industrial revolution, while halogenated molecules (CFCs) just entered the industrial use in around 1930, before which they found essentially no use in industry [Carlisle, 2004]. As shown in Figs. 8.3A and B, the results showed that despite the continued increase of atmospheric CO_2 in this pre-CFC era, global surface temperature [UK Met Office's HadCRUT4 dataset; see Morice *et al.*, 2012] remained nearly constant from 1850 to ~1930; when global surface temperature was plotted versus CO_2 concentration, a nearly zero correlation coefficient R (=−0.04) was found. In other words, the global temperature was independent of the rising CO_2 level (285 to 307 ppm) over this interval of 80 years [Lu, 2010b].

A more in-depth quantitative and statistical analysis of the data of CO_2, halogenated GHGs and global surface temperature from 1970 to 2012 (the post-CFC era) was further presented [Lu, 2010b, 2013]. This analysis was based on the fact that like TSI data, data from direct measurements of CFCs became available since the 1970s when considerable atmospheric impact (O_3 depletion) of CFCs started to be observed. Thus, the significant anthropogenic effect of CFCs on Earth's climate is also expected to begin around 1970, and reliable conclusions are achievable. The results are updated and re-plotted in Figs. 8.3C-F.

Time-series data of atmospheric CO_2 concentration, the total concentration of major atmospheric halogen-containing GHGs (CFCs, HCFCs, and CCl_4) and global surface temperature over the past 45 years from 1970 to 2014 are plotted in Figs. 8.3C and E, respectively. In Fig. 8.3C, it is interesting to show that both the CO_2 concentration and the total halogenated GHG concentration had a nearly identical growth shape during the three decades from 1970 to ~2000, while they have been drastically different since the beginning of this century. Namely, CO_2 has kept the identical rising rate, whereas the total halogenated GHG concentration has had a turnover since ~2002. Correspondingly, it is clearly shown in Fig. 8.3E that global surface temperature had a linear rise from 1975 to ~2002 and has had a slowly declining trend since 2002. The 3-point smoothed temperature data in Fig. 8.3E also exhibits small but visible 11-year cyclic modulations, which can be attributed to the solar effect or stratospheric cooling arising from O_3 loss caused by CFCs and CRs or the combination of the two effects [Lu, 2010a, 2010b, 2013].

Most interesting are the results shown in Figs. 8.3D and F for the period of 1970-2014, plotting observed global surface temperature as a function of CO_2 concentration and total concentration of halogen-containing GHGs, respectively. In Fig. 8.3F (and Fig. 8.3C), a 9-year delay in halogen-containing GHG concentrations in the stratosphere from surface-based measurements must be applied, otherwise, global surface temperature would show a sharp *rise* with high total halogenated GHG concentrations above 1100 ppt (1.1 ppb) [Lu, 2010b]. This delay of 9 years on the average in *global* stratospheric halogenated GHGs has been well justified by the observations of ozone loss reviewed in Chapter 5 (Fig. 5.11): The stratospheric EECl delays in the polar region (60°-90° S) and in the non-polar latitudes (65° S-65° N) are 1~2 and ≥10 years from surface-measured EECls, respectively [Lu, 2013]. Most strikingly, Fig. 8.3F shows that *global surface temperature has had a nearly perfect*

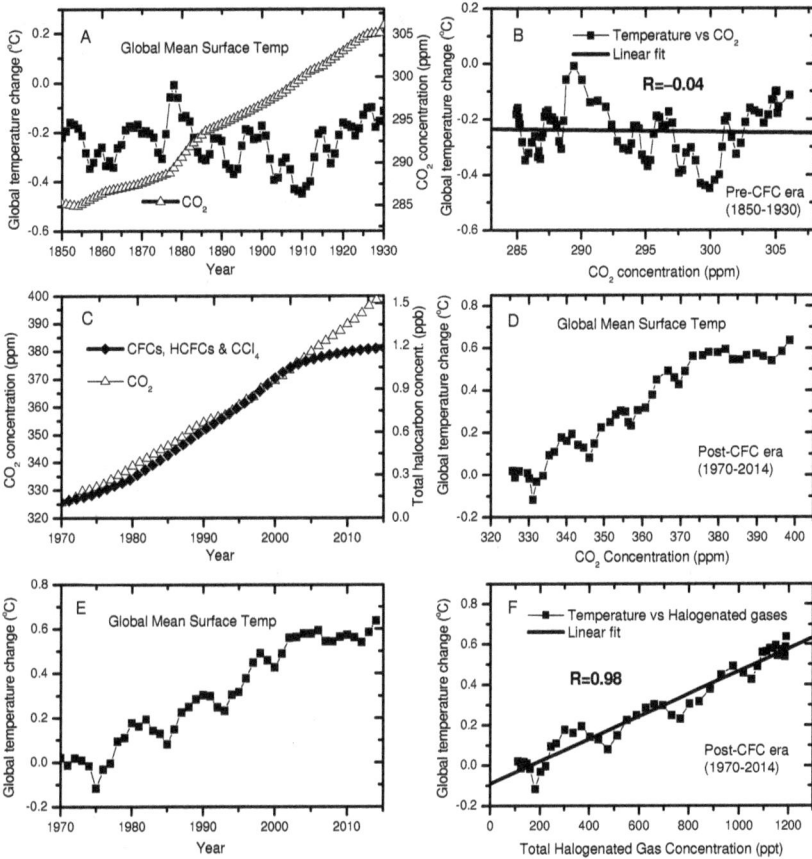

Fig. 8.3. Global surface temperature, CO_2 and halogen-containing greenhouse gases (GHGs) from 1850 to 2014. A and B for the *pre-CFC era* (1850-1930): A, time-series atmospheric CO_2 concentrations and global mean surface temperatures; B, observed global temperature versus atmospheric CO_2 concentration, and a linear fit to the observed data gives a nearly zero linear correlation coefficient R=−0.04. C-F for the *post-CFC era* (1970-2014): C, time-series atmospheric CO_2 concentrations and total concentrations of atmospheric halogen-containing GHGs (CFCs, HCFCs and CCl_4); D, observed global temperature versus CO_2 concentration; E, time-series global surface temperatures; F, observed global temperature versus total concentration of atmospheric halogen-containing GHGs, and a linear fit to the observed data gives a nearly perfect linear correlation coefficient R=0.98 with P<0.0001 for (R=0). For halogen-containing gas concentrations in the stratosphere, a 9-year delay from surface-based measurements was applied (see the text). Annual global surface temperatures were from the UK Met Office (HadCRUT4), relative to the temperature in 1970; only a minimum 3-point smoothing was applied to observed temperature data. Updated and adapted from Lu [2010b, 2013].

linear dependence on the total amount of atmospheric halogen-containing GHGs from the 1970 to the present. Statistically, the linear fit to observed global surface temperature data gives a statistical linear correlation coefficient R as high as 0.98 (close to unit) and P<0.0001 for R=0 [Lu, 2010b]. It is worthwhile to note that with the solar effect removed from the global surface temperature data, similarly high correlation coefficients of 0.96-0.97 between global temperature and total amount of halogen-containing GHGs was also obtained [Lu, 2013]. This result is consistent with the conclusion drawn in Chapter 7 that the natural (solar) effect has played a negligible role in climate change since 1970. Fig. 8.3D shows that at first glance, global temperature seemingly had a linear dependence on CO_2 concentration at 326-373 ppm (1970-2002), but it has entered a decreasing trend with rising CO_2 levels of ≥373 ppm (2002 - present).

In summary, the observed data shown in Figs. 8.3A-F strongly indicate that *neither the continued rises of CO_2 and other non-halogen GHGs nor the solar effect has played a considerable role in climate change; global surface temperature has been nearly perfectly controlled by the variation of atmospheric halogen-containing GHGs since 1970 up to date.*

8.3.3 *Observations versus IPCC-used radiative forcings of non-halogen greenhouse gases*

The interesting observed data shown in Fig. 8.3 deserve for a more in-depth and quantitative analysis. In Chapter 6.5.3, we introduce the *logarithmic* relationship between the radiative force ΔF and CO_2 concentration (Eq. 6.13) that has been used in IPCC Reports [2001, 2007, 2013] and has given nearly the same results as those from state-of-the-art AOGCMs [Dufresne and Bony, 2008; IPCC, 2013]. From Eq. 6.13, we obtain an identical radiative forcing $\Delta F \approx 0.72$ W.m^{-2} for the CO_2 concentration rises of 285 to 326 ppm in 1850-1970 and of 326 to 373 ppm in 1970-2002. This radiative forcing would cause a global surface temperature rise of about 0.6 °C, given the equilibrium climate sensitivity factor λ_c=0.8 K/(W/m^2). This is shown in Fig. 8.4. From the radiation physics aspect of the assumption that the radiative forcing of

CO_2 would arise from the in-complete saturation in absorption at the edge wings of the CO_2 IR line at 667 cm^{-1}, it is reasonably expected that the radiative forcing would tend to become lower with higher CO_2 concentrations [Lu, 2010b, 2013]. This is because the degree of saturation would increase with rising CO_2 concentration.

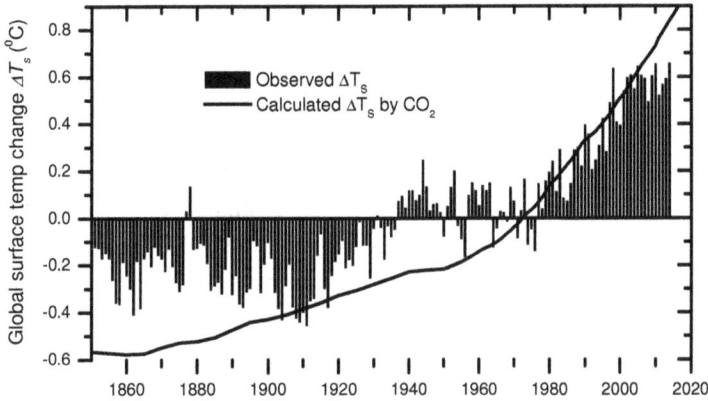

Fig. 8.4. Observed global mean surface temperature changes ΔT_s and calculated ΔT_s using the IPCC-used radiative forcing formula of CO_2 for 1850-2014, with respect to the temperature in 1970. ΔT_s are calculated by $\Delta T_s = \lambda_c \times \Delta F(CO_2)$ with $\lambda_c = 0.8$ K/(Wm^{-2}), where $\Delta F(CO_2)$ are calculated by Eq. 6.13. Annual global surface temperatures were from the UK Met Office's HadCRUT4 dataset.

The above calculated results actually contradicts the observed global surface temperature rise of about 0.2 °C during 1850-1970 and the sharp rise of ~0.6 °C during 1970-2002, not to mention the observed *negative* correlation with CO_2 concentrations at >373 ppm from 2002 to the present. It should also be noted that if the CO_2-warming model used by IPCC were correct, the current global temperature would be at least 0.2~0.3 °C higher than the observed value (see Fig. 8.4). Moreover, when the radiative forcings of CH_4 and N_2O calculated from the IPCC-used Eqs. 6.14 and 6.15 are added to that of CO_2, an even larger discrepancy between observed and calculated global surface temperatures is seen (see Fig. 6.10). In contrast, the observed global temperature has exhibited a nearly perfect linear positive correlation ($R \approx 1.0$) with the total concentration of halogenated GHGs since their considerable emission into the atmosphere in the 1970s (Fig. 8.3F). These observed and calculated results in Figs. 8.2-8.4 strongly indicate that the greenhouse effect of increasing non-halogen gases has been saturated (zero) and to a good approximation, halogenated gases (mainly CFCs) have completely governed the global climate change since 1970.

8.3.4 *Observations of outgoing longwave radiation (OLR) spectra*

Changes in the earth's greenhouse effect may be detected from variations in the radiance spectrum of outgoing longwave radiation (OLR) at outer space (at the top of atmosphere, TOA). If reliable data are available and careful analyses are performed properly, the OLR spectrum can be a measure of how the earth radiation emits to space and carries the signature of greenhouse gases that cause the warming effect [Goody *et al.*, 1998]. Harries *et al.* [2001] made the first investigation of the difference spectrum between the spectra of the OLR of the Earth as measured by the NASA Infrared Interferometric Spectrometer (IRIS) onboard the Nimbus 4 spacecraft in 1970 [Hanel *et al.*, 1972] and the interferometric monitor for greenhouse gases (IMG) onboard the ADEOS satellite in 1997 [Kobayashi *et al.*, 1999]. Harries *et al.* [2001] showed differences in the spectra associated with long-term changes in atmospheric CO_2, CH_4 and O_3 as well as CFC-11 and CFC-12, consistent with their theoretical calculations of the radiative forcing of increasing GHGs expected from the climate models, over the 27-year period (1970-1997) of the most rapid global warming. Particularly their main results showed that a negative-going brightness temperature difference was observed on the edge of the CO_2 v_2 band, between 710 and 740 cm^{-1}, in accord with the radiative force as a result of the known increase in atmospheric CO_2 concentrations between 1970 and 1997 expected from the IPCC-used climate models (GCMs).

However, the first careful analysis of the satellite data by Anderson *et al.* [2004] actually showed that to reliably sample the radiation field would require averages over ~10^4 independent radiance spectra. They achieved a more reliable analysis of the radiance spectra in the entire

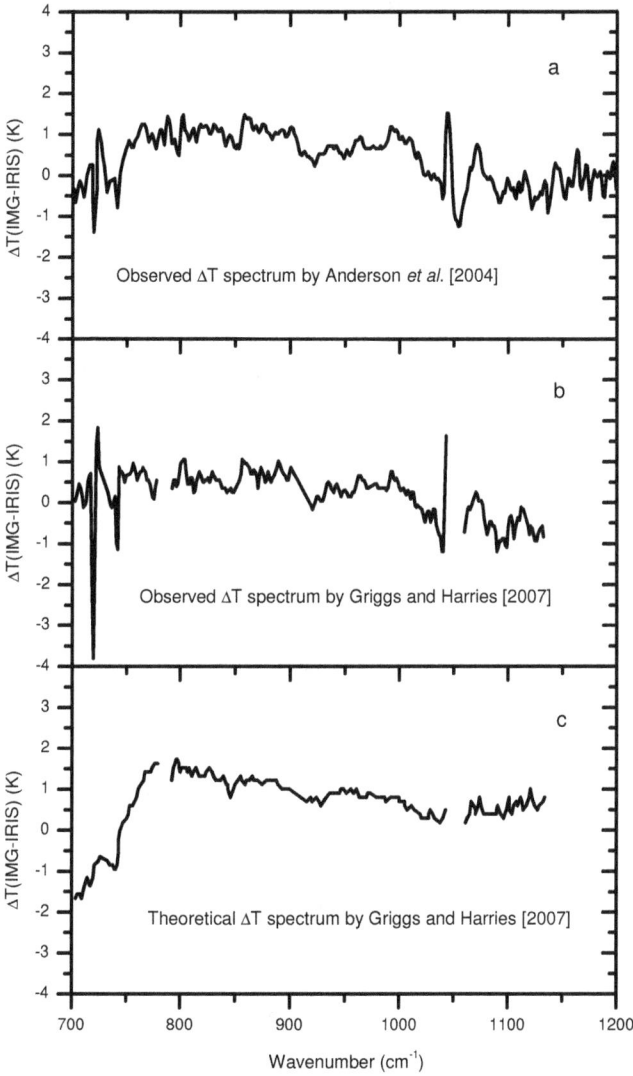

Fig. 8.5. Observed and theoretical difference spectrum of outgoing longwave radiation (OLR) for IMG-IRIS. *a*: Observations by Anderson *et al.* [2004]; *b:* Observations by Griggs and Harries [2007]; *c*: Simulated difference spectrum from Griggs and Harries [2007]. Based on data obtained from the cited references. Adapted from Lu [2010b, 2014b].

tropical belt between 30° N and 30° S measured by IRIS and IMG by increasing the data sample number to $2\sim3\times10^4$ spectra and applying an innovative method averaging the spectra. Their improvements largely reduced the random and system errors. As shown in Fig. 8.5a, Anderson *et al.* [2004] found the striking result that the strongly negative brightness temperature difference (by ~-1.5 K) at the wing from 800 to 600 cm^{-1} of the main 667 cm^{-1} CO_2 band, expected from the CO_2-warming theory, is absent in the observed difference spectrum. This is certainly an interesting and important observation, as this radiative forcing (of CO_2) lies at the heart of the debate on the cause of global warming observed in the late half of the 20th century.

Subsequently, Griggs and Harries [2007] re-analyzed the radiance spectra of IRIS and IMG. As shown in Figs. 8.5b, the results of Griggs and Harris [2007] are very similar to those obtained by Anderson *et al.* [2004]. These results show that the expected strong CO_2 absorption band in the 600 to 800 cm^{-1} region indeed disappears in the observed difference spectrum between 1970 and 1997. Griggs and Harries [2007] noted a negative brightness temperature difference observed in the CO_2 line at 720 cm^{-1} in the IMG-IRIS (1997-70) and the AIRS-IRIS (2003-1970) difference spectra, as well as a strongly negative brightness temperature difference in the CH_4 line at 1304 cm^{-1} for the AIRS–IRIS (2003–1970) difference spectra, indicating the increases of CO_2 and CH_4. However, they also noted that there was no signal at 720 cm^{-1} and there was a *positive* brightness temperature difference at 1304 cm^{-1} in the AIRS-IMG (2003–1997) OLR difference, despite the increases of CO_2 and CH_4 between these years. Note also that the theoretical OLR difference spectrum gives no such spike at 720 cm^{-1} either (Fig. 8.5c).

Nuccitelli *et al.* [2014] recently argued that the single negative spike at 720 cm^{-1} in Fig. 8.5a and b is evident of the unsaturated warming effect of CO_2 in 1970-1997. If this argument were correct, then one would have to account for the *positive* spike at 730 cm^{-1} as well. Even more obviously, it would also be difficult to explain why there was only a single negative spike in the interested wing at 600-800 cm^{-1} of the main CO_2 absorption line at 667cm^{-1}, as the absorption line is supposed to consist of many closely spaced spikes (lines). The latter is shown here by a high-resolution absorption (emission) spectrum of CO_2 plotted in

Fig. 8.6, given by a simulation using HITRAN2008 [Rothman *et al.*, 2009]. Thus, it is obvious that the single spike at 720 cm^{-1} in Figs. 8.5a and b cannot be attributed to the warming effect of increasing CO_2. Overall, the observed OLR difference spectra show no sign in warming effect of non-halogenated GHGs from 1971 to 1997 (2003) [Lu, 2010b, 2013, 2014b].

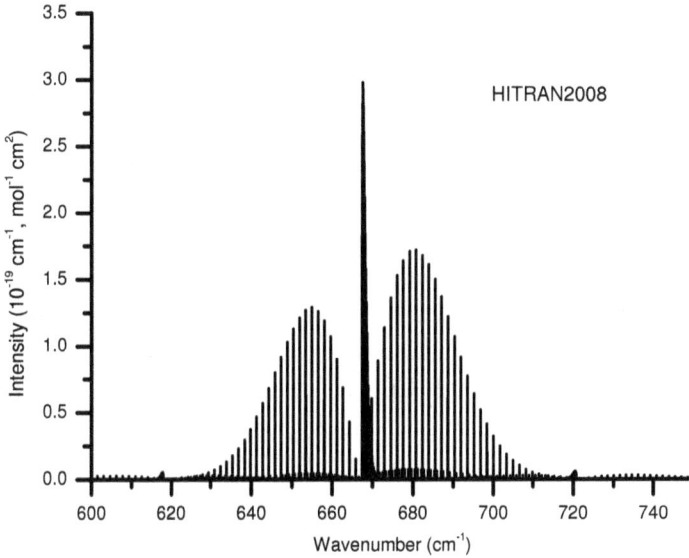

Fig. 8.6. Simulated (Theoretical) high-resolution absorption (emission) spectrum at 600-750 cm^{-1} by HITRAN2008 for CO_2. This shows that the very strong absorption line at 667cm^{-1} is actually a cluster of closely spaced but distinct spikes (lines).

8.3.5 Observations of paleoclimate

Some studies reviewed in the IPCC Reports [2001, 2007] showed that CO_2 co-varied with Antarctic temperature over glacial-interglacial cycles, suggesting a close link between CO_2 variation and temperature. However, linking paleoclimate studies to the modern anthropogenic warming has been very controversial, as there are also studies of ice core records that show the opposite: the rise in atmospheric CO_2 is the effect rather than the cause of surface warming in paleoclimate. For instance, a rapid rise by 5 °C in global average sea surface temperature occurred during transitions from the last Glacial Maximum to the onset of Holocene times, while detailed ice core studies by Smith *et al.* [1999] found that the concentration of atmosphere CO_2 increased by about 80 ppm (from ~190 to 270 ppm). But the authors found that the latter was due to the effect of the rise in global surface temperature on driving more CO_2 emission predominantly from the ocean. If one reversed the sequence and took the CO_2 increase as the cause of the 5 °C rise in global surface temperature, then it would be extremely difficult to understand why no global temperature increases were observed for the increases of atmospheric CO_2 from 285 to 308 ppm and 310 to 330 ppm in the periods of 1850 to 1930 and 1950 to 1975, respectively [Fig. 8.4, Lu, 2010b]. In fact, high-resolution ice core records of temperature proxies and CO_2 during deglaciations generally show that Antarctic temperature started to rise about one thousand years before the rise of atmospheric CO_2, as observed by Fischer *et al.* [1999], Veizer *et al.* [2000], Caillon *et al.* [2003] and Stott *et al.* [2007]. And despite strongly decreasing temperatures, high CO_2 concentrations can be sustained for thousands of years during glaciations [Fischer *et al.*, 1999]. These observations indicate that atmospheric CO_2 was not the main cause of these climate transitions. It also questions the application of the CO_2-paleoclimate relation to the recent anthropogenic warming.

In summary, *all the observations reviewed in this section have shown fairly strong and convincing evidence that increasing CO_2 and other non-halogen greenhouse gases have caused no significant warming effect since the 1950s.* Then, a question follows: Can the greenhouse effect of halogenated GHGs alone account for the observed surface temperature rise from the 1950s to ~2000? This will be addressed in the following sections.

8.4 Early calculations of surface temperature changes caused by halogenated greenhouse gases

In addition to their well-known role in ozone depletion, CFCs are also long-known greenhouse gases. Ramanathan [1975] made the first calculations that the greenhouse effect by CFCs and chlorocarbons could lead to a rise of ~0.9 K in global surface temperature *if* each atmospheric concentration of these compounds would increase to 2 parts per billion (ppb). This speculated concentration is far larger than the real values and has never been observed. Madden and Ramanathan [1980], Wang and Molnar [1985], and Ramanathan *et al.* [1985, 1987] also subsequently studied the potential climatic effects of halogen and non-halogen greenhouse gases (CO_2, CFCs, CH_4, N_2O, O_3 and others), showing that CFCs, through their indirect O_3-depleting effect, would have a potentially large stratospheric cooling effect, as large as that due to the CO_2 increase. Fisher *et al.* [1990] also calculated the greenhouse effect of another family of halogenated gases, HCFCs that have also strong absorption bands in the atmospheric window at 8-13 μm.

Since the 2000s, Ramanathan and co-workers [Ramanathan, 2006, 2007a, 2007b; Ramanathan et al., 2005, 2007a, 2007b, 2008] have turned to emphasize the importance of black carbon (BC) or atmospheric brown clouds (ABCs) in causing regional and global warming. They proposed that the warming effect of BC/ABC arises because it traps the solar radiation reflected by the earth's surface and clouds, which would have otherwise escaped to space. Ramanathan *et al.* propelled that BC plays a major role in the dimming of the surface and a correspondingly large solar heating of the atmosphere. They estimated that when globally averaged, the current BC radiative forcing at the TOA (the so-called radiative forcing as per IPCC) is as much as 60% of the current radiative forcing due to CO_2 greenhouse effect [Chung et al., 2005; Ramanathan, 2007a, 2007b]. They therefore concluded that BC in soot particles is potentially the second major contributor to the observed 20th century global warming, just next to CO_2. Thus, halogen-containing GHGs (CFCs and HCFCs) have become the third contributor to global warming observed in last century, according to their most recent estimates. Nevertheless, they also noted that the BC/ABC effect is subject to a threefold or larger uncertainty, compared with the CO_2 warming effect uncertainty of within ±15% [Ramanathan, 2007b].

Forster and Joshi [2005] used various climate models to examine the role of halogenated gases on stratospheric and tropospheric temperatures. They found that halogenated gases (mainly CFC-12) should have contributed a significant warming of ~0.4 K at the tropical tropopause since 1950, dominating the effect of other well-mixed GH gases. They also noted that the "disappearance" of such temperature increases would suggest that some other mechanism(s) such as stratospheric cooling due to O_3 loss are highly likely to be compensating for this, and as O_3 will likely recover in the next few decades, a slightly faster rate of warming would be expected from the net effect of halogenated gases. Interestingly, Wang *et al.* [1991, 1992] showed that the spatial distribution of atmospheric opacity which absorbs and emits the long-wave radiation for non-CO_2 gases (CFCs) is different from that for CO_2, for example, $CFCl_3$ is optically thin, whereas CO_2 is optically thick. Their simulations indicated that non-CO_2 greenhouse gases provide an important radiative energy source for the Earth climate system and different infrared opacities of CO_2 and non-CO_2 gases can lead to different climatic effects. They concluded that it is inappropriate to use an 'effective' CO_2 concentration to simulate the total greenhouse effect of CO_2, CFCs and other gases.

However, all these previous studies concluded that halogenated gases would play a certain but not dominant role in past and future surface temperature changes, compared with the greenhouse effect of non-halogen gases (CO_2, CH_4 and N_2O). It was generally concluded that CO_2 would play the dominant role in recent global warming and a warming trend will continue [IPCC, 2007; WMO, 2010; IPCC, 2013].

It should be noted that all the climate models mentioned in the previous paragraphs assumed no complete saturation in greenhouse effect of non-halogen GHGs but *continued* radiative forcings arising from their rising concentrations [Ramanathan *et al.*, 1985, 1987, 2008; Wang *et al.*, 1991, 1992; Forster and Joshi, 2005; Ramanathan 2007b; IPCC, 2001, 2007, 2013]. In view of the substantial observations reviewed in the previous Section, these models obviously overestimate the greenhouse effect of non-halogen gases including CO_2, BC, *etc.*

In striking contrast, the above assumption was recently found to be invalid, and the actual role of halogen-containing GHGs in global climate change was re-calculated [Lu, 2010a, 2010b]. Based on the strong observations that there has been a complete saturation, i.e., no greenhouse effect associated with the *increasing* of concentrations of non-halogen GHGs since the 1950s, it was shown that atmospheric halogenated gases (mainly CFCs) are most likely to cause the observed global surface temperature rise of ~0.6 °C from 1950 to 2002 [Lu, 2010b]. Also, it was predicted that global temperature will reverse slowly with the projected decrease of CFCs in coming five to seven decades [Lu, 2010a, 2010b]. This prediction seems to be surprising, but it is actually consistent with the WMO Report [2010]. The latter stated that "There have been no significant long-term trends in global-mean lower stratospheric temperatures since about 1995", and that following an apparent increase from 1980–2000, the stratospheric water vapor amount has decreased in the past decade. These observations are "not well understood" from photochemistry-climate models [WMO, 2010], but are actually consistent with the CFC warming mechanism [Lu, 2010a, 2010b]. Subsequently, Revadekar and Patil [2011] also found the positive correlation between surface temperature and CFCs over the region of India. A refined physical model of the warming effect of halogenated greenhouse gases will be presented in next section.

8.5 Refined physics model of global climate change

In Chapter 6, we consider that the earth-atmosphere system is imposed by a small energy perturbation, for example, by an increase of a greenhouse gas in the atmosphere. Before the surface temperature changes, this energy change translates into an initial radiative flux perturbation dF at TOA, called *radiative forcing*. A small concentration increment of a GHG initially *increases* the absorption of long-wave radiation in the atmosphere in the wavelength range of the GHG absorption, which leads to an initial *decrease* in outgoing long-wave radiation (OLR) at the TOA. To compensate this decrease in OLR at the TOA, the Earth's surface must raise its temperature to increase its upward long-wave radiation by dF_g. That is, the increase in F_g from the warming must balance the reduction in OLR at TOA due to the increased absorption of the GHG. Finally, a new equilibrium is restored at TOA. This is the well-known basic theory underlying global warming.

Thus, according to its classical definition, the radiative forcing due to a GHG rise can be written as

$$dF = d\left[\int_0^\infty [1 - \Gamma(\lambda)] B_\lambda(T_s) d\lambda\right] \neq (1 - \overline{\Gamma}) dF_g$$

(8.2)

where $B_\lambda(T_s)$ is the earth blackbody radiation energy flux in the wavelength interval $d\lambda$, given by Planck's formula, and $\Gamma(\lambda)$ is the wavelength-dependent transmittance of the atmosphere. From Eq. 8.2, one can see that it is misleading to find radiative forcings of various GHGs using an average transmittance Γ over the entire Earth blackbody spectrum. As shown in Fig. 8.1 and discussed in Sec. 8.2, a change in absorption in the atmospheric window sensitively influences the radiative process of the Earth; the atmospheric window plays a critical role in affecting the surface temperature. As a consequence, to a good approximation, Eq. 8.2 can be rewritten as

$$dF \approx (1 - \Gamma_{wd}) dF_g = (1 - \Gamma_{wd}) 4\sigma T_s^3 dT_s,$$

(8.3)

where Γ_{wd} is the (mean) transmittance in the atmospheric window at 8-13 μm. Hence, the climate sensitivity factor α is given by

$$\alpha \equiv \frac{dT_s}{dF} = \frac{1}{4(1 - \Gamma_{wd})\sigma T_s^3}.$$

(8.4)

In contrast to Eq. 6.19 given in Chapter 6, Eq. 8.4 now gives a climate sensitivity factor dependent on the (mean) transmittance Γ_{wd} in the atmospheric window, which can be determined from the OLR spectrum at the TOA measured by a satellite. Various infrared opacities of GHGs in the atmosphere can lead to quite different climatic effects. Below we will discuss the GH effects in the two extreme atmospheric limits: optically opaque and transparent limits.

8.5.1 *The optically opaque limit*

As mentioned in the above, CO_2 is a main GHG in the earth atmosphere because of its extremely high concentration and its main infrared absorption band at ~15 µm (667 cm^{-1}) lying in the spectral region of the surface blackbody radiation (Fig. 8.1). However, the CO_2 concentration is so high that the atmosphere has been completely opaque in the spectral region at 14-16 µm. As shown in Fig. 8.1, the OLR transmittance spectrum measured at the TOA by a satellite appears to give a mean transmittance Γ=~0.4 in this spectral region. But this non-zero transmittance is due to re-emission of CO_2 molecules in the tropopause or stratosphere. Indeed, the measured OLR flux in the CO_2 absorption band at 14-16 µm corresponds to a blackbody temperature ~220 K, which is about the temperature of the tropopause or stratosphere (Fig. 6.6). Moreover, the transmittance spectrum measured on the ground using the Sun as a radiation source also shows Γ=0 in this spectral region (Fig. 8.1).

It is worth noting that as revealed by substantial observations reviewed in Sec. 8.3, no further change in long-wave atmospheric absorption by CO_2 could occur and therefore the radiative forcing dF due to an increase of CO_2 is zero (dF=0). As a consequence, the continued rise of CO_2 contributes to no change in global surface temperature T_s.

8.5.2 *The optically transparent limit*

Halogenated GHGs such as CFCs and HCFCs have strong infrared absorption bands lying exactly in the atmospheric window at 8-13 µm (Fig. 8.1). Therefore CFCs and HCFCs are well-known extremely effective GHGs. From the satellite-measured atmospheric transmittance $\Gamma(\lambda)$ spectrum in Fig. 8.1, we directly obtain the mean transmittance Γ_{wd}=0.84 in the window of 8-13 µm. Substituting the measured Γ_{wd}=0.84 and T_s=288 K into Eq. 8.4, we obtain a climate sensitivity factor

$$\alpha = 1.16 \text{ K/(Wm}^{-2}) \text{ (for halogenated gases)}. \qquad (8.5)$$

Interestingly, this α factor is not far from the value of α=0.9 K/(Wm^{-2}) obtained in the best fit of calculated radiative forcings of halogenated greenhouse gases to the observed global mean surface temperature data [Lu, 2010b; 2013]. With the measured $\alpha = 1.16$ K/(Wm^{-2}) and the feedback amplification factor $\beta = 1.53$ measured from solar irradiance variability in solar cycles (Eq. 7.2 in Chapter 7), we now obtain the equilibrium climate sensitivity factor for halogenated greenhouse gases

$$\lambda_c^{halo} = \alpha\beta = 1.77 \text{ K/(Wm}^{-2}). \tag{8.6}$$

Amazingly, this equilibrium climate sensitivity factor is identical to the value $\lambda_c^{halo} = 1.8$ K/(Wm^{-2}) found in the best fits of the calculated radiative forcings arising from halogen-containing GHGs to the observed global surface temperatures in previous studies [Lu, 2010b, 2013].

8.6 Re-calculations of surface temperature changes caused by halogenated gases

With the above obtained λ_c^{halo}, we now calculate radiative forcing (ΔF) and global (equilibrium) surface temperature changes (ΔT_s) as a result of a rise of a halogenated greenhouse gas

$$\Delta T_s = \lambda_c^{halo} \times \Delta F^{halo}. \tag{8.7}$$

Here, the radiative forcings of halogenated gases can be calculated by Eq. 6.16, equivalently

$$\Delta F^{halo} = \chi C, \tag{8.8}$$

where C is the gas concentration and χ in Wm^{-2} ppb^{-1} is the radiative efficiency that refers to the change in net radiation at the TOA or at the climatological tropopause caused by a given change in gas concentration. In fact, Eq. 8.8 has been well proven by robust observations showing a nearly perfect linear dependence of global surface temperature on halogenated gas concentration with linear correlation coefficients as high as 0.98, as shown in Fig. 8.3 [Lu, 2010b, 2013]. The χ values for halogenated GHGs have been calculated using radiative transfer models of the atmosphere, taking into account of the strengths and spectral positions of a compound's absorption bands as well as atmospheric structure, surface temperature, and clouds, and are available in the WMO Reports [2010] or the IPCC Report [2013]. These χ values are readily used to calculate the radiative forcings with changing concentrations of halogenated GHGs.

Combining the solar radiative effect (Eq. 7.7) with the halogen GH effect (Eqs. 8.7 and 8.8), we obtain

$$\Delta T_s = \lambda_c^{halo} \times \Delta F^{halo} + \lambda_c^s \times \Delta F_0 \qquad (8.9)$$

where the second term refers to ΔT_s due to the change in solar radiative output F_S (total solar irradiance, TSI) with the solar equilibrium climate sensitivity factor λ_c^s=0.46 K/(Wm^{-2}) (Eq. 7.6). The total radiative forcing of fluorinated GHGs (HFCs, PFCs, and SF$_6$) controlled under the Kyoto Protocol has been negligibly small up to date [see Fig. 6.8; IPCC, 2013] and hence was not included in a recent study [Lu, 2013]. But the atmospheric concentrations of these gases are projected to increase significantly in the coming decades [IPCC, 2013]. Hence they are included in the current calculations. The observed and projected atmospheric abundances of 16 Montreal Protocol regulated halogen GHGs (CFCs, HCFCs, CCl$_4$, CH$_3$CCl$_3$, halons, CH$_3$Br and CH$_3$Cl) can be obtained from the WMO Report [2010], while the abundances of 12 fluorinated GHGs (HFCs, PFCs and SF$_6$) can be obtained from the averages of the 4 Representative Concentration Pathways (RCPs) given in the IPCC Report [2013]. Also, a global mean 9-year delay between surface measured and tropopause/stratospheric halogen levels is justified and applied [see Chapter 5; Lu, 2013].

For the solar constant (F_S) data, we use the 2013 IPCC-recommended TSI data up to 2012 [IPCC, 2013], which are shown in Fig. 7.3. For 2013-2070, the modulations in global surface temperature by the solar-cycle effect are simplified by $\lambda_c^s \times \Delta F_0 \approx 0.055 \times \cos[2\pi(i-2014)/11]$ (°C), which is similar to that used in Lu [2013], based on the observation that the temperature modulations are 0.11 °C from solar minima to solar maxima (see Sec. 7.6 in Chapter 7).

Using Eq. 8.9 and the above-mentioned data, we calculate global surface temperatures from 1950 to 2070. The results are shown in

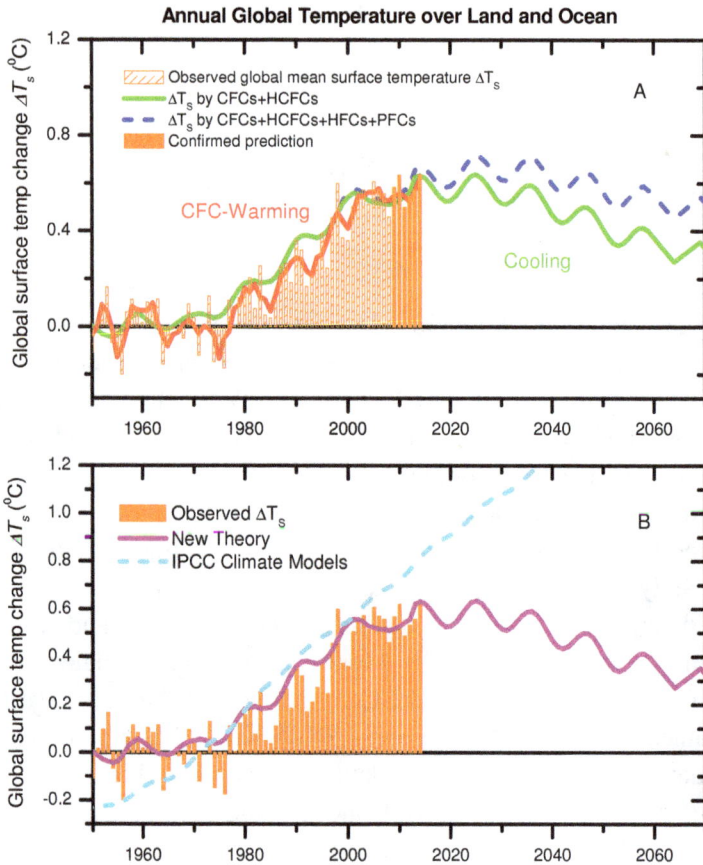

Fig. 8.7. Observed and calculated global mean surface temperature (GMST) changes ΔT_s for 1950-2014, and predicted future change, with respect to the mean T_s in 1950-1975. **A.** Observed GMSTs (bars in orange) were from the UK Met Office's HadCRUT4 time series data [Morice *et al.*, 2012]; the red curve is a 3-point smoothing of the observed data. Calculated ΔT_s were obtained by Eq. 8.9 with λ_c^{halo}=1.77 K/(Wm^{-2}) for the 16 halogen GHGs controlled under the Montreal Protocol (CFCs, HCFCs, Halons, *etc.*) without and with the addition of the 12 fluorinated GHGs controlled under the Kyoto Protocol (HFCs, PFCs, and SF$_6$), with the solar effect included. Data presented as filled bars in orange were observed *after* the prediction of a global cooling trend was made [Lu, 2010a, 2010b]. B. Solid line, calculated GMST changes ΔT_s caused by the 16 halogen GHGs controlled under the Montreal Protocol (CFCs, HCFCs, Halons, *etc.*) and the solar effect (same as in A). Dash line, transient temperature changes calculated by AOGCMs with $\Delta T_s^t = \lambda_c^t \times \Delta F$ with λ_c^t=0.51 K/(Wm^{-2}) for all halogen and non-halogen GHGs. Updated from Lu [2010b, 2013].

Table 8.1. Predicted/calculated and observed global mean surface temperatures as well as the deviations in percent between both since 2009

Year	Cal. ΔT_s [Lu, 2013]	Cal. ΔT_s Present[a]	Obs. ΔT_s wrt 1950-1975[b]	Obs. ΔT_s 3-yr average	Deviation (%)[c]
2009	0.518	0.518	0.570	0.550	-6.212
2010	0.535	0.531	0.621	0.559	-5.284
2011	0.562	0.547	0.487	0.547	-0.097
2012	0.591	0.560	0.533	0.526	6.016
2013	0.613	0.625	0.558	0.571	8.562
2014	0.620	0.633	0.623	0.623	1.639
2015	0.611	0.625			
2016	0.588	0.602			
2017	0.559	0.571			
2018	0.532	0.543			
2019	0.515	0.528			
2020	0.514	0.529			
2021	0.528	0.547			
2022	0.552	0.576			
2023	0.578	0.607			
2024	0.598	0.631			
2025	0.604	0.638			
2026	0.594	0.627			
2027	0.571	0.602			
2028	0.541	0.57			
2029	0.514	0.54			
2030	0.496	0.519			

*Note: a. The calculated data correspond to the green curve in Fig. 8.7A. *b*. The observed global surface temperature data were obtained from the UK Met Office's HadCRUT4, with respect to (wrt) the average temperature (-0.066 °C) for the period of 1950-1975. *b*. The deviation (%) is defined as [Cal. ΔT_s–Obs. ΔT_s (3-point average)]*100/Cal. ΔT_s.

Fig. 8.7A, together with the observed data since 1950. It is clearly seen that overall the calculated ΔT_s results show a surprisingly excellent agreement with the observed data during 1950-2014. These results strongly indicate that *the change in surface temperature since 1950 has been predominantly determined by halogenated GHGs (mainly CFCs).* Note that Eq. 8.9 includes no effects of natural drivers such as El Niño Southern Oscillation (ENSO) and volcanic eruptions, as well black carbon (BC), which themselves are subject to large uncertainties, because we by no means aim to make a perfect fit to the observed data in every detail. Instead, we are interested in comparing our calculations with the envelope of the global surface temperature change to access the success of our conceptual physical model to quantify the human contribution to the long-term global climate change.

It is also seen from Fig. 8.7A that the addition of the 12 fluorinated GHGs could produce an extra temperature increase up to ~0.20 °C by 2070, compared with the GH effect alone of all the 16 halogenated GHGs controlled under the Montreal Protocol. This addition will not change the secular cooling trend. Nevertheless, it is still important to phase out all halogenated GHGs globally; the control of fluorinated GHGs (HFCs, PFCs, and SF_6) will accelerate the reversal process of global temperature in the coming decades.

The digital data of the calculated global surface temperatures in Lu [2013] and the present Eq. 8.9, together with the observed data (the update UK Met office's HadCRUT4) [Morice *et al.*, 2012] with respective to the average temperature (−0.066 °C) in 1950-1975, as well as the deviations in percent between calculated and observed (3-point/year average) temperatures since 2009, are shown in Table 8.1. It can be seen that the deviations are within 10%. In particular, the predicted temperature of 0.620 °C for 2014 in Lu [2013] nearly perfectly agrees with the observed temperature of 0.623 °C in 2014. Nevertheless, it should be noted that it is always of risk to make a comparison with a particular year temperature, as the natural factor or climate dynamics can make a large temperature fluctuation. The latter can be reduced to a large extent by averaging/smoothing the observed data of neighbouring three years. Overall, the agreement between Eq. 8.9 and observed temperatures has been excellent.

In contrast to the well-established Eq. 8.8 (Eq. 6.16) for halogen GHGs, the assumed ΔF formulas for non-halogen GHGs (Eqs. 6.13-6.15 for CO_2, CH_4 and N_2O, respectively) used in the IPCC Reports do not pass the close examination by observed data, as clearly shown in Sec. 8.3. Nevertheless, researchers might still be attempted to use the ΔF formulas and λ_c given in climate models and IPCC Reports [2001, 2007, 2013] for the calculations of ΔT_s induced by all non-halogen (CO_2, CH_4 and N_2O) and halogen GHGs. Such calculated results are already shown in Fig. 6.10. Here, Fig. 8.7B gives a comparison with the results of the CFC-warming theory described in this Chapter. It is clearly seen that the assumption of an incomplete saturation in greenhouse effect of non-halogen GHGs inevitably leads to the result that a continued warming trend must have occurred over the past 16 years, obviously inconsistent with the observation, and would continue in the rest of this century. This conclusion has been drawn by IPCC [2013].

In contrast, the present model calculations show that the GH effect of halogenated GHGs has nearly perfectly accounted for the observed ΔT_s since the 1950s (Figs. 8.3F and 8.7A, and Table 8.1), without the need of introducing an additional parameter κ -- the ocean heat uptake efficiency as in conventional climate models. In view of the substantial and convincing observations reviewed in Sect. 8.3, there is good reason to expect a long-term cooling trend for the coming five to seven decades.

Most recently, Estrada *et al.* [2013] attributed the 'hiatus' in global warming since 1998 to the 'reductions' in the radiative forcings due to mainly the change of CH_4 and to a much less extent the changes of CFCs (~0.13, 0.05 and 0.08 Wm^{-2} for CH_4, CFC11 and CFC12, respectively), amounting to about 15% of the increase in total radiative forcing since 1880. However, such a small percent change in radiative forcing would only have a minor effect on surface temperature in the past decade, as shown in Fig. 8.7B. This is consistent with the newest IPCC Report [2013], which also states that the rate of increase in the (calculated) radiative forcing from the well-mixed GHGs over the last 15 years has been dominated by CO_2, and since AR4, CO_2 has accounted for over 80% of the ΔF increase. Even more evidently, it has been well documented that the growth rate of atmospheric CH_4 level slowed in the 1990s and was nearly constant from 1999 to 2006, but *since 2007 it has been strongly rising again* [IPCC, 2013; Nisbet *et al.*, 2014]. Thus, the 'hiatus' in global warming cannot be explained by the variation of mainly CH_4. The ceased surface warming might also be correlated with stratospheric aerosols from volcanic eruptions. However, the latter can only have a short-term effect, and if such effect were significant, it would be observed on the data of polar ozone depletion. This was not observed in the ozone data over the past 16 years, as shown in the observation reviewed in Chapter 5 and Fig. 8.2. Therefore, the reversal in global surface temperature cannot be attributed to volcanic eruptions either.

8.7 The potential nullifying effect of water

The missing of abundant non-halogen GHGs in contributing to recent warming is most likely due to the effect of water vapor as the most important GHG. Water has strong and extensive absorption bands covering the entire infrared spectral range where the extended wings of CO_2, CH_4 and N_2O absorption bands are located. Water is responsible for most of the absorption throughout the IR spectrum and may cause a nullifying effect on the weaker absorptions from other gases. It was actually shown previously that the effect of doubling CO_2 is small compared to the absorption due to water [Clark, 1999]. This probably leads to the failure of the IPCC-used relationships between radiative forcing and concentration for these non-halogen GHGs. In contrast, water vapor has some absorption at 11.5-13 µm (see Fig. 8.1), which can contribute to the positive feedback effect (β) on surface warming. But water has essentially no influence on the absorption of halogenated gases in the atmospheric window and hence has no interference on the GH effect of halogenated GHGs. Further studies on the potential erasing effect of water on non-halogen GHGs (CO_2, CH_4 and N_2O) will be interesting.

8.8 Summary

It has been generally agreed that the global surface temperature has stopped rising since around 2002.

There are substantial observations showing that the greenhouse effect of non-halogenated gases (CO_2, CH_4 and N_2O) has been completely saturated. Both the polar stratospheric temperature and the global surface temperature have predominantly been controlled by the presence of anthropogenic halogen-containing GHGs (mainly CFCs) in the atmosphere since 1950.

The global surface temperature exhibited zero correlation with atmospheric CO_2 prior to the industrial use of CFCs in 1930 with correlation coefficient R=−0.04. In striking contrast, the temperature has exhibited a nearly perfect linear correlation with the total amount of halogenated GHGs since the significant presence of CFCs in the atmosphere in 1970; an extremely high linear coefficient R=0.98 has been observed.

On the basis of radiation physics of the earth, we refine the physics of (equilibrium) climate sensitivity factor. It is shown that the climate sensitivity factor of a GHG depends strongly on its optical properties; there is a marked difference in climate effect between optically opaque and transparent GHGs, such as CO_2 and CFCs. From the refined expression of the climate sensitivity factor and the satellite-measured transmittance spectrum of the atmosphere, an equilibrium climate sensitivity factor (λ_c^{halo}) value of 1.77 K/(Wm^{-2}) for halogenated GH gases is now obtained. This climate sensitivity factor is unchanged from the value of 1.8 K/(Wm^{-2}) obtained by the best fits of the calculated radiative forcings caused by halogenated GHGs to the observed global surface temperature data in previous studies [Lu, 2010b, 2013].

With the obtained λ_c^{halo} value and the radiative efficiencies of halogenated GHGs given in IPCC and WMO Reports, the present physical model calculations with *no parameters* surprisingly well reproduce the observed data of global mean surface temperature since 1950, including the observed stopping in global surface temperature over the past 16 years. Thus, a continued cooling trend is expected in the coming decades under the international regulations to phase out CFCs, HCFCs, and HFCs.

In fact, the present calculated global mean surface temperatures since 2009 agree with the 3-year mean observed data nearly perfectly (within 10%). For example, the predicted temperature was 0.620 °C for 2014

[Lu, 2013], while the observed temperature was indeed 0.623 °C for 2014, with respective to the mean temperature in 1950-1975, according to the UK Met office HadCRUT4 dataset.

The substantial observations and the present calculations have shown strong evidence that humans are mainly responsible for significant climate change since the 1950s. The results strongly strengthen the conclusion that CFCs are the major culprit not only for ozone depletion but also for global warming by ~0.6 °C during 1970~2002. The successful execution of the Montreal Protocol and its amendments has shown its effectiveness in controlling the ozone hole in the polar regions and in reversing the global warming trend. It is therefore important to phase out all halogenated GHGs on the globe.

The conceptual physical model presented in this Chapter did not aim to make precise and perfect calculations of global temperature changes in every detail, unlike the sophisticated climate computer models including multiple parameters. But it has indeed shown that the warming effect of CO_2 and other non-halogen gases has most likely been completely saturated, and that halogenated gases (mainly CFCs and HCFCs) were responsible for the rapid global warming observed in the late 20th century. A long-term global cooling that started around 2002 is expected to continue for the next five to seven decades. This could be good news for humans. But we have to continue our efforts to phase out the production and use of halogenated greenhouse gases including CFCs, HCFCs, HFCs, PFCs, *etc.*

Finally, it is reiterated that the assumptions in any climate model must be validated with observations. Successful predictions are usually an indicator of the validity of a model. A model of greater simplicity and predictive capacity is likely the one closer to the truth.

Chapter 9

Impacts on Science, Policy and Economics

9.1 Summary of main results and conclusions and implications for the Montreal Protocol and the Kyoto Protocol

The main results and conclusions described in Chapters 3-8 are summarized in Figs. 9.1. and 9.2. Indeed, there exist persistent quantitative discrepancies between observations and photochemical models of ozone depletion, even though the models include a number of parameters. According to the photochemical models, an ozone recovery that would be attributable to the declining of the ozone-depleting substances regulated by the Montreal Protocol has not yet been observed [WMO, 2014]. More remarkably, the state-of-the-art photochemistry-climate models cannot reproduce the well-observed 11-year cyclic variations of polar ozone loss, nor can they capture the essential features of the climate in the polar stratosphere. These observations indicate that current atmospheric (photochemical) chemistry models have encountered challenges to place the Montreal Protocol on a firm scientific basis.

In contrast, the cosmic-ray-driven electron-induced reaction (CRE) theory of the ozone hole has been well confirmed by substantial and convincing observations from both laboratory and field measurements. According to this new theory, ozone loss in the polar O_3 hole can be calculated with the straightforward CRE equation: $-\Delta[O_3]=k[C]I^2$, where $[C]$ is the equivalent effective chlorine level in the polar stratosphere, I

253

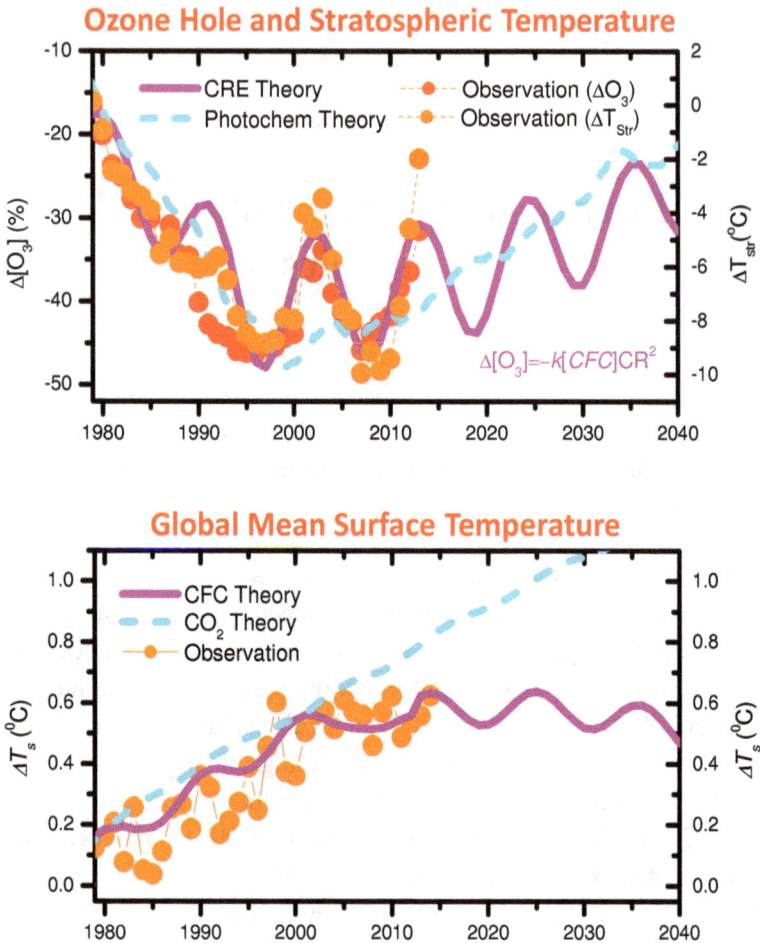

Fig. 9.1. Theoretical (conventional and new) and observed time-series data of the Antarctic ozone hole and global surface temperature changes. Upper panel: total ozone and lower stratospheric temperature changes in the Antarctic ozone hole, from the data presented in Fig. 3.13 and Fig. 5.14. Lower panel: global surface temperature changes, from the data presented in Fig. 6.10 and Fig. 8.7. *It is shown that neither the conventional photochemical theory of the O₃ hole (described in Chapter 3) nor the CO₂ theory of modern global warming (Chapter 6) can satisfactorily explain the observed data. In contrast, both the CRE theory of the ozone hole (Chapters 4 and 5) and the CFC theory of modern global warming (Chapter 8) give excellent agreements with the observed data.*

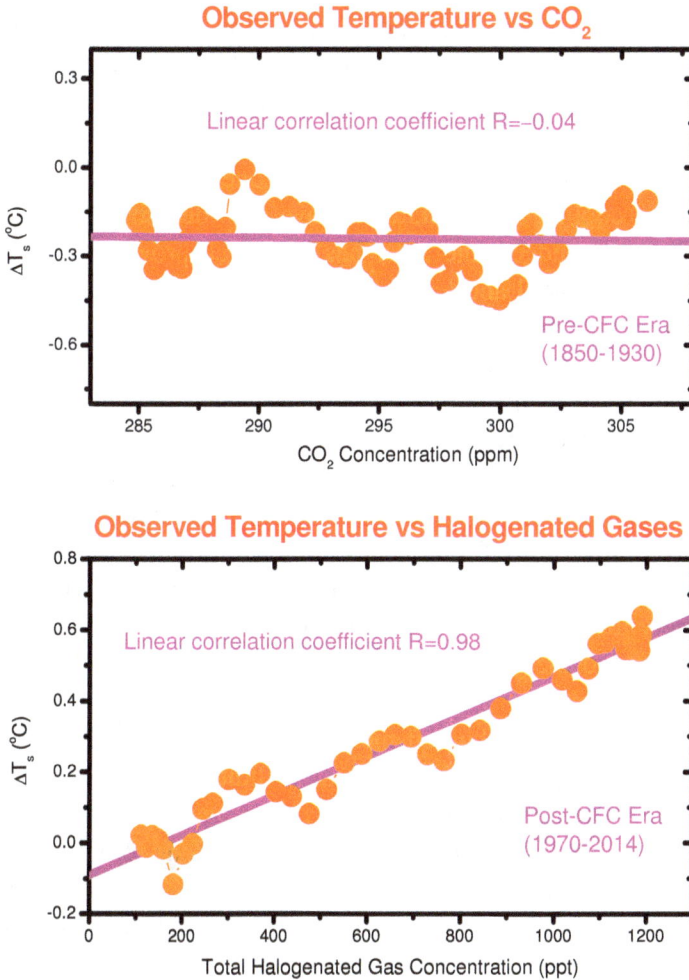

Fig. 9.2. Upper panel: Observed global surface temperature change (ΔT_s) vs CO_2 concentration for the pre-CFC era (1850-1930), a linear fit to the observed data gives a nearly zero linear correlation coefficient R=−0.04. Lower panel: Observed global surface temperature change (ΔT_s) vs total concentration of halogen-containing greenhouse gases for the post-CFC era (1970-2014), a linear fit to the observed data gives a nearly perfect linear correlation coefficient of R=0.98. Summarized from Fig. 8.3. *These observations, together with those in Fig. 9.1, show strong evidence that CO_2 has had a zero effect in modern climate change, whereas halogen-containing greenhouse gases (mainly CFCs) have exhibited a nearly perfect and complete control of global surface temperature.*

the CR intensity in the year, and k a constant {$[CFC]=[C]$ and $CR=I$ specifically in Fig. 9.1 (upper panel)}. The CRE mechanism can cause ozone depletion in both the winter and the springtime polar stratosphere.

The quantitative and statistical analyses of comprehensive datasets of halogenated molecules, CRs, total ozone and O_3-loss-induced stratospheric cooling over Antarctica have provided strong evidence of the CRE theory. Indeed, the latter has been well proven by substantial observations of spatial and temporal variations of CRs, CFCs, and total ozone. In particular, the CRE theory *with only one parameter* has predicted and well reproduced 11-year cyclic variations of the Antarctic O_3 hole and associated stratospheric cooling, and significantly improved our predictive capabilities for future polar ozone change.

A clear and steady recovery in the summer total ozone over Antarctica that started around 1995 has been observed. Consistently, after the removal of the CR effect, a pronounced ozone recovery in the Antarctic O_3 hole since ~1995 has been clearly discovered, while no sign in recovery of O_3 loss in mid-latitudes has been observed. These observations and the analyses enabled by the CRE theory have clearly unraveled the sensitive and quick response of the polar O_3 hole to the decline in total halogen burden measured in the low troposphere since 1994. This result has not only shown the validity of the CRE theory but placed the Montreal Protocol on a much firmer scientific ground.

Observed data have also shown that the long-term temperature variation in the lower Antarctic stratosphere is nearly completely controlled by the CRE equation with the equivalent effective chlorine level and CR intensity in the polar stratosphere as only two variables. Thus, the observed data have shown no sign in greenhouse effect of increasing non-halogen gases (CO_2, CH_4, N_2O) on the stratospheric climate of Antarctica over the past five to six decades.

Various climate models including AOGCMs have well established that the equilibrium surface temperature has a linear relationship with the radiative forcing. Radiative forcings ΔF calculated by the analytical (functional) expressions used in IPCC Reports are nearly identical to those given by the sophisticate GCMs. However, there exist large discrepancies between modeled results and observations, indicating that the assumptions in current climate models, especially in radiative forcings of various greenhouse gases, are invalid.

Observations have also shown strong evidence that natural climate drivers have played a negligibly small role in long-term climate change since 1950. Research on the potential climate impacts of solar variability should be focused on reliable measurements available after 1978. Based on the well-observed surface temperature variation during 11-year solar cycles, a total climate (feedback) amplification factor $\beta=1.53$ and a solar equilibrium climate sensitivity factor $\lambda_c^s=0.46$ K/(Wm^{-2}) have been determined. These results play an important role in quantifying human-made contribution to global climate change.

There are also substantial observations showing strong evidence that the greenhouse effect of non-halogenated gases (CO_2, CH_4 and N_2O) has been completely saturated. Both the polar stratospheric temperature and the global surface temperature have predominantly been controlled by the presence of anthropogenic halogenated greenhouse gases (mainly CFCs) in the atmosphere since 1950.

There are also very interesting observations that the global surface temperature exhibited zero correlation with atmospheric CO_2 prior to the industrial use of CFCs starting in 1930 with a correlation coefficient R=−0.04. In striking contrast, the temperature has exhibited a nearly perfect linear correlation with the total amount of halogenated greenhouse gases since the significant presence of atmospheric CFCs in 1970 with an extremely high linear coefficient R=0.98 (unit means perfect) observed.

According to the refined physics of climate sensitivity factor and the satellite-measured transmittance spectrum of the atmosphere, an equilibrium climate sensitivity factor (λ_c^{halo}) value of 1.77 K/(Wm^{-2}) for halogenated greenhouse gases that have strong absorption in the atmospheric window has been obtained.

With the obtained λ_c^{halo} value, the present *no-parameter* theoretical calculations surprisingly well reproduce the observed data of global mean surface temperature since 1950, including the observed reversal in global surface temperature over the past 16 years. Thus, a continued global cooling trend is expected in the coming decades under the international regulations to phase out CFCs, HCFCs, and HFCs.

In fact, the present calculated global mean surface temperatures since 2009 have agreed with the 3-year mean observed data excellently (within 10% uncertainty). For example, with respective to the mean temperature in 1950-1975, the predicted temperature was 0.620 °C for 2014 [Lu, 2013], while the observed temperature was indeed 0.623 °C for 2014, according to the UK Met office HadCRUT4 dataset.

The substantial and robust observations and the present calculations have shown solid and convincing evidence that humans are mainly responsible for significant climate change since the 1950s. The results have strongly strengthened the conclusion that CFCs are the major culprit not only for ozone depletion but also for global warming by ~0.6 °C during 1970~2002. The successful execution of the Montreal Protocol has shown its effectiveness in controlling the ozone hole in the polar regions and in reversing the global warming trend. It is therefore important to phase out all halogenated GHGs on the globe.

The observed data and reliable analyses have strongly shown that the warming effect of CO_2 and other non-halogen greenhouse gases has most likely been completely saturated, and that halogenated gases (mainly CFCs and HCFCs) were responsible for the rapid global warming observed in the late 20th century. The long-term global cooling that started around 2002 is expected to continue for the next five to seven decades. This could be good news for humans. But we have to continue our efforts to phase out the production and use of all halogenated greenhouse gases including CFCs, HCFCs, HFCs, PFCs, *etc.*

Researchers, policy-makers and the public should be aware of the fact that the IPCC-used climate models do not agree with the key observations of global surface temperatures not only for the period of the past 16 years but also for the period of 1850-1930 and of Antarctic stratospheric temperatures since the 1950s as well. From these reliable observations, *zero effects of rising CO_2* have been observed.

The new and succinct physics model for modern global warming reviewed in this book is far more comprehensible than the state-of-the-art three-dimensional general circulation models. It provides far more reliable and convincing results in excellent agreement with observations. The physics model has indeed great simplicity and predictive capability, and is therefore the one closer to the truth. The results and conclusions may be of great societal and global importance not only to the research community but also to the general public and the policy makers.

Since June 2013, the US, China and the European Union have subsequently reached agreements to go beyond the original Montreal Protocol for further control of the production and use of HFCs. There is no doubt that this is a correct step to continuously reverse the climate change humans have caused.

9.2 Jeopardizing politicized science

In the first and original sense, ozone depletion and climate change are two major scientific problems. They should first be treated more properly in a scientific way rather than a political approach. Although great global concerns about our environment and climate change are understandable and we should make our sustained efforts to protect the Earth, we should first make sure that our scientific understandings of the important issues are correct. An incorrect scientific understanding can mislead the public and the policy makers and lead to unwise decisions and policies. The latter might misuse our resources and jeopardize our industry and economics.

Furthermore, the over or improper interference and impacts on researchers from political parties can endanger the established and necessary procedures in solving a scientific problem.

Funding has been vital to research and has become increasingly competitive. On the other hand, researchers should also set the first priority to seek the truths and report real scientific data and findings, independent of political demanding and/or personal interests. In the long run, the truth prevails, and the trust from the public rewards.

9.3 What should we do first?

There are clear discrepancies between observations and atmospheric and climate models used by the WMO and IPCC. The conventional models have little success in reproducing the past ozone and climate changes and predicting future changes. Researchers, policy-makers and the public should be aware of this fact. An open mind about different models is required to access the impacts of human activities and atmospheric/climate change, and make decisions about future policies. Invalid assumptions in any model that disagree with solid observations should be abandoned. This is how science advances.

As far as the scientific evidence is concerned, there is no basis in claiming that CO_2 would be the main culprit of modern global warming observed in the period of 1970-2002.

It is most likely that researchers will have to make significant revisions of conventional (current-generation) atmospheric and climate models in order to better describe the ozone hole and climate change and more reliably and realistically predict future changes.

In approaching the big environmental and climate problems of major global and societal significance, the right strategy should perhaps be that science is science, and politics is politics. The real truth should be discovered by science first, and only afterwards should politics help to fix the problems. There has been some success in the Montreal Protocol, but it still has to be placed on a firmer scientific basis. The Kyoto Protocol is likely in a much more awkward situation for lack of scientific rationale that can stand from examinations by observations.

Bibliography

Abouaf, R., Paineau, R., and Fiquet-Fayard, F. (1976). Dissociative attachment in NO_2 and CO_2. *J. Phys. B* **9**, 3030-314.

Akbulut, M., Sack, N. J., and Madey, T. E. (1997). Elastic and inelastic process in the interaction of 1–10 eV ions with solids: ion transport through surface layers. *Surf. Sci. Rep.* **28**, 177-245.

Alge, E., Adams, N. G., and Smith, D. (1984). Rate coefficients for the attachment reactions of electrons with $c\text{-}C_7F_{14}$, CH_3Br, CF_3Br, CH_2Br_2, and CH_3I determined between 200 and 600 K using the FALP technique. *J. Phys. B: At. Mol. Phys.* **17**, 3827-3833.

Anderson, J. G., Toohey, D. W., and Brune, W. H. (1991). Free radicals within the Antarctic vortex: The role of CFCs in Antarctic ozone loss. *Science* **251**, 39-46.

Anderson, J. G., Dykema, J. A., Goody, R. M., Hua, H., and Kirk-Davidoff, D. B. (2004). Absolute, spectrally-resolved, thermal radiance: a benchmark for climate monitoring from space. *J. Quant. Spectrosc. & Radiat. Transfer* **85**, 367-383.

Andrews, D. G. (2000). *An Introduction to Atmospheric Physics*. (Cambridge University Press, Cambridge, UK).

Andronova, N. G., and Schlesinger, M. E. (2001). Objective estimation of the probability density function for climate sensitivity. *J. Geophys. Res.* **106**, 22605-22612.

Armbruster, M., Haberland, H., and Schindler, H.-G. (1981). Negatively Charged Water Clusters, or the First Observation of Free Hydrated Electrons. *Phys. Rev. Lett.* **47**, 323-326.

Austin, J., Butchart, N., and Shine, K. P. (1992). Possibility of an arctic ozone hole in a doubled-CO_2 climate. *Nature* **360**, 221-225.

Baletto, F., Cavazzoni, C., and Scandolo, S. (2005). Surface Trapped Excess Electrons on Ice. *Phys. Rev. Lett.* **95**, 176801(1-4).

Ball, W. T., Unruh, Y. C., Krivova, N. A., Solanki, S., Wenzler, T., Mortlock, D. J., and Jaffe, A. H. (2012). Reconstruction of total solar irradiance 1974–2009. *Astron. & Astrophys.* **541**, A27(1-15).

Bansal, K. M., and Fessenden, R. W. (1972). Electron disappearance in pulse irradiated CH_3Cl, C_2H_5Cl, CH_3Br, and C_2H_5Br. E. CH_3Cl, etc. *Chem. Phys. Lett.* **15**, 21-23.

Bass, A. D., Gamache, J., Parenteau, L., and Sanche, L. (1995) Absolute Cross Section for Dissociative Electron Attachment to CF_4 Condensed onto Multilayer Krypton. *J. Phys. Chem.* **99**, 6123-6127.

Bass, A. D., Gamache, J., Ayotte, P., and Sanche, L. (1996) Charge stabilization by chloromethane molecules on multilayer Kr films. *J. Chem. Phys.* **104**, 4258-4266.

Bates, D. R. and Nicolet, N. (1950). Atmospheric hydrogen. *Publ. Astron. Soc. Pacific* **62**, 106-110.

Becker, G., Müller, R., McKenna, D. S., Rex, M., and Carslaw, K. S. (1998). Ozone loss rates in the Arctic stratosphere in the winter 1991/92: Model calculations compared with Match results. *Geophys. Res. Lett.* **25**, 4325-4328.

Bertin, M., Meyer, M., Stähler, J., Gahl, C., Wolf, M., and Bovensiepen, U. (2009). Reactivity of water–electron complexes on crystalline ice surfaces. *Faraday Discuss.* **141**, 293-307.

Bhattacharya, S. K., Finn, J. M., Diep, V. P., Baletto, F., and Scandolo, S. (2010). CCl_4 dissociation on the ice Ih surface: an excess electron mediated process. *Phys. Chem. Chem. Phys.* **12**, 13034-13036.

Bloch, J., Mihaychuk, J. G., and van Driel, H. M. (1996). Electron Photoinjection from Silicon to Ultrathin SiO_2 Films via Ambient Oxygen. *Phys. Rev. Lett.* **77**, 920-923.

Boag, J. W., and Hart, E. J. (1963). Absorption Spectra of "Hydrated" Electron. *Nature* **197**, 45-47.

Böttcher, A., and Gießel, T. (1998). Dissociative chemisorption of N_2O molecules on Cs layers monitored via exoelectron emission. *Surf. Sci.* **408**, 212-222.

Bovensiepen, U., Gahl, C., Stähler, J., Bockstedte, M., Meyer, M., Baletto, F., Scandolo, S., Zhu, X.-Y., Rubio, A., and Wolf, M. (2009). A Dynamic Landscape from Femtoseconds to Minutes for Excess Electrons at Ice-Metal Interfaces. *J. Phys. Chem. C* **113**, 979–988.

Brasseur, G. P., Orlando, J. J., and Tyndall G. S. (Ed.) (1999). *Atmospheric Chemistry and Global Change*. (Oxford University Press, New York).

Brasseur, G. P., Orlando, J. J., and Tyndall G. S. (1999). Middle atmospheric ozone. In: G. P. Brasseur, J. J. Orlando and G. S. Tyndall (Ed.), *Atmospheric Chemistry and Global Change*. (Oxford University Press, New York) Chap. 14.

Brüning, F., Tegeder, P., Langer, J., and Illenberger, E. (2000). Low energy (0–14 eV) electron impact to CHF_2Cl at different phase conditions: Medium enhanced desorption of anions. *Int. J. Mass. Spectrom.* **195/196**, 507-516 .

Burch, J. L. (2001). The fury of space storms. *Sci. Am.* **284**, 86-94.

Burns, S. J., Matthews, J. M., and McFadden, D. L. (1996). Rate coefficients for dissociative electron attachment by halomethane compounds between 300 and 800 K. E, CF_2Cl_2, CCl_4, $CFCl_3$, CF_3Cl, $CHCl_3$, CH_2Cl_2, $CHFCl_2$, CH_3I, CH_2Br_2. *J. Phys. Chem.* **100**, 19436-19440.

Buxton, G. V., Greenstock, C. L., Helman, W. P., and Ross, A. B. (1988). Critical Review of Rate Constants for Reactions of Hydrated Electrons, Hydrogen Atoms and Hydroxyl Radicals (OH$^{\bullet}$/O$^{-\bullet}$) in Aqueous Solution. *J. Phys. Chem. Ref. Data* **17**, 513-886.

Caillon, N., Severinghaus, J. P., Jouzel, J., Barnola, J.-M., Kang, J., Lipenkov, V. Y. (2003). Timing of Atmospheric CO_2 and Antarctic Temperature Changes Across Termination III. *Science* **299**, 1728-1731.

Caplan, P. J., Poindexter, E. H., and Morrison, S. R. (1982). Ultraviolet bleaching and regeneration of Si≡Si$_3$ centers at the Si/SiO$_2$ interface of thinly oxidized silicon wafers. *J. Appl. Phys.* 53, 541-545.

Carlisle, R. (2004). *Scientific American Inventions and Discoveries.* (John Wiley & Sons, Inc., New Jersey) pp. 351.

Carslaw, K. S., Harrison, R. G., and Kirkby, J. (2002). Cosmic rays, clouds, and climate. *Science*, **298**, 1732-1737.

Chakarov, D., and Kasemo, B. (1998). Photoinduced Crystallization of Amorphous Ice Films on Graphite. *Phys. Rev. Lett.* **81**, 5181-5184.

Chaler, R., Vilanova, R., Santiago-Silva, M., Fernandez, P., and Grimalt, J. O. (1998) Enhanced sensitivity in the analysis of trace organochlorine compounds by negative-ion mass spectrometry with ammonia as reagent gas. *J. Chromatogr.* A**823**, 73-79.

Chapman, S. (1930). On ozone and atomic oxygen in the upper atmosphere. *Phil. Mag.* **10**, 369-383.

Christodoulides, A. A., Schumacher, R., and Schindler, R. N. (1975). Studies by the electron cyclotron resonance technique. X. Interactions of thermal-energy electrons with molecules of chlorine, hydrogen chloride, and methyl chloride. E, CH_3Cl, Cl_2, HCl; 293 - 513 K. *J. Phys. Chem.* **91**, 1904-1911.

Christodoulides, A. A., Schumacher, R., and Schindler, R. N. (1978). Studies by the electron cyclotron resonance (ECR) technique. XII. Interactions of thermal-energy electrons with the molecules CHF_3, $CHClF_2$, and $CHCl_2F$. *Int. J. Chem. Kinet.* **10**, 1215-1223.

Christophorou, L. G. (1976). Electron attachment to molecules in dense gases ("quasi-liquids"). *Chem. Rev.* **76**, 409-423.

Christophorou, L. G., Compton, R. N., and Dickson, H. W. (1968). Dissociative Electron Attachment to Hydrogen Halides and their Deuterated Analogs. *J. Chem. Phys.* **48**, 1949-1955.

Christophorou, L. G., McCorkle, D. L., and Christodoulides, A. A. (1984). In *Electron-Molecule Interactions and Their Applications*, Vol. 1, Christophorou, L. G. (Ed.) (Academic Press, Orlando) Chap. 6.

Christophorou, L. G., Compton, R. N., Hurst, G. S., and Reinhardt, P. W. (1965). Determination of Electron- Capture Cross Sections with Swarm-Beam Techniques. *J. Chem. Phys.* **43**, 4273-4281.

Chu, S. C. and Burrow, P. D. (1990). Dissociative attachment of electrons in the chloromethanes. *Chem. Phys. Lett.* **172**, 17-22.

Chung, C. E., Ramanathan, V., Kim, D., and Podgorny, I. A. (2005). Global Anthropogenic Aerosol Direct Forcing Derived from Satellite and Ground-Based Observations. *J. Geophys. Res.*, **110**, D24207, doi:10.1029/2005JD006356.

Chutjian, A., Garscadden, A., and Wadehra, J. M. (1996). Electron attachment to molecules at low electron energies. *Phys. Rep.* **264**, 393-470.

Clark, R. N. (1999). Chapter 1: Spectroscopy of Rocks and Minerals, and Principles of Spectroscopy. In *Manual of Remote Sensing*, Vol. 3, *Remote Sensing for the Earth Sciences* (Rencz, A. N., Ed.). (John Wiley and Sons, New York)pp.3-58.

Clerbaux, C., Hadji-Lazaro, J., Turquety, S., Mégie, G., and Coheur P.-F. (2003). Trace gas measurements from infrared satellite for chemistry and climate applications. *Atmos. Chem. Phys. Discuss.* **3**, 1495-1508.

Coe, J. V., Lee, G. H., Eaton, J. G., Arnold, S. T., Sarkas, H. W., Ludewigt, C., Haberland, H., Worsnop, D. R., and Bowen, K. H. (1990). Photoelectron Spectroscopy of Hydrated Electron Cluster Anions, $(H_2O)^-n=2-69$. *J. Chem. Phys.* **92**, 3980-3982.

Cole, Jr., R. K., and Pierce, E. T. (1965). Electrification in the Earth's atmosphere for altitudes between 0 and 100 kilometers. *J. Geophys. Res.* **70**, 2735-2749.

Crutzen, P. J. (1970). The influence of nitrogen oxide on the atmospheric ozone content. *Quart. J. Roy. Met. Soc.* **96**, 320-325.

Crutzen, P. J., and Arnold, F. (1986). Nitric acid cloud formation in the cold Antarctic stratosphere: a major cause for the springtime 'ozone hole'. *Nature* **324**, 651-655.

Crutzen, P. J., Isaksen, S. A., and Reid, G. C. (1975). Solar Proton Events: Stratospheric Sources of Nitric Oxide. *Science* **189**, 457-459.

Curran, R. K. (1961). Positive and Negative Ion Formation in CCl_3F. *J. Chem. Phys.* **34**, 2007-2010.

Currie, J. G., and Kallio, H. (1993). Triacylglycerols of human milk: Rapid analysis by ammonia negative ion tandem mass spectrometry. *Lipids* **28**, 217-222.

DeLand, M. T., and Cebula, R. P. (2008). Creation of a composite solar ultraviolet irradiance dataset. *J.Geophys. Res.* **113**, A11103.

Demuth, J. E., Schmeisser, D., and Avouris, Ph. (1981). Resonance Scattering of Electrons from N_2, CO, O_2, and H_2 Adsorbed on a Silver Surface. *Phys. Rev. Lett.* **47**, 1166-1169.

Dewitte, S, Crommelynck, D., Mekaoui, S., and Joukoff, A. (2004). Merasurement and Uncertainty of the long-term Total Solar Irradiance Trend. *Sol. Phys.* **224**, 209-216.

Dispert, H., and Lacmann, K. (1978). Negative Ion Formation in Collisions between Potassium and Fluoro- and Chloromethanes: Electron Affinities and Bond Dissociation Energies. *Int. J. Mass. Spectrom. Ion Phys.* **28**, 49-67.

Dixon-Warren, St.-J., Jensen, E. T., and Polanyi, J. C. (1991). Direct evidence for charge-transfer photodissociation at a metal surface: CCl_4/Ag(111). *Phys. Rev. Lett.* **67**, 2395-2398.

Dixon-Warren, St.-J., Jensen, E. T., and Polanyi, J. C. (1993). Photochemistry of adsorbed molecules. XI. Charge-transfer photodissociation and photoreaction in chloromethanes on Ag(111). *J. Chem. Phys.* **98**, 5938-5953.

Djamo, V., Teillet-Billy, D., and Gauyacq, J. P. (1993). Resonant vibrational excitation of adsorbed molecules by electron impact. *Phys. Rev. Lett.* **71**, 3267-3270.

Dobson, G. M. B. (1968). Forty Years' Research on Atmospheric Ozone at Oxford: a History. *App. Opt.* **7**, 387-405.

Douglass, D. H., and Clader, B. D. (2002), Climate sensitivity of the Earth to solar irradiance, *Geophys. Res. Lett.* **29**, 1786.

Douglass, D. H., Pearson, B. D., and Singer, S. F. (2004). Altitude dependence of atmospheric temperature trends: Climate models versus observation. *Geophys. Res. Lett.* **31**, L13208.

Dufresne, J.-L., and Bony, S. (2008). An Assessment of the Primary Sources of Spread of Global Warming Estimates from Coupled Atmosphere–Ocean Models. *J. Climate* **21**, 5135-5144.

Eddy, J. A. (1976). The Maunder minimum. *Science* **192,** 1189–1202.

Estrada, F., Perron, P., and Martínez-López, B. (2013). Statistically derived contributions of diverse human influences to twentieth-century temperature changes. *Nat. Geosci.* **6**, 1050-1055.

Fabrikant, I. I. (1990). Resonance processes in e-HCl collisions: Comparison of the R-matrix and the nonlocal-complex-potential methods. *Comments At. Mol. Phys.* **24**, 37–52.

Fabrikant, I. I. (1994). Semiempirical calculations of inelastic electron-methylchloride scattering. *J. Phys. B: At. Mol. Opt. Phys.* **27.** 4325-4336.

Fabrikant, I. I. (2007). Dissociative Electron Attachment on Surfaces and in Bulk Media. *Phys. Rev. A* **76**, 012902.

Fabrikant, I. I., Caprasecca, S., Gallup, G. A., and Gorfinkiel, J. D. (2012). Electron attachment to molecules in a cluster environment. *J. Chem. Phys.* **136**, 184301.

Fabrikant, I. I., Nagesha, K., Wilde, R., and Sanche, L. (1997). Dissociative electron attachment to CH_3Cl embedded into solid Krypton. *Phys. Rev. B* **56**, R5725.

Faradzhev, N. S., Perry, C. C., Kusmierek, D. O., Fairbrother, D. H. and Madey, T. E. (2004). Kinetics of electron-induced decomposition of CF_2Cl_2 coadsorbed with water (ice): a comparison with CCl_4. *J. Chem. Phys.* **121**, 8547-8561.

Farman, J. C., Gardiner, B. G., and Shanklin, J. D. (1985). Large losses of total ozone in Antarctica reveal seasonal ClO_x/NO_x interaction. *Nature* **315**, 207-210.

Fedor, J., May, O., and Allan, M. (2008). Absolute cross sections for dissociative electron attachment to HCl, HBr, and their deuterated analogs. *Phys. Rev. A* **78**, 032701.

Ferguson, E. E., Fehsenfeld, F. C., and Albritton, D. L. (1979). Ion chemistry of the earth's atmosphere. In *Gas Phase Ion Chemistry*, Vol. 1., Bowers, M. T. (Ed.). (Academic Press Inc., New York) pp.45-82.

Fehsenfeld, F. C., Crutzen, P. J., Schmeltekopf, A. L., Howard, C. J., Albritton, D. L., and Ferguson, E. E. (1976). Ion chemistry of chlorine compounds in the troposphere and stratosphere. *J. Geophys. Res.* **81**, 4454-4460.

Fischer, H., Wahlen, M., Smith, J., Mastroianni, D., and Deck, B. (1999). Ice Core Records of Atmospheric CO_2 Around the Last Three Glacial Terminations. *Science* **283**, 1712-1714.

Fisher, D. A., Hales, C. H., Wang, W.-C., Ko, M. K. W., and Sze, N. D. (1990). Model calculations of the relative effects of CFCs and their replacements on global warming. *Nature* **344**, 513-516.

Foltz, G. W., Latimer, C. J., Hildebrandt, G. F., Kellert, F. G., Smith, K. A., West, W. P., Dunning, F. B., and Stebbings, R. F. (1977). *J. Chem. Phys.* **67**, 1352-1359.

Forest, C. E., Stone, P. H., Sokolov, A. P., Allen, M. R., and Webster, M. D. (2002). Quantifying uncertainties in climate system properties with the use of recent observations. *Science* **295**, 113–117.

Forster, P.M. De F., and Joshi, M. (2005). The role of halocarbons in the climate change of the troposphere and stratosphere. *Climate Change* **71**, 249-266.

Foukal, P., Fröhlich, C., Spruit, H., and Wigley, T. M. L. (2006). Variations in solar luminosity and their effect on the Earth's climate. *Nature* **443**, 161-166.

Friis-Christensen, E., and Lassen, K. (1991). Length of the Solar Cycle: An Indicator of Solar Activity Closely Associated with Climate. *Science* **254**, 698–700.

Fröhlich, C. (2006). Solar Irradiance Variability since 1978: Revision of the {PMOD} Composite during Solar Cycle 21. *Space Sci. Rev.* **125**, 53–65.

Fröhlich, C. (2009). Evidence of a long-term trend in total solar irradiance. *Astron. & Astrophys.* **501**, L27–L30

Fröhlich, C. (2012). Total solar irradiance observations. *Surv. Geophys.* **33**, 453–473.

Fröhlich, C., and Lean, J. (1998). The Sun's total irradiance: Cycles and trends in the past two decades and associated climate change uncertainties. *Geophys. Res. Let.* **25**, 4377-4380.

Frolicher, T. L., Joos, F., and Raible, C. C. (2011). Sensitivity of atmospheric CO_2 and climate to explosive volcanic eruptions. *Biogeosciences* **8**, 2317-2339.

Fyfe, J. C., Gillett, N. P., and Zwiers, F. W. (2013). Overestimated global warming over the past 20 years. *Nat. Clim. Change* **3**, 767-768.

Gahl, C., Bovensiepen, U., Frischkorn, C., and Wolf, M. (2002). Ultrafast Dynamics of Electron Localization and Solvation in Ice Layers on Cu(111). *Phys. Rev. Lett.* **89** 107402.

Gast, P. R. (1961). Thermal Radiation, in Handbook of Geophysics, US Air Force Research Division, 2nd Ed. (Macmillian Publishing Co., New York).

Gilton, T. L., Dehnbostel, C. P., and Cowin, J. P. (1989). Electron transmission through layers of H_2O and Xe in the ultrahigh vacuum photoreduction of CH_3Cl on Ni(111). *J. Chem. Phys.* **91**, 1937.

Girardet, C., and Toubin, C. (2001). Molecular atmospheric pollutant adsorption on ice: a theoretical survey. *Surf. Sci. Rep.* **44**, 159-238.

Gole, J. L., and Zare, R. N. (1972). Determination of D_0^0(AlO) from Crossed-Beam Chemiluminescence of Al + O_3. *J. Chem. Phys.* **57**, 5331-5335.

Goody, R., Anderson, J., and North, G. (1998). Testing climate models: An approach. *Bull. Am. Meteorol. Soc.* **79**, 2541-2549.

Gregory, J. M., Stouffer, R. J., Raper, S. C. B., Stott, P. A., and Rayner, N. A. (2002). An observationally based estimate of the climate sensitivity. *J. Clim.* **15**, 3117-3121.

Griggs, J. A., and Harries, J. E. (2007). Comparison of Spectrally Resolved Outgoing Longwave Radiation over the Tropical Pacific between 1970 and 2003 Using IRIS, IMG, and AIRS. *J. Climate* **20**, 3982-4001.

Grooß, J.-W., and Müller, R. (2011). Do cosmic-ray-driven electron-induced reactions impact stratospheric ozone depletion and global climate change? *Atmos. Environ.* **45**, 3508-3514.

Grooß, J.-U., and Müller, R. (2013). Corrigendum to "Do cosmic-ray-driven electroninduced reactions impact stratospheric ozone depletion and global climate change?" [Atmos. Environ. 45 (2011), 3408–3514]. *Atmos. Environ.* **68**, 350.

Hanel, R. A. *et al.* (1972). The Nimbus4 Infrared Spectroscopy Experiment: 1. Calibrated Thermal Emission Spectra. *J. Geophys. Res.* **77**, 2629-2641.

Harries, J. E., Brindley, H. E., Sagoo, P. J., and Bantges, R. J. (2001). Increases in greenhouse forcing inferred from the outgoing longwave radiation spectra of the Earth in 1970 and 1997. *Nature* **410**, 355-357.

Harris, N. R. P., Farman, J. C., and Fahey, D. W. (2002). Comment on 'Effects of Cosmic Rays on Atmospheric Chlorofluorocarbon Dissociation and Ozone Depletion'. *Phys. Rev. Lett.* **89,** 219801.

Hart, E, and Anbar, M (1970). *The Hydrated Electron.* (John Wiley & Sons, Inc, New York).

Hayakawa, S. (1969). *Cosmic Ray Physics.* (Wiley-Interscience, New York) pp.8.

Hegerl, G. C., and Zwiers, F.W. (2011). Use of models in detection and attribution of climate change. Wiley Interdisc. Rev.: Clim. Change **2**, 570-591.

Hickam, W. M., and Berg, D. (1958). Negative Ion Formation and Electric Breakdown in Some Halogenated Gases. *J. Chem. Phys.* **29**, 517-523.

Horowitz, Y., and Asscher, M. (2012). Low energy charged particles interacting with amorphous solid water layers. *J. Chem. Phys.* **136**, 134701.

Hoyt, D. V. and Schatten, K. H. (1993). A Discussion of Plausible Solar Irradiance Variations, 1700-1992. *J. Geophys. Res.* **98**, 18895 -18906.

Huels, M. A., Parenteau, L., and Sanche, L. (1994). Substrate dependence of electron-stimulated O^- yields from dissociative electron attachment to physisorbed O_2. *J. Chem. Phys.* **100**, 3940-3956.

Huetz, H., Gresteau, F., ad Mazeau, J. (1980). 'Dissociative attachment' in N_2. *J. Phys. B.* **13**, 3275-3284.

Idso, S. B. (1980). The climatological significance of a doubling of earth's atmospheric carbon dioxide concentration. *Science* **207**, 1462-1463.

Idso, S. B. (1998). CO_2-induced global warming: a skeptic's view of potential climate change. *Clim. Res.* **10**, 69-82.

Illenberger, E. (1982). Energetics of negative ion formation in dissociative electron attachment to CCl_4, $CFCl_3$, CF_2Cl_2 and CF_3Cl. E, CF_2Cl_2, CCl_4, $CFCl_3$, CF_3Cl; 0 – 10 eV. *Ber. Bunsenges. Phys. Chem.* **86**, 252-261.

Illenberger, E. (1992). Electron-attachment reactions in molecular clusters. *Chem. Rev.* **92**, 1589-1609.

Illenberger, E., Scheunemann, H.-U., and Baumgärtel, H. (1979). Negative-ion formation in CF_2Cl_2, CF_3Cl and $CFCl_3$ following low-energy (0-10 eV) impact with near monoenergetic electrons. *Chem. Phys.* **37**, 21-31.

IPCC (2001). Third Assessment Report: Climate Change (TAR): The Physical Science Basis. (Cambridge University Press, Cambridge, UK).

IPCC (2007). Fourth Assessment Report: Climate Change (AR4): The Physical Science Basis. (Cambridge University Press, Cambridge, UK).

IPCC (2013). *Fifth Assessment Report: Climate Change (AR5)*: The Physical Science Basis. (Cambridge University Press, Cambridge, UK) Chapters 8 and 9.

Jackman, C. H. (1991). Effects of energetic particles on minor constituents of the middle atmosphere. *J. Geomagnet. Geoelectri.* **43**, S637-646.

Jacob, D. J. (1999). *Introduction to Atmospheric Chemistry.* (Princeton University Press, Princeton) Chapter 7.

Jacobi, K., Bertolo, M., Geng, P., Hansen, W., and Astaldi C. (1990). H_2O-induced quenching of the negative-ion resonance scattering for N_2 physisorbed on Al(111). *Chem. Phys. Lett.* **173**, 97-102.

Jarvis, G. K., Mayhew, C. A., Singleton, L., and Spyrou, S. M. (1997). An investigation of electron attachment to $CHCl_2F$, $CHClF_2$ and CHF_3 using an electron-swarm mass spectrometric technique. *Int. J. Mass Spectrom. Ion Proc.* **164**, 207-223.

Johnson, R. E. (1990) Energetic Charge-Particles Interactions with Atmospheres and Surfaces, Physics and Chemistry in Space Planetology, Vol. 19 (Springer-Verlag, Berlin).

Johnston, H. S. (1971). Reduction of stratospheric ozone by nitrogen oxide catalysis from SST exhaust. *Science* **173**, 517-522.

Kasatkina, E. A., and Shumilov, O. I. (2005). Cosmic ray-induced stratospheric aerosols: A possible connection to polar ozone depletions. *Ann. Geophys.* **23**, 675-679.

Kee, T. K., Son, D. H., Kambhampati, P., and Barbara, P. F. (2001). A unified electron transfer model for the different precursors and excited states of the hydrated electron. *J. Phys. Chem.* A**105**, 8434-8439.

Kerr, R. A. (2009). What Happened to Global Warming? Scientists Say Just Wait a Bit. *Science* **326**, 28-29.

Kidder, S. Q., and Vonder Haar, T. H. (1995). *Satellite Meteorology: An Introduction.* (Academic Press, San Diego).

Kirkby J. *et al.* (2011). Role of sulphuric acid, ammonia and galactic cosmic rays in atmospheric aerosol nucleation. *Nature* **476**, 429-433 .

Kiss, J., Lennon, D., Jo, S. K., and White, J. M. (1991). Photoinduced dissociation and desorption of nitrous oxide on a platinum(111) surface. *J. Phys. Chem.* **95**, 8054-8059.

Klar, D., Ruf, M.-W., Fabrikant, I. I., and Hotop, H. (2001). Dissociative electron attachment to dipolar molecules at low energies with meV resolution: $CFCl_3$, 1,1,1-$C_2Cl_3F_3$, and HI. *J. Phys. B* **34**, 3855-3878 .

Klots, C. E. (1976). Rate constants for unimolecular decomposition at threshold. *Chem. Phys. Lett.* **38**, 61-64.

Knight, J., Kennedy, J. J., Folland, C., Harris, G., Jones, G. S., Palmer, M., Parker, D., Scaife, A., and Stott, P. (2009). Do global temperature trends over the last decade falsify climate predictions? in *State of the Climate in 2008. Bull. Amer. Meteor. Soc.* **90**, S22-S23.

Klein, M. J. (1962). Max Planck and the beginnings of the quantum theory. *Archive for History of Exact Sciences* **1**, 459-479.

Kobayashi, H., Shimota, A., Kondo, K., Okumura, E., Kameda, Y., Shimoda, H., and Ogawa, T. (1999). Development and evaluation of the interferometric monitor for greenhouse gases: a high-throughput Fourier-transform infrared radiometer for nadir Earth observation. *Appl. Opt.* **38**, 6801-6807.

Kopp, G., and Lean, J. L. (2011). A new, lower value of total solar irradiance: evidence and climate significance. *Geophys. Res. Lett.* **38**, L01706.

Kosaka, Y., and Xie, S.-P. (2013). Recent global-warming hiatus tied to equatorial Pacific surface cooling. *Nature* **501**, 403-407.

Kragh, H. (1999). *Quantum Generations. A History of Physics in the Twentieth Century* (Princeton University Press, Princeton, New Jersey).

Kreuzer, H. J., and Wang, R. L. C. (1994). Physics and chemistry in high electric fields. *Philos. Mag. B* **69**, 945-955.

Krivova, N. A., Balmaceda, L., and Solanki, S. K. (2007). Reconstruction of solar total irradiance since 1700 from the surface magnetic flux. *Astron. & Astrophys.* **467**, 335–346.

Krivova, N. A., and Solanki, S. K. (2013). Models of Solar Total and Spectral Irradiance Variability of Relevance for Climate Studies. In *Climate and Weather of the Sun-Earth System (CAWSES)*, Lübken F.-J. (ed.) (Springer, Berlin) Chapter 2.

Krivova, N. A., Solanki, S. K., & Wenzler, T. (2009a). ACRIM-gap and total solar irradiance revisited: is there a secular trend between 1986 and 1996? *Geophys. Res. Lett.* **36**, L20101.

Krivova, N. A., Solanki, S. K., Wenzler, T., & Podlipnik, B. (2009b). Reconstruction of solar UV irradiance since 1974. *J. Geophys. Res.* **114**, D00I04.

Krivova, N. A., Vieira, L. E. A., and Solanki, S. K. (2010). Reconstruction of solar spectral irradiance since the Maunder minimum. *J. Geophys. Res.* **115**, A12112.

Laenen, R., Roth, T., and Laubereau, A. (2000). Novel Precursors of Solvated Electrons in Water: Evidence for a Charge Transfer Process. *Phys. Rev. Lett.* **85**, 50-53.

Langer, J., Matt, S., Meinke, M., Tegeder, P., Stamatovic, A., and Illenberger, E. (2000) Negative ion formation from low energy 0–15 eV electron impact to CF_2Cl_2 under different phase conditions. *J. Chem. Phys.* **113,** 11063-11070.

Le Garrec, J. L., Sidko, O., Queffelec, J. L., Hamon, S., Mitchell, J. B. A., and Rowe, B. R. (1997). Experimental studies of cold electron attachment to SF_6, CF_3Br, and CCl_2F_2. *J. Chem. Phys.* **107,** 54-63.

Lean, J. L., and Rind, D. H. (2008). How natural and anthropogenic influences alter global and regional surface temperatures: 1889 to 2006. *Geophys. Res. Lett.* **35,** L18701(1-6).

Lean, J., Skumanich, A., and White, O. (1992). Estimating the sun's radiative output during the Maunder minimum. *Geophys. Res. Lett.* **19,** 1595–1598.

LeDourneuf, M., Schneider, B. I., and Burke, P. G. (1979). Theory of vibrational excitation and dissociative attachment: an R-matrix approach. *J. Phys. B.* **12,** L365-L370.

Lehnert, S. (2008) *Biomolecular Action of Ionizing Radiation* (Taylor & Francis Ltd.) pp.14.

Levine, R. D., and Bernstein, R. B. (1987). *Molecular Reaction Dynamics and Chemical Reactivity* (Oxford University Press, Oxford) pp. 134.

Lewerenz, M., Nestmann, B., Bruna, P. J., and Peyerimhoff, S. D. (1985). The electronic-spectrum, photodecomposition and dissociative electron-attachment of CF_2Cl_2 - an abinitio configuration-interaction study. *Theochem-J. Mol. Struct.* **24,** 329-342.

Lindzen, R. S. (1997). Can increasing carbon dioxide cause climate change? *Proc. Natl. Acad. Sci. USA* **94,** 8335-8342.

Liu, H. (2012). *Electron-induced CCl_4 Adsorption on Ice*. MSc. Thesis, University of Waterloo, Waterloo, Ontario, Canada.

Lockwood, M., and Fröhlich, C. (2007). Recent oppositely directed trends in solar climate forcings and the global mean surface air temperature. *Proc. R. Soc.* A**463,** 2447-2460.

Long, F. H., Lu, H., and Eisenthal, K. B. (1990). Femtosecond studies of the presolvated electron--an excited-state of the solvated electron. *Phys. Rev. Lett.* **64,** 1469-1472.

Lu, Q.-B. (2009). Correlation between Cosmic Rays and Ozone Depletion. *Phys. Rev. Lett.* **102,** 118501 (1-4).

Lu, Q.-B. (2010a). Cosmic-Ray-Driven Electron-Induced Reactions of Halogenated Molecules Adsorbed on Ice Surfaces: Implications for Atmospheric Ozone Depletion and Global Climate Change. *Phys. Rep.* 487, 141-167(published online in 2009).

Lu, Q.-B. (2010b). What is the Major Culprit for Global Warming: CFCs or CO_2? *J. Cosmology* **8,** 1846-1862.

Lu, Q.-B. (2010c). Effects of Ultrashort-Lived Prehydrated Electrons in Radiation Biology and Their Applications for Radiotherapy of Cancer, *Mutat. Res.-Rev. Mutat. Res.* **704,** 190-199.

Lu, Q.-B. (2012a). On Cosmic-Ray-Driven Electron Reaction Mechanism for Ozone Hole and Chlorofluorocarbon Mechanism for Global Climate Change. arXiv:1210.1498 [physics.ao-ph].

Lu, Q.-B. (2012b). Cosmic Rays, CFCs, Ozone Hole and Global Climate Change: Understandings from a Physicist. arXiv:1210.6844v1 [physics.ao-ph].

Lu, Q.-B. (2013). Cosmic-Ray-Driven Reaction and Greenhouse Effect of Halogenated Molecules: Culprits for Atmospheric Ozone Depletion and Global Climate Change. *Int. J. Mod. Phys. B27*, 1350073 (1-38).

Lu, Q.-B. (2014a). Reply to "Comment on 'Cosmic-ray-driven reaction and greenhouse effect of halogenated molecules: Culprits for atmospheric ozone depletion and global climate change' by Rolf Müller and Jens-Uwe Grooß". *Int. J. Mod. Phys. B* **28**, 1482002.

Lu, Q.-B. (2014b). Reply to "Comment on 'Cosmic-ray-driven reaction and greenhouse effect of halogenated molecules: Culprits for atmospheric ozone depletion and global climate change' by Dana Nuccitelli et al". *Int. J. Mod. Phys. B* **28**, 1482004.

Lu, Q.-B. (2014c). Author's reply to a Note on the Reply to Comment on "Cosmic-ray-driven reaction and greenhouse effect of halogenated molecules: Culprits for atmospheric ozone depletion and global climate change". *Int. J. Mod. Phys. B* **28**, 1475003.

Lu, Q.-B., Baskin, J. S., and Zewail, A. H. (2004). The Presolvated Electron in Water: Can It Be Scavenged at Long Range? *J. Phys. Chem. B108*, 10509-10514.

Lu, Q.-B., Bass, A., and Sanche, L. (2002). Superinelastic Electron Transfer: Electron Trapping in H_2O Ice via the $N_2*(^2\Pi_g)$ Resonance. *Phys. Rev. Lett.* **89**, 219804.

Lu, Q.-B., Ma, Z., and Madey, T. E. (1998) Negative-ion formation in electron-stimulated desorption of CF_2Cl_2 adsorbed on Ru(0001). *Phys. Rev. B58*, 16446-16454.

Lu, Q.-B., and Madey, T. E. (1999a). Negative-ion enhancements in electron-stimulated desorption of CF_2Cl_2 coadsorbed with nonpolar and polar gases on Ru(0001). *Phys. Rev. Lett.* **82**, 4122-4125.

Lu, Q.-B., and Madey, T. E. (1999b). Giant enhancement of electron-induced dissociation of chlorofluorocarbons coadsorbed with water or ammonia ices: Implications for the atmospheric ozone depletion. *J. Chem. Phys.* **111**, 2861-2864.

Lu, Q.-B., and Madey, T. E. (2000). Mechanism for giant Cl^- and F^- enhancements in electron-induced dissociation of CF_2Cl_2 coadsorbed with water or ammonia ices. *Surf. Sci.* **451**, 238 -342.

Lu, Q.-B., and Madey, T. E. (2001). Factors influencing Cl^- and F^- enhancements in electron-stimulated desorption of CF_2Cl_2 coadsorbed with other gases. *J. Phys. Chem. B105*, 2779-2784 .

Lu, Q.-B., and Madey, T. E., Parenteau, L., Weik, F., and Sanche, L. (2001). Structural and Temperature Effects on Cl^- Yields in Electron-Induced Dissociation of CF_2Cl_2 Adsorbed on Water Ice. *Chem. Phys. Lett.* **342**, 1-6.

Lu, Q.-B., and Sanche, L. (2001a). Effects of Cosmic Rays on Atmospheric Chlorofluorocarbon Dissociation and Ozone Depletion. *Phys. Rev. Lett.* **87**, 078501 (1-4).

Lu, Q.-B., and Sanche, L. (2001b). Enhanced Dissociative Electron Attachment to CF_2Cl_2 by Transfer of Electrons Localized in Preexisting Traps of Water and Ammonia Ice. *Phys. Rev.* **B63**, 153403 (1-4).

Lu, Q.-B., and Sanche, L. (2001c). Large Enhancement in Dissociative Electron Attachment to HCl adsorbed on H_2O Ice via Transfer of Presolvated Electrons. *J. Chem. Phys.* **115**, 5711-5714.

Lu, Q.-B., and Sanche, L. (2003). Condensed-Phase Effects on Absolute Cross Sections for Dissociative Electron Attachment to CFCs and HCFCs Adsorbed on Kr. *J. Chem. Phys.* **119**, 2658-2662.

Lu, Q.-B., and Sanche, L. (2004). Dissociative Electron Attachment to CF_4, CFCs and HCFCs adsorbed on H_2O Ice. *J. Chem. Phys.* **120**, 2434-2438.

Luo, C. W., Wang, Y. T., Chen, F. W., Shih, H. C., and Kobayashi, T. (2009). Eliminate coherence spike in reflection-type pump-probe measurements. *Opt. Express* **17**, 11321-11327.

Mackenzie, K., Kopinke, F.-D., and Remmler, M. (1996) Reductive destruction of halogenated hydrocarbons in liquids and solids with solvated electrons. *Chemosphere* **33**, 1495-1513.

Madden, R. A., and Ramanathan, V. (1980). Detecting Climate Change Due to Increasing CO_2 in the Atmosphere. *Science* **209**, 763-768.

Madey T. E., and Yates, Jr., J. T. (1971). Electron-Stimulated Desorption as a Tool for Studies of Chemisorption: A Review. *J. Vac. Sci. Technol.* **8**, 525-555.

Manney, G. L. *et al.* (2011). Unprecedented Arctic ozone loss in 2011. *Nature* **478**, 469-475.

Marsh, E. P., Gilton, T. L., Meier, W., Schneider, M. R., and Cowin, J. P. (1988). Electron-transfer–mediated and direct surface photochemistry: CH_3Cl on Ni(111). *Phys. Rev. Lett.* **61**, 2725-2728.

Marsh, N. D. and Svensmark, H. (2000). Low Cloud Properties Influenced by Cosmic Rays. *Phys. Rev. Lett. 85, 5004* .

Marsolais, R., Deschênes, M., and Sanche, L. (1989). Low-energy electron transmission method for measuring charge trapping in dielectric films. *Rev. Sci. Instrum.* **60**, 2724-2732.

Michaud, M., and Sanche, L. (1990). The $^2\Pi_g$ shape resonance of N_2 near a metal surface and in rare gas solids. *J. Electron. Spectrosc. Relat. Phenom.* **51**, 237-248.

Migus, A., Gauduel, Y., Martin, J. L., and Antonetti, A. (1987) Excess electron in liquid water: First evidence of a prehydrated state with femtosecond lifetime. *Phys. Rev. Lett.* **58**, 1559-1563.

Miller, T. M. (2003). Electron Affinities. In *CRC Handbrook of Chemistry and Physics*, 84th Ed, Lide, D. R. (Ed) (CRC Press, Boston) pp. 10-147 to 10-162

Molina, L. T., and Molina, M. J. (1987). Production of Cl_2O_2 from the self-reaction of the ClO radical. *J. Phys. Chem.* **91**, 433-436.

Molina, M. J., and Rowland, F. S. (1974) Stratospheric sink for chlorofluoromethanes: chlorine atomic-atalysed destruction of ozone. *Nature* **249**, 810-812.

Morice, C. P., Kennedy, J. J., Rayner, N. A., and Jones, P. D. (2012). Quantifying uncertainties in global and regional temperature change using an ensemble of observational estimates: The HadCRUT4 dataset. *J. Geophys. Res.* **117**, D08101.

Mothers, K. G., Schultes, E., and Schindler, R. N. (1972). Application of electron cyclotron resonance technique in studies of electron capture processes in the thermal energy range. E, CCl_2F_2, NF_3, CCl_4, SF_6, C_4F_8, HBr, etc. *J. Phys. Chem.* **76**, 3578-3764.

Müller, R. (2003). Impact of Cosmic Rays on Stratospheric Chlorine Chemistry and Ozone Depletion. *Phys. Rev. Lett.* **91**, 058502.

Müller, R. (2008). Comment on "Resonant dissociative electron transfer of the presolvated electron to CCl_4 in liquid: Direct observation and lifetime of the CCl_4^{*-} transition state'[*J. Chem. Phys.* **128**, 041102]". *J. Chem. Phys.* **129**, 027101.

Müller, R., and Grooß, J. U. (2009). Does cosmic-ray-induced heterogeneous chemistry influence stratospheric polar ozone loss? *Phys. Rev. Lett.* **103**, 228501(1-4).

Müller, R., and Grooß, J.-U. (2014a). Comment on "Cosmic-ray-driven reaction and greenhouse effect of halogenated molecules: Culprits for atmospheric ozone depletion and global climate change". *Int. J. Mod. Phys. B* **28**, 1482001.

Müller, R., and Grooß, J.-U. (2014b). A note on the 'Reply to 'Comment on "Cosmic-ray-driven reaction and greenhouse effect of halogenated molecules: Culprits for atmospheric ozone depletion and global climate change" by Rolf Müller and Jens-Uwe Grooß' by Q.-B. Lu'. *Int. J. Mod. Phys. B* **28**, 1475002.

Myhre, G., Highwood, E. J., Shine, K. P., and Stordal, F. (1998). New estimates of radiative forcing due to well mixed greenhouse gases. *Geophys. Res. Lett.* **25**, 2715–2718.

Nagesha, K. and Sanche, L. (1998). Effects of Band Structure on Electron Attachment to Adsorbed Molecules: Cross Section Enhancements via Coupling to Image States. *Phys. Rev. Lett.* **81**, 5892-5895.

Nightingale, R. W., Roche, A. E. *et al.* (1996). Global CF_2Cl_2 measurements by UARS cryogenic limb array etalon spectrometer: Validation by correlative data and a model. *J. Geophys. Res.* **101**, 9711-9736.

Nisbet, E. G., Dlugokencky, E. J., and Bousquet, P. (2014). Methane on the Rise—Again. *Science* **343**, 493-495.

Nuccitelli, D., Cowtan, K., Jacobs, P., Richardson, M., Way, R. G., Blackburn, A.-M., Stolpe, M. B. and Cook, J. (2014). Comment on "Cosmic-ray-driven reaction and greenhouse effect of halogenated molecules: Culprits for atmospheric ozone depletion and global climate change". *Int. J. Mod. Phys. B* **28**, 1482003.

Oku, A., Kimura, K., and Sato, M. (1988). Chemical decomposition of chlorofluorocarbons by reductive dehalogenation using sodium naphthalenide. *Chem. Lett.*, 1789-1792.

O'Malley, T. F. (1966). Theory of Dissociative Attachment. *Phys. Rev.* **150**, 14-29.

O'Malley, T. F. (1967). Theory of Dissociative Attachment. *Phys. Rev.* **155**, 59-63.

Onda, K., Li, B., Zhao, J., Jordan, K. D., Yang, J., and Petek, H. (2005). Wet Electrons at the $H_2O/TiO_2(110)$ Surface. *Science* **308**, 1154-1158.

Orlando, J. J., and Schauffler S. (1999). Halogen Compounds. In *Atmospheric Chemistry and Global Change*, edited by Brasseur, G. P., Orlando, J. J., and Tyndall G. S. (Oxford University Press, New York)chap.8.

Oster, T., Kühn, A., and Illenberger, E. (1989). Gas phase negative ion chemistry. *Int. J. Mass Spectrosc. Ion. Proc.* **89**, 1-72.

Oum, K. W., Lakin, M. J., DeHaan, D. O., Brauers, T., and Finlayson-Pitts, B. J. (1998). Formation of Molecular Chlorine from the Photolysis of Ozone and Aqueous Sea-Salt Particles. *Science* **279**, 74-76.

Overduin, J. M. (2003). Eyesight and the solar Wien peak. *Am. J. Phys.* **71**, 216-219.

Parker, E. N. (1965). The passage of energetic charged particles through interplanetary space. *Planet. Space Sci.* **13**, 9-49.

Parnis, J. M., Hoover, L. E., Pedersen, D. B., and Paterson, D. D. (1995). Methanol Production from Methane in Lithium-Doped Argon Matrixes by Photoassisted, Dissociative Electron Attachment to N_2O. *J. Phys. Chem.* **99**, 13528-13536.

Patra, P. K. and Santhanam, M. (2002). Comment on 'Effects of Cosmic Rays on Atmospheric Chlorofluorocarbon Dissociation and Ozone Depletion'. *Phys. Rev. Lett.* **89**, 219803.

Petrovié, Z. L., Wang, W. C., and Lee, L. C. (1988). Attachment of Low Energy Electrons to HCl. *J. Appl. Phys.* **64**, 1625-1631.

Peyerimhoff, S. D., and Buenker, R. J. (1979). Potential curves for dissociative electron-attachment of $CFCl_3$. *Chem. Phys. Lett.* **65**, 434-439.

Phelps, A. V. (1969). Laboratory studies of electron attachment and detachment processes of aeronomic interest. *Can. J. Chem.* **47**, 1783-1793.

Pielke Jr., R. A. (2001). Room for doubt. *Nature* **410**, 151.

Pierce, J. R., and Adams, P. J. (2009). Can cosmic rays affect cloud condensation nuclei by altering new particle formation rates? *Geophys. Res. Lett.* **36**, L09820(1-6).

Pimblott, S. M., Laverne, J. A. (1998). On the radiation chemical kinetics of the precursor to the hydrated electron. *J. Phys. Chem.* A**102**, 2967-2975.

Polanyi, J. C., and Zewail, A. H. (1995). Direct Observation of the Transition State. *Acc. Chem. Res.* **28**, 119-132.

Pomerantz, M. A. (1971). *Cosmic Physics.* (Van Nostrand Reinhold Company, New York)pp.66.

Ramanathan, V. (1975). Greenhouse Effect Due to Chlorofluorocarbons: Climatic Implications. *Science* **190**, 50-52.

Ramanathan, V. (2006). Atmospheric Brown Clouds: Health, Climate and Agriculture Impacts. In the Pontifical Academy of Sciences Scripta Varia 106 Interactions Between Global Change and Human Health, (Pontifica Academia Scientiarvm 2006) pp. 47-60.

Ramanathan, V. (2007a). Global Dimming by Air Pollution and Global Warming by Greenhouse Gases: Global and Regional Perspectives; Extended Abstracts of the Plenary lecture presented at the 17th International Conference on Nucleation andAtmospheric Aerosols, Galway, Ireland, August 13th-17th, 2007

Ramanathan, V. (2007b). Role of Black Carbon in Global and Regional Climate Change, Testimonial to the House Committee on Oversight and Government Reform, October 18, 2007, and references therein.

Ramanathan, V., Callis, L., Cess, R., Hansen, J., Isaksen, I., Kuhn, W., Lacis, A., Luther, F., Mahlamn, J., Reck, R., and Schlesinger, M. (1987). Climate-Chemical Interactions and Effects of Changing Atmospheric Trace Gases. *Rev. Geophys.* **25**, 1441-1482.

Ramanathan, V., Chung, C., Kim, D., Bettge, T., Buja, L., Kiehl, J. T., Washington, W. M., Fu, Q., Sikka, D. R., and Wild, M. (2005), Atmospheric Brown Clouds: Impacts on South Asian Climate and Hydrological Cycle. *Proc. Natl. Acad. Sci. USA* **102**, 5326-5333.

Ramanathan, V., Cicerone, R. J., Singh, H. B., and Kiehl, J. T. (1985). Trace Gas Trends and Their Potential Role in Climate Change. *J. Geophys. Res.* **90**, 5547-5566.

Ramanathan, V., Li, F., Ramana, M. V., Praveen, P. S., Kim, D., Corrigan, C. E., Nguyen, H. (2007a). Atmospheric Brown Clouds: Hemispherical and regional variations in long range transport, absorption, and radiative forcing. *J. Geophys. Res.* **112**, D22, S21, doi:10.1029/2006JD008124.

Ramanathan, V., Ramana, M. V., Roberts, G., Kim, D., Corrigan, C. E., Chung, C. E., and Winker, D. (2007b). Warming trends in Asia amplified by brown cloud solar absorption. *Nature* **448**, 575-578.

Ramanathan, V. et al. (2008). Atmospheric Brown Clouds: Regional Assessment Report with Focus on Asia, *published by the United Nations Environment Program, Nairobi, Kenya*, p. 1-360.

Ramaswamy, V., Schwarzkopf, M. D., and Randel, W. J. (1996). Fingerprint of ozone depletion in the spatial and temporal pattern of recent lower-stratospheric cooling. *Nature* **382**, 616-618.

Ramaswamy, V., Schwarzkopf, M. D., Randel, W. J., Santer, B. D., Soden, B. J., and Stenchikov, G. L. (2006). Anthropogenic and natural influences in the evolution of lower stratospheric cooling. *Science* **311**, 1138-1141.

Randel, W. J., Stolarski, R. S., Cunnold, D. M., Logan, J. A., Newchurch, M. J., and Zawodny, J. M. (1999). Trends in the vertical distribution of ozone. *Science* **285**, 1689-1692.

Ratkowski, A. J., Anderson, G. P., Chetwynd, J. H., Nadile, R. M., Devir, A. D., Conley, T. D. (1998). Paper presented at the RTO SET Symposium on "E-O Propagation, Signature and System Performance under Adverse Meteorological Conditions Considering Out-of-Area Operations" (Italian Air Force Academy, Naples, Italy) and published in RTO MP-I.

Reid, G. C. (1987). Influence of solar variability on global sea surface temperatures. *Nature* **329**, 142–143.

Revadekar, J. V., and Patil, S. D. (2011). On the surface air temperature variation in relation to chlorofluorocarbons over the Indian region. *Atmos. Environ.* **45**, 6658-6668.

Ring, M. J., Lindner, D., Cross, E. F., and Schlesinger, M. E. (2012). Causes of the Global Warming Observed since the 19th Century. *Atmos. Clim. Sci.* **2**, 401-415.

Roche, A. E. *et al.* (1994). Observations of Lower-Stratospheric $ClONO_2$, HNO_3, and Aerosol by the UARS CLAES Experiment between January 1992 and April 1993. *J. Atmos. Sci.* **51**, 2877-2902.

Rossky, P. J., and Schnitker, J. (1988). The hydrated electron--quantum simulation of structure, spectroscopy, and dynamics. *J. Phys. Chem.* **92**, 4277-4285.

Rothman, L. *et al.* (2009). The *HITRAN* 2008 molecular spectroscopic database. *J. Quant. Spectrosc. & Radiat. Transfer* **110**, 533-572.

Rowland, F. S., and Molina, M. J. (1975). Chlorofluoromethanes in the environment, *Rev. Geophys. & Space Phys.* **13**, 1-35.

Rowntree, P., Sambe, H., Parenteau, L., and Sanche, L. (1993). Formation of anionic excitations in the rare-gas solids and their coupling to dissociative states of adsorbed molecules. *Phys. Rev. B***47**, 4537-4554.

Ruderman, M. A., Foley, H. M., and Chamberlain, J. W. (1976). 11-year variation in polar ozone and stratospheric-ion chemistry. *Science* **192**, 555-557.

Ryu, S., Chang, J., Kwon, H., and Kim, S. K. (2006). Dynamics of solvated electron transfer in thin ice film leading to a large enhancement in photodissociation of $CFCl_3$. *J. Am. Chem. Soc.* **128**, 3500-3501.

Salby, M. L. (2012). *Physics of the Atmosphere and Climate*. Cambridge University Press, Cambridge, New York.

Sanche, L. (1995). Interactions of low-energy electrons with atomic and molecular solids. *Scanning Micros.* **9**, 619-656.

Sanche, L., Bass, A. D., Ayotte, P., and Fabrikant, I. I. (1995) Effect of the condensed phase on dissociative electron attachment: CH_3Cl condensed on a Kr surface. *Phys. Rev. Lett.* **75**, 3568-3571.

Sander, R., and Crutzen, P. J. (1996). Model study indicating halogen activation and ozone destruction in polluted air masses transported to the sea. *J. Geophys. Res.* **101**, 9121-9138.

Schneider, C. W., Kucerovsky, Z., and Brannen, E. (1989). Carbon dioxide absorption of He-Ne laser radiation at 4.2 μm: characteristics of self and nitrogen broadened cases. *Appl. Opt.* **28**, 959-966.

Schultes, E., Christodoulides, A. A., and Schindler, R. N. (1975). Studies by the electron cyclotron resonance (ECR) technique. VIII. Interations of low-energy electrons with the chlorine-containing molecules CCl_4, $CHCl_3$, CH_2Cl_2, $C_nH_{2n+1}Cl$ (n = 1 to 4), C_2H_3Cl, $COCl_2$, $NOCl$, $CNCl$ and Cl_2. E, CH_3Cl, C_2H_5Cl, CCl_4, Cl_2, $CHCl_3$, CH_2Cl_2, etc. *Chem. Phys.* **8**, 354-365.

Schulz, G. J. (1973a). Resonances in Electron Impact on Atoms. *Rev. Mod. Phys.* **45**, 378-422.

Schulz, G. J. (1973b). Resonances in Electron Impact on Diatomic Molecules. *Rev. Mod. Phys.* **45**, 423-486.

Seeley, J. V., Miller, T. M., and Viggiano, A. A. (1996). *Ab initio* study of the chlorine nitrate anion. *J. Chem. Phys.* **105**, 2127-2128.

Seliger, H. H. (1977). Environmental photobiology. In *The Science of Photobiology*. Smith, K. C. (Ed). (Plenum Press, New York and London) pp.143-174.

Shamir, N., Mihaychuk, J. G., van Driel, H. M., and Kreuzer, H. J. (1999). Universal mechanism for gas adsorption and electron trapping on oxidized silicon. *Phys. Rev. Lett.* **82**, 359-361.

Shamir, N., and van Driel, H. M. (2000). Trapping and detrapping of electrons photoinjected from silicon to ultrathin SiO_2 overlayers. II. In He, Ar, H_2, N_2, CO, and N_2O. *J. Appl. Phys.* **88**, 909-917 .

Shindell, D. T., Rind, D., and Lonergan, P. (1998). Increased polar stratospheric ozone losses and delayed eventual recovery owing to increasing greenhouse-gas concentrations. *Nature* **392**, 589-593.

Silva, C., Walhout, P. K., Yokoyama, K., and Barbara, P. F. (1998). Femtosecond Solvation Dynamics of the Hydrated Electron. *Phys. Rev. Lett.* **80**, 1086-1089.

Smart, D. F., Shea, M. A. (1997). Solar radiation. In: *Encyclopedia of applied physics*, Vol. 18. (New York, NY: VCH Publishers,) pp. 393-429.

Smith, D., and Adams, N. G. (1980). Elementary plasma reactions of environmental interest. *Top. Curr. Chem.* **89**, 1-43.

Smith, D., Adams, N. G. and Alge, E. (1984). Attachment coefficients for the reactions of electrons with CCl_4, CCl_3F, CCl_2F_2, $CHCl_3$, Cl_2 and SF_6 determined between 200 and 600 K using the FALP technique. *J. Phys. B* **17**, 461-472 .

Smith, H. J., Fischer, H., Wahlen, M., Mastroianni, D., and Deck, B. (1999). Dual modes of the carbon cycle since the Last Glacial Maximum. *Nature* **400**, 248-250.

Soffer, B. H., and Lynch, D. K. (1999). Some paradoxes, errors, and resolutions concerning the spectral optimization of human vision. *Am. J. Phys.* **67**, 946-953.

Solanki, S. K., and Krivova, N. A. (2003). Can solar variability explain global warming since 1970? *J. Geophys. Res.* **108**, 1200.

Solanki, S. K., Krivova, N. A., and Haigh, J. D. (2013). Solar Irradiance Variability and Climate. *Annu. Rev. Astron. Astrophys.* **51**, 311–351.

Solomon, S. (1990). Progress towards a quantitative understanding of Antarctic ozone depletion. *Nature* **347**, 347-354.

Bibliography

Solovev, S., Kusmierek, D. O., and Madey, T. E. (2004). Negative ion formation in electron-stimulated desorption of CF_2Cl_2 coadsorbed with polar NH_3 on Ru(0001). *J. Chem. Phys.* **120**, 968-978.

Soon, W., Baliunas, S., Idso, S. B., Kondratyev, K. Y., and Posmentier, E. S. (2001). Modeling climatic effects of anthropogenic carbon dioxide emissions: unknowns and uncertainties. *Clim. Res.* **18**, 259-275.

Souda, R. (2009). Glass transition and crystallization dynamics of thin CF_2Cl_2 films deposited on Ni(111) graphite and water-ice films. *J. Chem. Phys.* **131**, 164501.

Spence, D., and Burrow, P. D. (1979). Resonant dissociation of N_2 by electron impact . *J. Phys. B.* **12**, L179-L184.

Stähler, J., Gahl, C., and Wolf, M. (2012). Trapped Electrons on Supported Ice Crystallites. *Acc. Chem. Res.* **45**, 131-138.

Stolarski, R. S., and Cicerone, R. J. (1974). Stratospheric chlorine: A possible sink for ozone. *Can. J. Chem.* **52**, 1610-1615.

Stott, L., Timmermann, A., and Thunell, R. (2007). Southern Hemisphere and Deep-Sea Warming Led Deglacial Atmospheric CO_2 Rise and Tropical Warming. *Science* **318**, 435-438.

Svensmark, H. (1998). Influence of Cosmic Rays on Earth's Climate. *Phys. Rev. Lett.* **81**, 5027-*5030.*

Svensmark, H., Bondo, T., and Svensmark, J. (2009). Cosmic ray decreases affect atmospheric aerosols and clouds. *Geophys. Res. Lett.* **36**, L15101(1-4).

Svensmark, H., and Friis-Christensen, E. (1997). Variation of cosmic ray flux and global cloud coverage - A missing link in solar-climate relationships. *J. Atmos. Sol.-Terr. Phys.* **59**, 1225-1232.

Tachikawa, H. (2008). Dissociative electron capture of halocarbon caused by the internal electron transfer from water trimer anion. *Phys. Chem. Chem. Phys.* **10**, 2200-2206.

Tachikawa, H., and Abe, S. (2007). Reaction dynamics following electron capture of chlorofluorocarbon adsorbed on water cluster: A direct density functional theory molecular dynamics study. *J. Chem. Phys.* **126**, 194310 .

Thorne, R. M. (1977). Energetic radiation belt electron-precipitation - natural depletion mechanism for stratospheric ozone. *Science* **195**, 287-289.

Toohey, D. W. *et al* (1993). The seasonal evolution of reactive chlorine in the northern hemisphere stratosphere. *Science* **261**, 1134-1136.

Toon, O. B., and Turco, R. P. (1991). Polar stratospheric clouds and ozone depletion. *Sci. Am.* **264**, 68-74.

Torr, D. G. (1985). In *The Photochemistry of Atmospheres*, Levine, J. S. (Ed). (Academic Press, Orlando) Chap. 5.

Trenberth, K. E., and Fasullo, J. T. (2013). An apparent hiatus in global warming? *Earth's Future* **1**, 19-32.

Tronc, M., Figuet-Fayard, F., Schermann, C., and Hall, R. I. (1977). Angular distributions of O^- from dissociative electron attachment to N_2O between 1.9 to 2.9 eV. *J. Phys. B.* **10**, L459-L462.

Van Allen, J. A., and Singer, S. F. (1950). On the primary cosmic-ray spectrum. *Phys. Rev.* **78**, 819.

Van Doren, J. M., McClellan, J., Miller, T. M., Paulson, J. F., and Viggiano, A. A. (1996). Electron attachment to $ClONO_2$ at 300 K. *J. Chem. Phys.* **105**, 104-110.

Van Doren, J. M., Miller, T. M., Williams, S., and Viggiano, A. A. (2003). Direct Measurement of the Thermal Rate Coefficient for Electron Attachment to Ozone in the Gas Phase, 300–550 K: Implications for the Ionosphere. *Phys. Rev. Lett.* **91**, 223201(1-4).

Veizer, J., Godderis, Y., and Francois, L. M. (2000). Evidence for decoupling of atmospheric CO_2 and global climate during the Phanerozoiceon. *Nature* **408**, 698-701.

Viggiano, A. A., and Arnold, F. (1995). Chemistry and composition of the atmosphere. In *Handbook of atmospheric electrodynamics*. Volland, H. (Ed). (CRC Press, Boca Raton) Chapter 1. pp.1–25.

von Storch, H. (2010). Climate models and modeling: an editorial essay. *Wiley Interdisc. Rev.: Clim. Change* **1**, 305-310.

von Storch, H., and Stehr, N. (2000). Climate change in perspective. *Nature* **405**, 615.

Vondrak, T., Plane, J. M. C., and Meech, S. R. (2006). Photoemission from sodium on ice: A mechanism for positive and negative charge coexistence in the mesosphere. *J. Phys. Chem. B* **110**, 3860-3863.

Vondrak, T., Plane, J. M. C., and Meech, S. R. (2006). Influence of submonolayer sodium adsorption on the photoemission of the Cu(111)/water ice surface. *J. Chem. Phys.* **125**, 224702.

Vondrak, T., Meech, S. R. and Plane, J. M. C. (2009). Photoelectric emission from the alkali metal doped vacuum-ice interface. *J. Chem. Phys.* **130**, 054702.

Wang, C.-R., Hu, A., and Lu, Q.-B. (2006). Direct Observation of the Transition State of Ultrafast Electron Transfer Reaction of a Radiosensitizing Drug Bromodeoxyuridine. *J. Chem. Phys.* **124**, 241102(1-4).

Wang, C.-R., and Lu, Q.-B. (2007). Real-Time Observation of a Molecular Reaction Mechanism of Aqueous 5-Halo-2'-deoxyuridines under UV/Ionizing Radiation. *Angew. Chem. Intl. Ed.* **46**, 6316-6320.

Wang, C.-R., Drew, K., Luo, T., Lu, M.-J., and Lu, Q.-B. (2008a). Resonant Dissociative Electron Transfer of the Presolvated Electron to CCl4 in Liquid: Direct Observation and Lifetime of the CCl4*⁻ Intermediate. *J. Chem. Phys.* **128**, 041102 (1-4).

Wang, C. R., Luo, T., and Lu, Q.-B. (2008b). On the Lifetimes and Physical Nature of Prehydrated Electrons in Liquid Water. *Phys. Chem. Chem. Phys.* **10**, 4463-4470.

Wang, C.-R., Drew, K., Luo, T., Lu, M.-J., and Lu, Q.-B. (2008c). Response to "Comment on 'Resonant Dissociative Electron Transfer of the Presolvated

Electron to CCl_4 in Liquid: Direct Observation and Lifetime of the CCl_4^{*-} Intermediate' [*J. Chem. Phys.128, 041102*(2008)]". *J. Chem. Phys.* **129**, 027102 (2008).

Wang, W. C., Dudek, M. P., Liang, X. Z., and Kiehl, J. T. (1991). Inadequacy of effective CO_2 as a proxy in simulating the greenhouse effect of other radioactively active gases. *Nature* **350,** 573-577.

Wang, W.-C., Dudek, M. P., and Liang, X.-Z. (1992). Inadequacy of effective CO_2 as a proxy in assessing the regional climate change due to other radiatively active gases. *Geophys. Res. Lett.* **19**, 1375-1378.

Wang, W.-C., and Molnar G. (1985). A Model Study of the Greenhouse Effects Due to Increasing Atmospheric CH_4, N_2O, CF_2Cl_2, and $CFCl_3$. *J. Geophys. Res.* **90**, 12971-12980.

Wang, Y., Christophorou, L. G., and Verbrugge, J. K. (1998). Effect of temperature on electron attachment to and negative ion states of CCl_2F_2. E. CCl_2F_2; 298-500 K, 0.04 - 1 eV. *J. Chem. Phys.* **109**, 8304-5310.

Waters, J. W., Froidevaux, L., Read, W. G., Manney, G. L., Elson, L. S., Flower, D. A., Jarnot, R. F., and Harwood, R. S. (1993). Stratospheric CIO and ozone from the Microwave Limb Sounder on the Upper Atmosphere Research Satellite. *Nature* **362**, 597-602.

Webster, C. R. *et al.* (1993). Chlorine chemistry on polar stratospheric cloud particles in the Arctic winter. *Science* **261**, 1130-1134.

Weyl, W. (1864). Uber Metall-Ammonium-Verbindungen. *Ann. Phys. (Leipzig)* **197**, 601-612.

Wilde, R. S., Gallup, G. A., and Fabrikant, I. I. (1999). Semiempirical *R*-matrix theory of low energy electron–CF_3Cl inelastic scattering. *J. Phys. B* **32**, 663–673.

Williams, S., Campos, M. F., Midey, A. J., Arnold, S. T., Morris, R. A., and Viggiano, A. A. (2002). Negative Ion Chemistry of Ozone in the Gas Phase. *J. Phys. Chem. A* **106**, 997-1003.

Willson, R. C. (1997). Total solar irradiance trend during solar cycles 21 and 22. *Science*, **277**, 1963-1965.

Willson, R. C., and Mordinov, A.V. (2003). Secular total solar irradiance trend during solar cycles 21 and 22. *Geophys. Res. Lett.* **30**, 1199-1202.

Woods, T. N., Tobiska, W. K., Rottman, G. J., & Worden, J. R. (2000). Improved solar Lyman-α irradiance modeling from 1947 through 1999 based on UARS observations. *J. Geophys. Res.* **105**, 27195–27215.

WMO (World Meteorological Organization) (1994). *Scientific Assessment of Ozone Depletion: 1994* (Global Ozone Research and Monitoring Project-Report No. 47, Geneva, Switzerland).

WMO (World Meteorological Organization) (1998). Scientific Assessment of Ozone Depletion: 1998 (Global Ozone Research and Monitoring Project-Report No. 44, Geneva) Chap 12.

WMO (World Meteorological Organization) (2006). *Scientific Assessment of Ozone Depletion: 2006* (Global Ozone Research and Monitoring Project-Report No. 50, Geneva, Switzerland).

WMO (World Meteorological Organization) (2010). *Scientific Assessment of Ozone Depletion: 2010* (Global Ozone Research and Monitoring Project-Report No. 52, Geneva, Switzerland).

WMO (World Meteorological Organization) (2014). *Scientific Assessment of Ozone Depletion: 2014* (Global Ozone Research and Monitoring Project-Report No. 55, Geneva, Switzerland).

Wyatt, M. G., and Curry, J. A. (2014). Role for Eurasian Arctic shelf sea ice in a secularly varying hemispheric climate signal during the 20th century. *Clim. Dynamics* **42**, 2763-2782.

Yu, F. Q. (2004). Formation of large NAT particles and denitrification in polar stratosphere: possible role of cosmic rays and effect of solar activity. *Atmos. Chem. Phys.* **4**, 2273-2283.

Yu, F. Q., and Turco, R. P. (2001). From molecular clusters to nanoparticles: Role of ambient ionization in tropospheric aerosol formation. *J. Geophys. Res.* **106**, 4797-4814.

Zewail, A. H. (2000). Femtochemistry: Atomic-Scale Dynamics of the Chemical bond using ultrafast lasers (Nobel lecture). *Angew. Chem. Int. Ed.* **39**, 2587-2631.

Index